D082447

Free-Radical Chain Reactions

Earl S. Huyser

Department of Chemistry

University of Kansas

WILEY-INTERSCIENCE, a Division of John Wiley & Sons
New York · London · Sydney · Toronto

Library of Congress Catalogue Card Number: 77-106013

SBN 471   42596   6

Printed in the United States of America

10 9 8 7 6 5 4 3 2 1

Preface

This book is intended primarily for chemists who are not experts in free-radical chemistry but wish to become acquainted with some of the kinetic and synthetic aspects of reactions that involve free-radical intermediates. The material presented here requires a working knowledge of organic chemistry and familiarity with the elementary aspects of kinetics and thermodynamics.

The discussion is limited to free-radical chain reactions. I am aware that several other aspects of free-radical chemistry are undergoing rapid development, and in a more comprehensive treatment of the subject these areas would warrant extensive discussion. However, the practicing organic chemist most often finds himself in need of information about and an appreciation of the concepts encountered in the chain reactions.

The book is divided into two parts. Chapters 1 through 4 introduce the principles of free-radical chain reactions and serve as a basis for the remaining chapters which discuss various types of free-radical reactions of interest to organic chemists. Chapter 1 mainly examines the subject in terms of its historical development and its place in organic chemistry at the present time. Pertinent definitions of free radicals and their reactions are also given. Chapter 2 provides a background of descriptive information in free-radical reactions and establishes the concept of the chain reaction. Some basic kinetic principles of free-radical chain reactions are introduced in Chapter 3, and the relationship of structure and reactivity of both free radicals and substrates in chain-propagating reactions is discussed in Chapter 4.

Various reaction types examined in terms of the principles developed in the earlier chapters make up the major portion of this book. *No attempt was made to be comprehensive in the survey of the literature on any subject in these chapters.* The reactions employed to illustrate the principles were chosen on the basis of my familiarity with the reaction, as well as of my evaluation of the reaction's appropriateness as an illustration. Both recent work and work of long standing are used for this purpose.

Chapter 5 is concerned with displacement reactions and, for the most part, deals with halogenation reactions. Free-radical additions are discussed in Chapters 6 and 7, the first being devoted largely to the free-radical chemistry of the adding reagent and the second to the chemistry of the unsaturated species that participate in these reactions. In Chapter 8 some free-radical reactions that include a β-elimination as a chain-propagating step are discussed, and Chapter 9 deals with reactions in which a rearrangement occurs as a chain-propagating reaction. Some of the more common chemical initiators and their free-radical chemistry are the subjects of Chapter 10, and Chapter 11 examines certain free-radical chain oxidation and reduction reactions.

Chapter 12 introduces some kinetic aspects of free-radical vinyl polymerization reactions. The background given in earlier chapters is used to develop the kinetic principles of vinyl polymerization, although the reverse order of presentation is somewhat more traditional. I have found, however, that presenting free-radical chain reactions initially in terms of polymerization reactions tends to "turn off" many organic chemists before they realize that the subject may be of interest to them. An appreciation of the principles of free-radical chain kinetics in terms of reactions that are more familiar to the reader may develop a desire to become acquainted with the chemistry of vinyl polymerizations. Although the subject cannot be treated in a thorough manner, some aspects of vinyl polymerization that pertain to the principles of chain reactions can readily become a part of the reader's working knowledge of organic chemistry.

The reader who is well versed in free-radical reactions may on occasion find too much repetition of both the structures and reactions. It should be kept in mind, however, that the book was written primarily for those who are not so familiar with these concepts. A certain amount of repetition may serve the reader better than being subjected to the problem of determining what the author is assuming he should be able to deduce from the text and previous reactions. The same is true for the derivations of some of the kinetic expressions. I have spent many "happy hours" trying to discover how an equation in a derivation was arrived at after being told that it was obvious that it followed from the preceding equation. Including the details of some of these derivations may seem unnecessary to some readers already familiar with them, but the reader not so well acquainted with such derivations may develop an appreciation for the concepts involved.

I am deeply indebted to Professors Cheves Walling and J. C. Martin for reading the entire manuscript in its original form and making many helpful and necessary suggestions which have been incorporated in the book. I also acknowledge with appreciation the help of Mrs. Connie F. Yother in preparing the manuscript and Miss Nancy L. Huyser in assembling the indexes.

EARL S. HUYSER

Lawrence, Kansas
November 1969

Contents

Free-Radical Chain Reactions

Chapter 1

Introduction

I. Definition of 'Free Radical'

Free radicals may be defined as species having one or more unpaired electrons and, therefore, a large variety of chemical substances can be classified as free radicals. In some cases, free radicals are monatomic species such as the halogen atoms, the alkali metal atoms and certain metallic ions having an

$$Br\cdot \quad Cl\cdot \quad I\cdot \quad F\cdot \quad Li\cdot \quad Na\cdot \quad K\cdot \quad Co^{+2} \quad Cu^{+2} \quad Ce^{+3} \quad Fe^{+3} \quad Ag^{2+}$$

unpaired electron. Much of the chemistry of these monatomic species indicates that they are indeed free radicals.

More familiar are polyatomic aggregates that have an unpaired electron. The original meaning of the word radical, a meaning still retained in modern chemistry, denotes a group of atoms that retain their identity through a series of chemical reactions. A free radical would be such a species with an unpaired

1

electron. In some free radicals the unpaired electron is on a carbon atom as in the case of the alkyl radicals. Other functionalities, e.g. hydroxy, carboxy or

$$CH_3 \cdot \qquad CH_3CH_2 \cdot \qquad (CH_3)_3C \cdot$$

halogen, may be bonded to the carbon atom having the unpaired electron.

$$\cdot CH_2OH \qquad \cdot CH(CO_2C_2H_5)_2 \qquad Cl_3C \cdot$$

The unpaired electron is found on oxygen or on sulfur in alkoxy, peroxy, and

$$(CH_3)_3CO \cdot \qquad C_6H_5\overset{\displaystyle CH_3}{\underset{\displaystyle CH_3}{C}}OO \cdot \qquad n\text{-}C_4H_9S \cdot$$

thiyl radicals. Free radicals with the unpaired electron on phosphorus are also

$$\cdot P(OC_2H_5)_2 \qquad Cl_3Si \cdot \qquad Cl_3Ge \cdot$$

known, as well as free radicals with the unpaired electron on silicon and germanium.

The above examples are merely illustrative of the variety of species that could be classified as free radicals. For the most part, free radicals are chemically reactive showing a tendency to do something to accommodate the unpaired electron in some manner. In the case of some free radicals this desire to react is greater than it is for others and, although we might classify free radicals as generally reactive, some are unreactive enough to be isolated. Examples of such free radicals are "galvinoxyl" (1), di-t-butyl nitroxide (2), and diphenyl-

1   2

3

$$R = (CH_3)_3C$$

picryhydrazyl (**3**). Some stable isolable molecules can also be classified, due to the bookkeeping of the electrons, as free radicals. Such is the situation for nitric

$$\cdot \ddot{N} :: \overset{\cdot\cdot}{\underset{\cdot\cdot}{O}} \qquad\qquad : \overset{\cdot\cdot}{\underset{\cdot\cdot}{O}} : \ddot{N} : \overset{\cdot\cdot}{\underset{\cdot\cdot}{O}} :$$

$$\textbf{4} \qquad\qquad\qquad\qquad \textbf{5}$$

oxide (**4**) and nitrogen dioxide (**5**). In the case of oxygen, the electronic configuration of oxygen in the ground state is such that two electrons are unpaired and have parallel spins (triplet state); the molecule exists as a diradical. Although unreactive compared to many free radicals, the behavior of molecular oxygen in many reactions can be attributed to its free-radical character. Indeed, the importance of molecular oxygen for maintaining life on this planet is cause for some reflection on the possible free-radical character of many biochemical processes.

Except for some of the unreactive free radicals just mentioned, most free radicals are too reactive to be isolated. They exist only as intermediates in various chemical reactions. Any reaction that involves one or more free radicals as reaction intermediates can be regarded as a free-radical reaction. A very interesting characteristic of most free-radical reactions, except, of course, those that may involve an unreactive radical as an intermediate or produce one as a product, is that the free radicals are truly reaction intermediates since neither the reactants nor the products are free radicals. The main concern in this book is to learn how free radicals are produced and how they react as reaction intermediates in the process of converting reactants to products.

II. Historical

It is only in the past two decades that free-radical chemistry has emerged in the minds of many organic chemists from the realm of the somewhat mysterious to become an acceptable aspect of reaction mechanisms worthy of serious consideration. Very few modern textbooks in organic chemistry are without some sections devoted to free-radical reactions whereas older texts often failed to mention free radicals or, if they did, pointed out the existence of free radicals as an interesting observation but of somewhat nebulous value to organic chemistry.

Interest in free-radical chemistry is by no means recent. As far back as the earliest days of the development of modern structural theory, organic chemists recognized the possible existence of such species and attempted to isolate free radicals. In the 1850's, Wurtz found that reactions of sodium with methyl iodide and ethyl iodide produced compounds with empirical formulas corresponding

$$CH_3I + Na \rightarrow NaI + CH_3\cdot \tag{1-1}$$

$$C_2H_5I + Na \rightarrow NaI + C_2H_5\cdot \tag{1-2}$$

to methyl and ethyl radicals. These gases were not the free radicals, but the dimeric species ethane and *n*-butane (mixed with some ethane and ethylene), respectively, as was deduced subsequently from their vapor densities. (Concurrent with the writing of this chapter there is activity in the investigation of the Wurtz reaction and evidence is accruing for free-radical intermediates in the reaction.)

With the advent of the valence theory around 1860 and its success in explaining many of the then-known aspects of organic chemistry in terms of tetravalent carbon, little, if any, effort was made to isolate or even to detect organic free radicals until the turn of the century. In 1900, Gomberg reported the existence of triphenylmethyl radicals (**7**), free radicals produced in solution by the dissociation of hexaphenylethane (**6**), a hydrocarbon formed by reaction of

$$2(C_6H_5)_3CCl \xrightarrow[\text{(or Ag)}]{\text{Zn}} (C_6H_5)_3CC(C_6H_5)_3 \rightleftharpoons 2(C_6H_5)_3C\cdot \qquad (1\text{-}3)$$
$$\textbf{6} \qquad\qquad\qquad \textbf{7}$$

triphenylmethyl chloride with either silver or zinc dust.[1] Although accepted with some reluctance, the existence of such free radicals became a part of organic chemistry. However, it was almost another forty years before the importance of free radicals as reaction intermediates began to be fully appreciated. This does not mean that there was no effort by some rather enlightened and prestigious (although in some instances the prestige came considerably after the attempts to enlighten the chemical world) investigators to postulate free radicals as reaction intermediates. In 1920, Staudinger suggested that a trivalent carbon species might be involved in the polymerization of styrene (see Chapter 12).[2] A free-radical chain process was postulated by Bäckstrom in 1934 to explain the autoxidation of benzaldehyde (see Chapter 11).[3] Free-radical chain mechanisms were suggested by Haber and Willstätter for other organic reactions[4] although many of their proposed mechanisms did not stand after further investigation.

In 1937, three significant publications appeared that might be considered to have launched modern free-radical chemistry. A review article by Hey and Waters outlined free-radical mechanisms for several organic reactions.[5] Among these reactions was the abnormal addition of hydrogen bromide to alkenes, a reaction reported by Kharasch and Mayo in 1934.[6] In 1937, Kharasch postulated the presently accepted free-radical chain process for the abnormal addition of hydrogen bromide to alkenes.[7] The same year, Flory published a most significant paper on the kinetics of vinyl polymerization in terms of a free-radical chain reaction.[8] It is probably not surprising that the suggestions made in these publications were not greeted with immediate acceptance. The ideas proposed in these publications did, however, form the groundwork for much significant research in many industrial and academic laboratories.

III. Formation of Free Radicals

The following sections serve only as an outline of some of the general principles of free-radical chemistry and many of the concepts in this chapter will be discussed in greater detail in subsequent chapters.

A. THERMOLYTIC REACTIONS

Possibly the most familiar radical-producing reaction is the thermal decomposition of an organic molecule containing a peroxide linkage. For example, t-butyl peroxide (**8**) decomposes at temperatures around 125°C yielding two

$$(CH_3)_3COOC(CH_3)_3 \xrightarrow{\sim 120°} 2(CH_3)_3CO\cdot \qquad (1\text{-}4)$$
$$\mathbf{8} \qquad\qquad\qquad \mathbf{9}$$

t-butoxy radicals (**9**). Such a reaction introduces free radicals into a reaction mixture and the importance of such a process will soon become apparent. Other peroxidic compounds that are often used to produce free radicals on thermolysis are diacetyl peroxide (**10**), benzoyl peroxide (**11**), and t-butyl

$$\underset{\mathbf{10}}{CH_3\overset{\overset{\displaystyle O}{\|}}{C}OO\overset{\overset{\displaystyle O}{\|}}{C}CH_3} \xrightarrow{\sim 75°} 2CH_3\cdot + 2CO_2 \qquad (1\text{-}5)$$

$$\underset{\mathbf{11}}{C_6H_5\overset{\overset{\displaystyle O}{\|}}{C}OO\overset{\overset{\displaystyle O}{\|}}{C}C_6H_5} \xrightarrow{\sim 80°} 2C_6H_5\overset{\overset{\displaystyle O}{\|}}{C}O\cdot \qquad (1\text{-}6)$$

$$\underset{\mathbf{12}}{CH_3\overset{\overset{\displaystyle O}{\|}}{C}OOC(CH_3)_3} \xrightarrow{\sim 100°} CH_3\cdot + CO_2 + \cdot OC(CH_3)_3 \qquad (1\text{-}7)$$

peracetate (**12**). The chemistry of these peroxides as well as others is treated in greater detail in Chapter 10.

Thermolysis of azobisisobutyronitril (**13**) at about 70°C yields a molecule of nitrogen and two isobutyronitrile radicals (**14**). Although **13** is the most widely

$$(CH_3)_2\underset{\underset{\mathbf{13}}{CN}}{\overset{\displaystyle |}{C}}-N{=}N-\underset{CN}{\overset{\displaystyle |}{C}}(CH_3)_2 \xrightarrow{\sim 70°} 2(CH_3)_2\underset{\underset{\mathbf{14}}{CN}}{\overset{\displaystyle |}{C}}\cdot + N_2 \qquad (1\text{-}8)$$

used of the azonitriles, similar compounds have been prepared and their decomposition reactions have been investigated.

Peroxides and azonitriles are the most familiar compounds for producing free radicals by a thermal decomposition reaction. Other compounds have been studied but none have proven to be as practical as either the peroxides or azonitriles in producing free radicals.

B. PHOTOLYTIC REACTIONS

Many compounds are capable of absorbing light and undergoing photochemical decomposition yielding free radicals. Probably the most familiar examples of such processes are the photochemical decompositions of molecular

$$Cl_2 \xrightarrow{h\nu} 2Cl\cdot \qquad (1\text{-}9)$$

$$Br_2 \xrightarrow{h\nu} 2Br\cdot \qquad (1\text{-}10)$$

chlorine and bromine. Although comparatively high temperatures are required for the thermolysis of t-butyl peroxide, this peroxide decomposes at almost any temperature when exposed to ultraviolet light. The azonitriles also undergo temperature-independent photolysis reactions. The photolysis of bromotri-

$$(CH_3)_3COOC(CH_3)_3 \xrightarrow{h\nu} 2(CH_3)_3CO\cdot \qquad (1\text{-}11)$$

$$\underset{\overset{|}{CN}}{(CH_3)_2C} - N{=}N - \underset{\overset{|}{CN}}{C(CH_3)_2} \xrightarrow{h\nu} 2\underset{\overset{|}{CN}}{(CH_3)_2C}\cdot + N_2 \qquad (1\text{-}12)$$

chloromethane (**15**) is a facile method of producing a bromine atom and a

$$\underset{\textbf{15}}{BrCCl_3} \xrightarrow{h\nu} Br\cdot + \underset{\textbf{16}}{Cl_3C}\cdot \qquad (1\text{-}13)$$

trichloromethyl free radical (**16**). Several other similar photolysis reactions yielding free radicals will be encountered in later chapters.

Much work has been done in the area of photochemistry and many fascinating transformations have been brought about by the action of light on certain organic compounds. The formation of an electronically excited state of a molecule by the absorption of a quantum of energy produces a reaction intermediate that behaves differently in many cases from the ground state of the molecule. Dissociation of the molecule, as shown in the previous paragraph, is just one mode of reaction. There are others, some of which indicate that the excited state has definite free-radical character. For example, illumination of benzophenone (**17**) ultimately produces a species that behaves very much as if it were the diradical radical **18**. It should not be inferred, however, that all

$$\underset{\textbf{17}}{\overset{\overset{O}{\|}}{C_6H_5CC_6H_5}} \xrightarrow{h\nu} \left(\overset{\overset{O}{\|}}{C_6H_5CC_6H_5} \right)^* \rightarrow \underset{\textbf{18}}{\overset{\overset{O\cdot}{|}}{C_6H_5\underset{\bullet}{C}C_6H_5}} \qquad (1\text{-}14)$$

photochemical transformations proceed *via* radical intermediates. In many instances, the chemical transformation may occur in an excited state in which the electron spins are still paired. Radical character will result if there is intersystem crossover from this excited singlet state with the paired spins to a triplet state in which the electron spins are parallel. It should be kept in mind that a triplet species may possibly differ both in reactivity and in the nature of its reactions from ground state radicals such as those produced by thermolysis reactions. Not only is the triplet species a diradical but it is a species in an excited state.

C. HIGH-ENERGY RADIATION

Much attention has been given to the formation of free radicals by the action of either γ-rays (generally from a Co^{60} source), X-rays, and high-energy electrons (from a Van de Graaff generator). The free-radical producing process in each case is somewhat similar in that it involves absorption of energy by the molecule which results in the loss of an electron by the molecule yielding a radical-ion. Free radicals can be produced either by recombination of the radical-cation with a low-energy electron yielding a molecule in a high-energy state which decomposes into free radicals, or by capture of an electron by a molecule producing a radical-anion which decomposes into an anion and a free radical (e.g. γ-radiation of $BrCCl_3$).

$$
BrCCl_3 \xrightarrow{\gamma\text{-radiation}} BrCCl_3^+ + e^-
\begin{array}{l}
\xrightarrow{BrCCl_3} BrCCl_3^{*(-)} \longrightarrow Br^- + Cl_3C \cdot \\
\xrightarrow{e^-} BrCCl_3^* \longrightarrow Br \cdot + Cl_3C \cdot
\end{array}
\tag{1-15}
$$

These processes are by no means the only ones that occur when molecules are subjected to high-energy radiation. Indeed, the more interesting aspects of high-energy radiation chemistry lie in the chemistry of the radical-ions themselves or that of the excited molecules that are produced in these processes.

When we consider the equipment requirements for producing the radicals by γ-radiation, disadvantages of the process become evident. There are, however, some interesting industrial possibilities for γ-radiation as well as for the Van de Graaff generator. For example, continuous use of the ultraviolet radiation sources used to induce certain chain reactions results in a considerable amount of coating on the surface of the lamp, and periodic shutdown of the reactor is required to allow for cleaning of the apparatus. A continuous reaction can be maintained with γ-radiation without the necessity of shutdown for cleaning. Indeed, the γ-radiation induced addition of hydrogen bromide to ethylene yielding ethyl bromide (see Chapter 7) has proven commercially feasible. Other applications of high-energy radiation lie in the possibility of introducing free radicals in places where it might prove difficult (or impossible) to do so with chemical initiators or photochemically, for example, in solids or on the surface of a film.

D. REDOX REACTIONS

Certain oxidation-reduction reactions produce, among other products, free radicals. A good example is the reaction of hydrogen peroxide with the ferrous

$$H_2O_2 + Fe^{+2} \rightarrow Fe^{+3} + OH^- + HO \cdot \tag{1-16}$$

ion (Fenton reaction). A similar reaction occurs between alkyl hydroperoxides and metal ions capable of oxidation by loss of a single electron. Such reactions

$$ROOH + M^{+n} \rightarrow RO \cdot + OH^- + M^{+(n+1)} \tag{1-17}$$

may well be far more general than is generally appreciated and may occur between many pairs of species provided one is an oxidizing agent and the other a reducing agent.

IV. Reactions of Free Radicals

Interestingly, the kinds of reactions in which free radicals actually participate are quite limited. They can be classified either as radical-propagating reactions or radical-destroying reactions. In radical-propagating reactions, free radicals react in either a bimolecular reaction with a substrate molecule or in a unimolecular process producing other free radicals. Radical-destroying reactions are bimolecular reactions of two free radicals yielding products that are not free radicals. The significance of the terms "chain propagating" and "chain terminating" will become evident in the next chapter in which the mechanisms of some free-radical reactions are examined in terms of the free-radical producing reactions and the reactions of free radicals as reaction intermediates.

A. RADICAL PROPAGATING REACTIONS

1. Displacement (or Abstraction). Free radicals are capable of participating in displacement reactions that are similar to the S_N2 reaction encountered in ionic organic reactions. A free radical can attack an atom displacing a different free radical from this atom as shown in 1-18 where $X \cdot$ is the attacking radical

$$X \cdot + YZ \rightarrow XY + Z \cdot \tag{1-18}$$

that displaces the radical $Z \cdot$ from Y. Displacement is a radical-propagating reaction since one radical reacts with a substrate producing another radical.

As in the case of S_N2 displacements, steric effects play a significant role in radical-displacement reactions. Although almost any organic free radical is capable of participating in displacement reactions, the different atoms on which the radical displacement reaction can occur is limited. Except for a few special cases cited below, the displacement reactions are observed to occur only on hydrogen or halogen atoms. Several familiar examples are listed in equations 1-19 through 1-23. Reactions that involve displacement on carbon are not

$$Cl\cdot + H\!-\!CH_3 \;\rightarrow\; ClH + \cdot CH_3 \qquad (1\text{-}19)$$

$$CH_3\cdot + Cl\!-\!Cl \;\rightarrow\; CH_3Cl + Cl\cdot \qquad (1\text{-}20)$$

$$CH_3\cdot + H\!-\!CH_2C_6H_5 \;\rightarrow\; CH_4 + \cdot CH_2C_6H_5 \qquad (1\text{-}21)$$

$$Cl_3CCH_2\dot{C}H_2 + ClCCl_3 \;\rightarrow\; Cl_3CCH_2CH_2Cl + Cl_3C\cdot \qquad (1\text{-}22)$$

$$Br\cdot + H\!-\!CH_2CH\!\!=\!\!CH_2 \;\rightarrow\; BrH + \cdot CH_2CH\!\!=\!\!CH_2 \qquad (1\text{-}23)$$

normally observed if displacement can occur on another atom. An alkyl radical $R\cdot$ attacks a molecule such as Cl_4C on chlorine yielding an alkyl chloride and a trichloromethyl radical rather than on carbon yielding a tri-

$$R\cdot + Cl_4C \;\rightarrow\; RCl + Cl_3C\cdot \qquad (1\text{-}24)$$

$$R\cdot + CCl_4 \;\nleftrightarrow\; RCCl_3 + Cl\cdot \qquad (1\text{-}25)$$

chloroalkane and a chlorine atom. The chlorine atoms in carbon tetrachloride are exposed to attack by the alkyl radical, whereas attack at the carbon atom is sterically hindered by the chlorines. Any univalent atom will be more accessible to attack by a free radical than a polyvalent atom since the former will generally be on the periphery of the molecule and, therefore, more approachable by the attacking free radical.

Some interesting exceptions are known. Certain peroxy-oxygen atoms are labile to attack by free radicals as is evident in the behavior of acyl peroxides with ether-derived free radicals yielding acylals (1-26). In this case there is

$$\underset{R'}{ROCH}\cdot + CH_3\overset{O}{\overset{\|}{C}}OO\overset{O}{\overset{\|}{C}}CH_3 \;\rightarrow\; \underset{R}{ROCH}O\overset{O}{\overset{\|}{C}}CH_3 + CH_3\cdot + CO_2 \qquad (1\text{-}26)$$

displacement on a bivalent atom in a chemically reactive peroxide linkage. Dialkyl disulfides undergo attack by free radicals on a sulfur atom (1-27).

$$R\cdot + R'SSR' \;\rightarrow\; RSR' + R'S\cdot \qquad (1\text{-}27)$$

However, no reports of displacement on divalent oxygen in alcohols, ethers, or esters have been reported nor on the sulfur in mercaptans or sulfides. The hydrogen atoms in these compounds are more reactive toward displacement than are divalent oxygen and sulfur.

Most organic compounds have several different types of hydrogens, and displacement reactions on the various hydrogens by free radicals are by no means necessarily equally facile. The character of the radical being displaced and the attacking radical play significant roles in determining which hydrogen is attacked. The different halogens also are not equally prone toward attack by various free radicals. Situations may be encountered where the attacking free radical may have to choose between displacing a free radical from either a hydrogen or a halogen.

2. Addition and Elimination. Free radicals add to the π-electron system of many unsaturated compounds yielding an adduct free radical as illustrated for

$$X \cdot + \begin{array}{c} R \\ \diagdown \\ \diagup \\ R \end{array} C{=}C \begin{array}{c} R \\ \diagup \\ \diagdown \\ R \end{array} \rightleftharpoons X{-}\overset{\displaystyle R}{\underset{\displaystyle R}{C}}{-}\overset{\displaystyle }{C} \begin{array}{c} R \\ \diagup \\ \diagdown \\ R \end{array} \cdot \qquad (1\text{-}28)$$

the reaction of a free radical $X \cdot$ with an alkene. It is often not fully appreciated that this addition reaction is reversible. The fragmentation reaction of an adduct radical is called a β-elimination (or β-scission) and yields the unsaturated compound and the adding free radical. The equilibrium of this reversible reaction is well on the side of the adduct radical in the temperature range at which most free-radical reactions involving addition are performed ($25\text{--}150°\text{C}$) and hence, the reversibility of the addition reaction is of little concern. At higher temperatures, β-elimination in such reactions can, of course, become important. Some free radicals undergo the β-elimination reaction at room temperature or lower and the course of reaction of such radicals is strongly influenced by the reversibility of the addition reaction.

Additions to alkynes yielding vinyl radicals as the adduct occur somewhat

$$X \cdot + R{-}C{\equiv}C{-}R \rightleftharpoons \begin{array}{c} X \\ \diagdown \\ \diagup \\ R \end{array} C{=}C \begin{array}{c} R \\ \diagup \\ \\ \cdot \end{array} \qquad (1\text{-}29)$$

less readily than additions to olefinic linkage. Aromatic rings are also susceptible to addition of reactive-free radicals. Addition to carbonyl functions can yield

$$X \cdot + \bigcirc \rightarrow \bigcirc \overset{H \quad X}{\cdot} \leftrightarrow \underset{\cdot}{\bigcirc} \overset{H \quad X}{} \leftrightarrow \bigcirc \overset{H \quad X}{\cdot} \qquad (1\text{-}30)$$

either an α-alkoxyalkyl radical (1-31) or an alkoxy radical (1-32).

$$XOC\cdot \quad\quad\quad (1\text{-}31)$$

$$X\cdot + O{=}C \quad\quad\quad$$

$$\cdot O{-}C{-}X \quad\quad\quad (1\text{-}32)$$

Both alkoxy radicals and α-alkoxyalkyl radicals fragment readily yielding the carbonyl function and a free radical.

Many factors concerning both the adding free radical and the unsaturated compound influence the addition reaction. Questions that require answers concern not only the reactivities of various free radicals and unsaturated compounds in such reactions but also such matters as the direction of addition to an unsymmetrically substituted unsaturated linkage (i.e., 1-31 and 1-32).

3. Rearrangement. Certain free radicals undergo a rearrangement reaction in which a group Y migrates from one atom to an adjacent atom which is the

$$Y{-}C{-}C\cdot \quad \rightarrow \quad \cdot C{-}C{-}Y \quad\quad\quad (1\text{-}33)$$

radical site. Such 1,2-shifts are not encountered as often in free-radical reactions as they are in reactions involving cationic intermediates. The process is generally (but not necessarily) exothermic, that is, the substituents R stabilize a radical site on the atom to which they are bonded more effectively than do the substituents R′. A most significant aspect of free-radical rearrangements is that only certain groups are capable of migration. Whereas in carbonium ion reactions extensive skeletal rearrangements are often encountered resulting from alkyl group migrations, alkyl groups do not undergo a 1,2-shift in free-radical reactions. Hydride ion migrations are also prevalent in carbonium ion chemistry but hydrogen atoms do not migrate in free radicals even if the process is exothermic. The only groups that have been observed to participate in such 1,2-shifts in free radicals are chlorine atoms (possibly bromine and iodine), aryl groups, vinyl groups, and acyl groups. What we generally find is that formation of a free radical that fulfills both of these requirements, namely that the 1,2-shift would be an exothermic reaction and that a suitable group be bonded to an atom beta to the radical site, is comparatively rare. However, several such systems have been studied and some of the significant aspects of rearrangement reactions are discussed in Chapter 9.

4. Reduction Reactions. Some free radicals have the ability of reducing a substrate by transfer of either a hydrogen atom or an electron to the substrate. Although the former case, illustrated by the reduction of peroxides (1-34) and by reduction of the carbonyl function of a ketone (1-35) by an α-hydroxyalkyl radical, could be regarded as a combination of a β-elimination reaction and either a displacement or addition, it is classified here with the electron-transfer

$$\begin{matrix} R \\ \diagdown \\ \cdot COH + \\ \diagup \\ R \end{matrix} \quad \begin{matrix} R' \\ \diagup \\ O\!-\!O \\ \diagdown \\ R' \end{matrix} \quad \rightarrow \quad \begin{matrix} R \\ \diagdown \\ C\!=\!O + HOR' + R'O \cdot \\ \diagup \\ R \end{matrix} \qquad (1\text{-}34)$$

$$\begin{matrix} R \\ \diagdown \\ \cdot COH + O\!=\!C \\ \diagup \\ R \end{matrix} \begin{matrix} R' \\ \diagup \\ \diagdown \\ R' \end{matrix} \quad \rightleftharpoons \quad \begin{matrix} R \\ \diagdown \\ C\!=\!O + HOC \cdot \\ \diagup \\ R \end{matrix} \begin{matrix} R' \\ \diagup \\ \diagdown \\ R' \end{matrix} \qquad (1\text{-}35)$$

reaction. The transfer of a hydrogen atom from an α-hydroxyalkyl radical to the carbonyl function of a ketone is a reversible process and the position of the equilibrium depends on the stabilities of the two free radicals involved. Hydrogen atom transfer from α-aminoalkyl radicals to peroxide linkages also occurs readily.

An acyl radical is oxidized by persulfate ion to the acylium ion in a reaction that probably involves transfer of an electron to the peroxide linkage of the

$$R\dot{C}\!=\!O + S_2O_8^{2-} \rightarrow R\overset{(+)}{C}\!=\!O + SO_4^- \cdot + SO_4^{2-} \qquad (1\text{-}36)$$

persulfate. Other carbon radicals that may possibly participate in such reactions are α-alkoxyalkyl radicals and α-aminoalkyl radicals having no nitrogen-bonded hydrogen atoms. Some of the reactions of metal ions with peroxides cited earlier as initiation reactions might well be regarded as electron-transfer reactions since they also involve reduction by transfer of a single electron.

B. RADICAL-DESTROYING REACTIONS

1. Radical-Coupling Reactions. Two free radicals can react with each other in such a way that the two unpaired electrons form a covalent sigma bond as illustrated for the reaction of two trichloromethyl radicals coupling to form

$$2Cl_3C \cdot \rightarrow Cl_3CCCl_3 \qquad (1\text{-}37)$$
$$\mathbf{19}$$

hexachloroethane (**19**). Cross-coupling can occur if two different kinds of free radicals are present in the same system.

$$C_6H_5\dot{C}H_2 + Cl_3C \cdot \rightarrow C_6H_5CH_2CCl_3 \qquad (1\text{-}38)$$

Radical-destroying processes contrast with the chain-propagating reactions outlined in the previous section in that coupling is bimolecular in free radicals whereas the radical-propagating reactions are unimolecular in free radicals. The latter aspect is quite significant because, as might be expected, the radical-coupling reaction has a large rate constant and can be a fast reaction. By maintaining free-radical concentrations that are small, coupling reactions (and the bimolecular disproportionation reactions discussed in the next section) do not occur to the complete exclusion of the chain-propagating reactions. Although they have smaller rate constants than the termination reactions, chain-propagating reactions are kinetically first order in free radicals and, in the cases of displacement, addition and reduction, first order in the substrate as well. The substrate may be present in very high concentrations compared to radical concentrations making the propagating reaction rate fast enough to compete with the coupling reaction. This makes it possible to have many free-radical reactions that involve a chain sequence of propagating reactions without having the fast coupling reaction remove the free radicals before any significant amount of reaction in the chain sequence has taken place.

2. Disproportionation Reactions. Some free radicals participate in a bimolecular reaction resulting in the oxidation of one with the concurrent reduction of the other. The reaction of two ethyl radicals yielding ethylene and

$$2CH_3CH_2 \cdot \ \rightarrow \ CH_2{=}CH_2 + CH_3CH_3 \qquad (1\text{-}39)$$

ethane is a good example of this radical-destroying process. Free radicals that are not capable of such reactions are removed from reaction mixtures only by

$$2CH_3 \cdot \ \rightarrow \ CH_3CH_3 \qquad (1\text{-}40)$$

$$2C_6H_5CH_2 \cdot \ \rightarrow \ C_6H_5CH_2CH_2C_6H_5 \qquad (1\text{-}41)$$

coupling reactions as would be the case with methyl and benzyl radicals. It is possible that such radicals could react with other radicals in a disproportionation reaction but, in every case, they would have to be the oxidizing radical, e.g. the case in a reaction between a methyl and an ethyl radical.

$$CH_3 \cdot + CH_3CH_2 \cdot \ \rightarrow \ CH_4 + CH_2{=}CH_2 \qquad (1\text{-}42)$$

It should not be inferred that radicals capable of participation in disproportionation do not couple. The coupling reaction is not only energetically favorable but, when conditions are suitable, predominates over the redox reaction. These conditions cannot be precisely outlined but some generalities can be inferred. The coupling process evolves a considerable amount of energy owing to the formation of a covalent bond. If the molecule formed in the reaction is large enough and possesses enough atoms to allow for the dissipation of this energy by various modes of vibration, the molecule will remain intact. If not, it may fragment again into the free radicals. Another important method of losing the energy produced in the coupling reaction is by allowing the molecule to collide either with another molecule or with the walls of the vessel. The situation for free radicals in solution is such that, except for some of the very simple radicals or some with special chemical features, the coupling reaction generally occurs readily and probably to the almost complete exclusion of the redox reaction. On the other hand, in gas phase reactions of many of the simpler free radicals, the disproportionation reaction is often a very important process.

Certain metal ions capable of a change of a single oxidation state oxidize free radicals. Such reactions, as illustrated for the oxidation of an α-hydroxyalkyl radical by a ferric ion, do occur readily in solution because the formation of the

$$Fe^{+3} + \underset{R}{\overset{R}{\diagdown}} \cdot COH \rightarrow Fe^{+2} + \underset{R}{\overset{R}{\diagdown}} C{=}O + H^+ \tag{1-43}$$

metal-carbon bond is unlikely. Such a redox reaction differs from the others shown before in that an electron transfer rather than a hydrogen atom transfer is involved in the reaction.

C. MECHANISTIC AND KINETIC SIGNIFICANCE OF REACTION TYPES

It is pertinent to point out that each of the reactions discussed should be regarded as a kinetically observable event and not necessarily as a rigorous description of the mechanism of the reaction. Since this book is centered on the free-radical chain concept and is based on the reaction types just described, it may be well to explain in more detail at this point what is meant here by a kinetically observable event.

Although any one of the chain-propagating or chain-terminating reactions described previously may involve the formation of one or more intermediates, the overall process is regarded kinetically as a single reaction. For example, addition of a free radical such as $X \cdot$ to $CH_2{=}CHY$ yielding $XCH_2\dot{C}HY$ may not be a single-step process but may proceed by formation of a π-complex intermediate which is converted in turn to the σ-complex adduct radical as shown in 1-44. Conversely, the β-elimination reaction must also involve the

$$X \cdot + CH_2{=}CHY \rightleftharpoons \underset{\underset{X}{\uparrow}}{CH_2{=}CHY} \rightleftharpoons \underset{\underset{X}{|}}{CH_2\dot{C}HY} \tag{1-44}$$

intermediate π-complex. If formation of the σ-complex is rate determining, the kinetically observable fact will be that the addition (or the β-elimination) reaction has occurred and the existence of the π-complex can only be inferred.

Its existence could be demonstrated, however, if formation of the π-complex were rate determining.

Formation of an intermediate may be inferred for the rearrangement reaction. In these reactions an intermediate adduct radical that can fragment yielding the rearranged radical might be postulated. If this indeed is the mechanism

$$
\begin{array}{ccc}
\underset{\underset{\displaystyle X}{|}}{\overset{\overset{\displaystyle R}{|}}{R{-}C{-}C}}\overset{\displaystyle R'}{\underset{\displaystyle R'}{\diagdown}}\cdot & \rightarrow & R{-}\underset{\overset{\displaystyle R}{|}}{\underset{\displaystyle \dot{X}}{C}}\underset{}{\qquad}\overset{\overset{\displaystyle R'}{|}}{C{-}R'} \rightarrow \quad \underset{R}{\overset{R}{\diagup}}\overset{\overset{\displaystyle R'}{|}}{\underset{\underset{\displaystyle X}{|}}{\cdot C{-}C{-}R'}}
\end{array}
\tag{1-45}
$$

for the 1,2-shifts, the rearrangement reaction could be excluded as a discrete reaction type. The event of a radical undergoing a 1,2-shift is of kinetic value and, although it may be a series of two events, it can kinetically be regarded as a single event. The same kind of reasoning could be applied to displacement reactions, redox processes, and even the coupling reaction.

As interesting as it may be, speculation about the existence of intermediates in such reactions is often of limited value at this point. Although several intermediates may be involved in a reaction of a free radical, it appears that the rate-limiting step generally is dictated by some process that involves the formation of the product radical. Except where speculation concerning such intermediates may be fruitful, the reactions of free radicals just described will be regarded as single kinetically observable events.

References

1. M. Gomberg, *J. Amer. Chem. Soc.*, **22**, 757 (1900); *Ber.*, **33**, 3150 (1900).
2. H. Staudinger, *ibid.*, **53**, 1073 (1920).
3. H. L. J. Bäckström, *Z. physik. Chem.*, **25B**, 99 (1934).
4. F. Haber and R. Willstätter, *Ber.*, **64**, 2844 (1931).
5. D. H. Hey and W. A. Waters, *Chem. Rev.*, **21**, 169 (1937).
6. M. S. Kharasch and F. R. Mayo, *J. Amer. Chem. Soc.*, **55**, 2468 (1933).
7. M. S. Kharasch, H. Engelmann, and F. R. Mayo, *J. Org. Chem.*, **2**, 288 (1937). See also F. R. Mayo, *Vistas in Free-Radical Chemistry*, W. A. Waters, Ed. Pergamon Press, New York, 1959, pp. 139–142.
8. P. J. Flory, *J. Amer. Chem. Soc.*, **59**, 241 (1937).

Chapter 2

Mechanisms of Free-Radical Reactions

In the previous chapter it was shown how free radicals were produced and how they reacted in both radical-propagating reactions and radical-destroying reactions. Any reaction proceeding by a free-radical mechanism must include some radical-producing reaction, which is generally referred to as an initiation reaction, and these initiator-derived radicals react with the reactants producing reactant-derived radicals. The reaction products are formed either in a chain sequence of propagating reactions of the reactants with the reactant-derived radicals or in a termination reaction.

This chapter consists of examples that illustrate the general principles of free-radical reactions. The reactions will be divided into two groups, the first being those in which the products are formed in a chain sequence of propagating reactions followed by those in which the products are formed in a termination reaction. Little will be said of these reactions at this point other than to show the general mechanistic characteristics in terms of the reactions of the free radicals. Many of the reactions given will be presented in greater detail in subsequent sections and appropriately referenced.

I. Product Formation in a Chain Sequence

A. CHAIN SEQUENCE INVOLVING DISPLACEMENT REACTIONS

Free-radical halogenations provide a good example of a reaction in which only displacement reactions are involved as chain-propagating reactions in the chain sequence. Illustrative is the light-induced reaction of bromine with toluene (1) yielding benzyl bromine (2) and hydrogen bromide. The initiation reaction is the photolysis of bromine yielding two bromine atoms each starting

$$C_6H_5CH_3 + Br_2 \xrightarrow{h\nu} C_6H_5CH_2Br + HBr \qquad (2\text{-}1)$$
$$\mathbf{1} \qquad\qquad\qquad \mathbf{2}$$

via   *Mechanism*

$$Br_2 \xrightarrow{h\nu} 2Br\cdot \qquad (2\text{-}2)$$

$$Br\cdot + C_6H_5CH_3 \longrightarrow HBr + C_6H_5CH_2\cdot \qquad (2\text{-}3)$$

$$C_6H_5CH_2\cdot + Br_2 \longrightarrow C_6H_5CH_2Br + Br\cdot \qquad (2\text{-}4)$$

$$C_6H_5CH_2\cdot + Br\cdot \longrightarrow C_6H_5CH_2Br \qquad (2\text{-}5)$$

$$2C_6H_5CH_2\cdot \longrightarrow C_6H_5CH_2CH_2C_6H_5 \qquad (2\text{-}6)$$

$$2Br\cdot \longrightarrow Br_2 \qquad (2\text{-}7)$$

a chain sequence. The conversion of the reactants takes place in the chain sequence of displacement reactions 2-3 and 2-4. The algebraic sum of the reactions comprising the chain sequence is the stoichiometric equation for the reaction provided the chain sequence is long enough. Any chain sequence having two different chain-carrying radicals has three possible termination processes. Two of these (2-6 and 2-7) are simple dimerizations of chain-carrying radicals whereas 2-5 is a cross-coupling reaction yielding a single molecule of one of the reaction products. The amounts of termination product formed in a reaction such as this will be small if the chain sequence is allowed to repeat itself many times before termination occurs.

Free-radical halogenations that involve agents other than the molecular halogens proceed in a very similar manner. For example, carbon tetrachloride reacts with alkanes in a benzoyl peroxide (3) induced reaction yielding an alkyl chloride and chloroform. The chain sequence 2-11 and 2-12 accounts for the

$$RH + Cl_4C \xrightarrow{Bz_2O_2} RCl + HCCl_3 \qquad (2\text{-}8)$$
Mechanism

$$Bz_2O_2 \rightarrow 2BzO\cdot \qquad (2\text{-}9)$$
$$\mathbf{3}$$

$$BzO\cdot + RH \rightarrow BzOH + R\cdot \qquad (2\text{-}10)$$

$$R \cdot + Cl_4C \rightarrow RCl + Cl_3C \cdot \tag{2-11}$$

$$Cl_3C \cdot + RH \rightarrow HCCl_3 + R \cdot \tag{2-12}$$

$$2Cl_3C \cdot \rightarrow C_2Cl_6 \tag{2-13}$$

products observed in this reaction. The initiation reaction 2-9 does not produce a free radical that is involved in the chain sequence, but the propagating reaction 2-10 involving the initiator-derived benzoyloxy radical and the alkane does produce the chain-carrying free radical $R \cdot$ and benzoic acid. If the chain sequence 2-11 and 2-12 is long, the amounts of benzoic acid and the proposed termination product hexachloroethane will be small relative to the products produced in the chain sequence. The chain sequence is long if a low concentration of chain-carrying radicals is maintained. This can best be accomplished by having only a small amount of benzoyl peroxide present and allowing it to decompose at a comparatively slow rate. Experience shows that best results are obtained if the initiator is present in small amounts. A reaction temperature should be maintained such that the decomposition of the initiator proceeds at a rate fast enough to produce free radicals to start chains but not at such a rate that the concentrations of free radicals becomes high enough to have the termination reaction compete with the chain-propagating reactions. An "average kinetic chain length" of a reaction such as this may be defined as the ratio between the amount of either product formed in the chain sequence and the amount of benzoic acid produced, since each molecule of benzoic acid produced indicates the initiation of a chain sequence.

The bromination of toluene with bromotrichloromethane (**4**) in a light-induced reaction is similar. The products of this reaction are formed in the chain sequence 2-17 and 2-18. Hydrogen bromide is also formed (2-16) as a result

$$C_6H_5CH_3 + BrCCl_3 \xrightarrow{h\nu} C_6H_5CH_2Br + HCCl_3 \tag{2-14}$$
$$\mathbf{4}$$

Mechanism

$$BrCCl_3 \xrightarrow{h\nu} Br \cdot + Cl_3C \cdot \tag{2-15}$$

$$Br \cdot + C_6H_5CH_3 \longrightarrow HBr + C_6H_5CH_2 \cdot \tag{2-16}$$

$$C_6H_5CH_2 \cdot + BrCCl_3 \longrightarrow C_6H_5CH_2Br + Cl_3C \cdot \tag{2-17}$$

$$Cl_3C \cdot + C_6H_5CH_3 \longrightarrow HCCl_3 + C_6H_5CH_2 \cdot \tag{2-18}$$

$$2Cl_3C \cdot \longrightarrow Cl_3CCCl_3 \tag{2-19}$$

of reaction of a radical other than a chain-carrying radical being formed in the initiation reaction 2-15. If the initiation reaction is effective in starting long kinetic chains, very little hydrogen bromide will be formed in comparison to the amounts of chloroform and benzyl bromide.

Chlorination of alkanes with t-butyl hypochlorite (5) proceeds by a similar chain sequence of radical displacement reactions. The stoichiometry of the

$$(CH_3)_3COCl + RH \xrightarrow{h\nu} (CH_3)_3COH + RCl \qquad (2\text{-}20)$$
$$\mathbf{5}$$

Mechanism

$$(CH_3)_3COCl \xrightarrow{h\nu} (CH_3)_3CO\cdot + Cl\cdot \qquad (2\text{-}21)$$

$$Cl\cdot + RH \longrightarrow HCl + R\cdot \qquad (2\text{-}22)$$

$$R\cdot + (CH_3)_3COCl \longrightarrow RCl + (CH_3)_3CO\cdot \qquad (2\text{-}23)$$

$$(CH_3)_3CO\cdot + RH \longrightarrow (CH_3)_3COH + R\cdot \qquad (2\text{-}24)$$

$$(CH_3)_3CO\cdot + R\cdot \longrightarrow (CH_3)_3COR(?) \qquad (2\text{-}25)$$

reaction is determined by the chain sequence 2-23 and 2-24. The hydrogen chloride formed in reaction 2-22 will be only a minor by-product as will the termination product produced in 2-25 if the kinetic chain length is long.

These examples should suffice to illustrate that the chain sequence is the significant part of such reactions. Although the initiation and termination processes are essential parts of the reaction, the main course of the reaction in which reactants are converted to products occurs in the chain sequence. For this reason, most of the following reactions will be treated only from the standpoint of the chain sequence. It must be kept in mind that some suitable means of initiating the chain sequence by the formation of one of the chain-carrying radicals is necessary in each case and that the chain sequence is terminated by either coupling or disproportionation of chain-carrying radicals.

The following examples involve a subsequent reaction of one of the reaction products of the chain sequence and therefore it is not obvious from the stoichiometry of the reaction that such a free-radical chain sequence was actually operative. The chlorination of cyclohexane with trichloromethanesulfonyl chloride (6) yields, in addition to the chlorocyclohexane, sulfur dioxide and

$$Cl_3CSO_2Cl + C_6H_{12} \rightarrow HCCl_3 + SO_2 + C_6H_{11}Cl \qquad (2\text{-}26)$$
$$\mathbf{6}$$

chloroform. There is ample evidence that the chain sequence involved in this reaction is one in which the trichloromethanesulfonyl radical participates in a displacement reaction (2-28) yielding a cyclohexyl radical and trichloromethanesulfinic acid (7) which decomposes yielding the observed reaction products.

$$Cl_3CC_6H_{11}\cdot + SO_2Cl \rightarrow C_6H_{11}Cl + Cl_3CSO_2\cdot \qquad (2\text{-}27)$$

$$Cl_3CSO_2 \cdot + C_6H_{12} \rightarrow Cl_3CSO_2H + C_6H_{11} \cdot \qquad (2\text{-}28)$$
$$7$$

$$Cl_3CSO_2H \rightarrow HCCl_3 + SO_2 \qquad (2\text{-}29)$$
$$7$$

Allylic brominations of alkenes with N-bromosuccinimide (8) yield succinimide (9) as a reaction product. Investigation (see Chapter 5) has shown that succinimide is not produced in the chain sequence but rather by reaction of hydrogen bromide, one of the products of the chain sequence, with N-bromo-

$$(2\text{-}30)$$

succinimide This reaction also produces molecular bromine, one of the necessary reactants in the chain sequence 2-31 and 2-32.

$$(2\text{-}31)$$

$$(2\text{-}32)$$

$$(2\text{-}33)$$

Formation of trialkyl phosphate (10) by reaction of diethyl phosphite (11) with bromotrichloromethane, an alcohol and a tertiary amine, may involve a free-radical chain halogenation. The stoichiometry of reaction 2-34 shows that the polyhalomethane is reduced to chloroform and that bromine appears as

$$HP(OC_2H_5)_2 + BrCCl_3 + ROH + R_3'N \rightarrow$$
$$11$$
$$(2\text{-}34)$$
$$ROP(OC_2H_5)_2 + HCCl_3 + R_3'N \cdot HBr$$
$$10$$

an amine hydrobromide. A mechanism for this reaction that accounts for these products consists, in part, of a free-radical halogenation of the dialkyl phosphite by the chain sequence 2-35 and 2-36. One of the observed reaction products,

$$Cl_3C \cdot + \overset{\displaystyle O}{\underset{\displaystyle |}{HP(OC_2H_5)_2}} \rightarrow HCCl_3 + \overset{\displaystyle O}{\underset{\displaystyle |}{\cdot P(OC_2H_5)_2}} \qquad (2\text{-}35)$$

$$\overset{\displaystyle O}{\underset{\displaystyle |}{\cdot P(OC_2H_5)_2}} + BrCCl_3 \rightarrow \overset{\displaystyle O}{\underset{\displaystyle |}{BrP(OC_2H_5)_2}} + Cl_3C \cdot \qquad (2\text{-}36)$$
$$\mathbf{12}$$

chloroform, is produced in this chain sequence. Diethyl phosphorobromidate (**12**), the other product of the chain sequence, reacts readily with the alcohol in an ionic reaction yielding the trialkyl phosphate and hydrogen bromide. The latter combines with the amine forming the amine salt.

$$\overset{\displaystyle O}{\underset{\displaystyle |}{BrP(OC_2H_5)_2}} + ROH \rightarrow \overset{\displaystyle O}{\underset{\displaystyle |}{ROP(OC_2H_5)_2}} + HBr \qquad (2\text{-}37)$$

$$HBr + R_3'N \rightarrow R_3'N \cdot HBr \qquad (2\text{-}38)$$

B. CHAIN SEQUENCE INVOLVING ADDITION AND DISPLACEMENT

Addition of a reagent across unsaturated linkages can be accomplished in a free-radical chain sequence consisting of addition of a free radical to the unsaturated linkage yielding an adduct-free radical which reacts with the adding reagent in a displacement reaction yielding the addition product and the free radical that adds to the unsaturated linkage. Illustrative of such a reaction is the addition of hydrogen bromide to a terminal alkene which yields

$$HBr + CH_2{=}CHR \xrightarrow[\text{or } O_2]{ROOH} BrCH_2CH_2R \qquad (2\text{-}39)$$

a product different from that obtained in "Markownikoff addition" of the hydrogen halide to an unsymmetrical alkene. The anti-Markownikoff addition of hydrogen bromide to an alkene is accomplished by the free-radical chain sequence which, in the simplest terms, can be represented by 2-41 and 2-42.

$$HBr + ROOH \text{ (or } O_2) \rightarrow Br \cdot + RO \cdot + H_2O \text{ (or } \cdot O_2H) \qquad (2\text{-}40)$$

$$Br \cdot + CH_2{=}CHR \rightarrow BrCH_2\dot{C}HR \qquad (2\text{-}41)$$

$$BrCH_2\dot{C}HR + HBr \rightarrow BrCH_2CH_2R + Br \cdot \qquad (2\text{-}42)$$

The direction of the addition appears to be determined by the mode of addition of the bromine atom to the alkene rather than by that of the proton as in the case of Markownikoff addition. (However, see Chapter 6 for further discussion of HBr addition.) In the former case, a comparatively stable secondary alkyl radical is produced when the addition involves bonding to the terminal carbon whereas bonding with the nonterminal carbon yields a less stable primary alkyl radical. In the Markownikoff addition, the adding proton prefers to bond to the terminal carbon yielding a secondary carbonium ion rather than with the nonterminal carbon yielding a primary carbonium ion.

$$H^+ + CH_2{=}CHR \rightarrow CH_3\overset{+}{C}HR \qquad (2\text{-}43)$$

$$CH_3\overset{+}{C}HR + Br^- \rightarrow CH_3CHBrR \qquad (2\text{-}44)$$

Many different classes of compounds, both organic and inorganic, add to unsaturated linkages by a free-radical mechanism. At this time a few examples that are typical and illustrative of the chain sequence involved in these additions will be examined. A significant feature of these additions is that the addition product is produced in a displacement reaction of the adduct radical on the adding reagent. Since displacements occur readily only on hydrogen and on halogens, it might seem that this limitation may affect the generality of the Kharasch addition reaction. However, when we realize how prevalent hydrogen containing compounds are, this limitation does not damage the potential versatility of the reaction.

Another factor to consider is that the adduct radical may add to another molecule of the unsaturated species. Consider, for example, the chain sequence

$$Cl\cdot + CH_2{=}CH_2 \rightarrow ClCH_2CH_2\cdot \qquad (2\text{-}45)$$

$$ClCH_2CH_2\cdot + HCl \rightarrow Cl \mid CH_2CH_2 {+} H \qquad (2\text{-}46)$$

$$ClCH_2CH_2\cdot + CH_2{=}CH_2 \rightarrow ClCH_2CH_2CH_2CH_2\cdot \qquad (2\text{-}47)$$

$$ClCH_2CH_2CH_2CH_2\cdot + HCl \rightarrow Cl \mid CH_2CH_2 {+} CH_2CH_2 {+} H \quad (2\text{-}48)$$

13

$$ClCH_2CH_2CH_2CH_2\cdot + CH_2{=}CH_2 \rightarrow Cl(CH_2CH_2)_2CH_2CH_2\cdot \qquad (2\text{-}49)$$

$$Cl(CH_2CH_2)_2CH_2CH_2\cdot + HCl \rightarrow Cl \mid (CH_2CH_2)_2 \mid CH_2CH_2 {+} H + Cl\cdot \qquad (2\text{-}50)$$

for the addition of hydrogen chloride to ethylene. Reactions 2-45 and 2-46 comprise a chain sequence leading to the formation of ethyl chloride, a chain sequence the same as that shown for the addition of hydrogen bromide to an alkene. However, hydrogen chloride is not as reactive as hydrogen bromide in displacement reactions (see bond-dissociation energies in Chapter 4) and, consequently, the adduct radical—rather than react with hydrogen chloride in a displacement reaction—can add to another molecule of alkene as shown in 2-47 yielding an adduct radical with two ethylene units. The 2:1 adduct radical is presented with the same choice and, rather than react with hydrogen chloride yielding a 2:1 addition product (**13**), it may react with another molecule of ethylene yielding a 3:1 adduct radical. The result is the formation of a series of alkyl chlorides (ethyl, *n*-butyl, *n*-hexyl, etc.). A reaction such as this is called a telomerization reaction and the products formed are called telomers.

The factors that determine the amounts of telomerization in these addition reactions will be covered in detail in Chapter 6. In the remainder of this section we will be concerned with other examples of free-radical addition reactions with the purpose of illustrating the character of the chain sequence of free-radical propagating reactions yielding the addition product.

The additions of the polyhalomethanes to alkenes are illustrative of reactions that occur readily by a free-radical mechanism but not by ionic mechanisms. The synthesis of some of the addition products produced in the free-radical reactions by other means presents a challenge. The addition of carbon tetrachloride to a terminal alkene such as 1-octene (**14**) yields a tetrachlorononane (**15**) by the chain sequence 2-52 and 2-53. The displacement reaction in this case occurs on a chlorine atom, the only atom available on the adding reagent.

$$Cl_4C + CH_2=CHC_6H_{13}\text{-}n \xrightarrow{\text{Bz}_2O_2} Cl_3CCH_2CHClC_6H_{13}\text{-}n \qquad (2\text{-}51)$$

$$\qquad\qquad\qquad \textbf{14} \qquad\qquad\qquad\qquad\qquad \textbf{15}$$

via

$$Cl_3C\cdot + CH_2=CHC_6H_{13}\text{-}n \longrightarrow Cl_3CCH_2\dot{C}HC_6H_{13}\text{-}n \qquad (2\text{-}52)$$

$$Cl_3CCH_2\dot{C}HC_6H_{13}\text{-}n + Cl_4C \longrightarrow Cl_3CCH_2CHClC_6H_{13}\text{-}n + Cl_3C\cdot \quad (2\text{-}53)$$

The additions of bromotrichloromethane and chloroform to the same alkene, however, show that even though chlorine may be the predominant atom of the adding reagent, other factors apparently play important roles in determining the course of the displacement reaction. It should be noted that the radical

$$BrCCl_3 + CH_2=CHR \rightarrow Cl_3CCH_2CHBrR \qquad (2\text{-}54)$$

via

$$Cl_3C\cdot + CH_2=CHR \rightarrow Cl_3CCH_2\dot{C}HR \qquad (2\text{-}55)$$

$$Cl_3CCH_2\dot{C}HR + BrCCl_3 \rightarrow Cl_3CCH_2CHBrR + Cl_3C\cdot \qquad (2\text{-}56)$$

$$HCCl_3 + CH_2=CHR \rightarrow Cl_3CCH_2CH_2R \qquad (2\text{-}57)$$

via

$$Cl_3C \cdot + CH_2 =\!\!= CHR \ \rightarrow \ Cl_3CCH_2\dot{C}HR \tag{2-58}$$

$$Cl_3CCH_2\dot{C}HR + HCCl_3 \ \rightarrow \ Cl_3CCH_2CH_2R + Cl_3C \cdot \tag{2-59}$$

addition reactions (2-52, 2-55, and 2-58) are the same and produce the same adduct radical but that the chemical reactivities of the three polyhalomethanes toward reaction with this adduct radical differ. As evidenced by the amounts of telomeric products produced in these reactions, the reactivities of the poly-halomethanes toward reaction with the adduct radical are $BrCCl_3 > Cl_4C > HCCl_3$.

Aldehydes add to alkenes in a free-radical reaction yielding ketones as 1:1 addition products. Although several different kinds of hydrogens are available

$$CH_3CH_2CH_2CHO + CH_2 =\!\!= CHR \ \rightarrow \ CH_3CH_2CH_2\overset{\overset{\displaystyle O}{\displaystyle \|}}{C}CH_2CH_2R \tag{2-60}$$
$$\mathbf{16}$$

via

$$n\text{-}C_3H_7\dot{C}O + CH_2 =\!\!= CHR \ \rightarrow \ n\text{-}C_3H_7\overset{\overset{\displaystyle O}{\displaystyle \|}}{C}CH_2\dot{C}HR \tag{2-61}$$

$$n\text{-}C_3H_7\overset{\overset{\displaystyle O}{\displaystyle \|}}{C}CH_2\dot{C}HR + n\text{-}C_3H_7CHO \ \rightarrow \ n\text{-}C_3H_7\overset{\overset{\displaystyle O}{\displaystyle \|}}{C}CH_2CH_2R + n\text{-}C_3H_7\dot{C}O$$
$$\tag{2-62}$$

in n-butyraldehyde (**16**), the displacement occurs only on the aldehydic hydrogen giving rise to an acyl radical.

Primary and secondary alcohols add to alkenes yielding secondary and tertiary alcohols, respectively. The displacement reaction occurs on a hydrogen

$$R_2CHOH + CH_2 =\!\!= CHR \ \xrightarrow{\ \text{Per.}\ } \ \underset{\underset{\displaystyle OH}{\displaystyle |}}{R_2C}CH_2CH_2R \tag{2-63}$$

via

$$R_2\dot{C}OH + CH_2 =\!\!= CHR \ \longrightarrow \ \underset{\underset{\displaystyle OH}{\displaystyle |}}{R_2C}CH_2\dot{C}HR \tag{2-64}$$

$$\underset{\underset{\displaystyle OH}{\displaystyle |}}{R_2C}CH_2\dot{C}HR + R_2CHOH \ \longrightarrow \ \underset{\underset{\displaystyle OH}{\displaystyle |}}{R_2C}CH_2CH_2R + R_2\dot{C}OH \tag{2-65}$$

R = alkyl or hydrogen

bonded to the carbon bearing the hydroxy group yielding an α-hydroxyalkyl radical. The rate of the displacement reaction is slow compared to the rate of addition of the adduct radical to another molecule of alkene as evidenced by the formation of considerable amounts of the telomeric products in these reactions. Although methanol also adds to alkenes by a free-radical mechanism yielding primary alcohols as the 1:1 addition products, the amount of telomer formation is extensive.

Certain esters add to alkenes in a free-radical reaction that involves a displacement reaction on an α-hydrogen relative to the carbonyl function. Although this displacement reaction is slow with simple esters and telomeric products result, the reaction does proceed readily with active methylene compounds such as malonic ester, acetoacetic ester and the cyanoacetates, compounds in which the hydrogens are activated by two activating groups. For example, malonic ester (17) can be alkylated with 1-octene (14) yielding diethyl n-octylmalonic ester (18), in high yields in a di-t-butyl peroxide-induced reaction.

$$CH_2(CO_2C_2H_5)_2 + n\text{-}C_6H_{13}CH{=}CH_2 \rightarrow n\text{-}C_8H_{17}CH(CO_2C_2H_5)_2 \quad (2\text{-}66)$$
$$\qquad \textbf{17} \qquad\qquad\qquad \textbf{14} \qquad\qquad\qquad\qquad \textbf{18}$$

via

$$\cdot CH(CO_2C_2H_5)_2 + n\text{-}C_6H_{13}CH{=}CH_2 \rightarrow n\text{-}C_6H_{13}\overset{\cdot}{C}HCH_2CH(CO_2C_2H_5)_2$$
$$\qquad\qquad\qquad\qquad \textbf{14}$$
$$\qquad\qquad\qquad\qquad\qquad\qquad\qquad\qquad (2\text{-}67)$$

$$n\text{-}C_6H_{13}\overset{\cdot}{C}HCH_2CH(CO_2C_2H_5)_2 + CH_2(CO_2C_2H_5)_2 \rightarrow$$
$$n\text{-}C_6H_{13}CH_2CH_2CH(CO_2C_2H_5)_2 + \cdot CH(CO_2C_2H_5)_2 \quad (2\text{-}68)$$

The preceding examples illustrate Kharasch additions that involve the formation of a carbon-carbon bond. Kharasch additions that, like the hydrogen halide additions, involve the formation of a carbon-hetero-atom bond are also known. Mercaptans, for example, add readily to alkenes in a free-radical reaction yielding sulfides as the simple addition products as illustrated by addition of n-butyl mercaptan (19) to 1-octene. In this reaction the displacement

$$n\text{-}C_4H_9SH + CH_2{=}CHR \rightarrow n\text{-}C_4H_9SCH_2CH_2R \qquad\qquad (2\text{-}69)$$
$$\quad \textbf{19}$$

via

$$n\text{-}C_4H_9S\cdot + CH_2{=}CHR \rightarrow n\text{-}C_4H_9SCH_2\overset{\cdot}{C}HR \qquad\qquad (2\text{-}70)$$
$$\quad \textbf{21} \qquad\qquad\qquad\qquad\qquad \textbf{20}$$

$$n\text{-}C_4H_9SCH_2\overset{\cdot}{C}HR + n\text{-}C_4H_9SH \rightarrow n\text{-}C_4H_9SCH_2CH_2R + n\text{-}C_4H_9S\cdot \quad (2\text{-}71)$$
$$\textbf{20}$$

reaction by the adduct radical 20 occurs on the sulfur-bonded hydrogen producing a thiyl radical 21 which is the adding radical. The displacement reactions on a mercaptan contrast markedly with those of primary and secondary alcohols in which a carbon-bonded hydrogen undergoes displacement. There is no reason to believe that the carbon-bonded hydrogen α to the sulfhydryl group of a mercaptan is not as reactive as the α-hydrogen of an alcohol. The sulfur-bonded hydrogen apparently is far more reactive toward displacement by the adduct radical as evidenced both by the nature of the product and by the relatively small amount of telomer formation in these reactions.

The bisulfite ion (22) adds to alkenes yielding alkane sulfonates (23). The addition is an anti-Markownikoff addition induced by oxygen or peroxides suggesting the same sort of chain sequence proposed for the preceding Kharasch

$$HSO_3^- + RCH{=}CH_2 \;\rightarrow\; RCH_2CH_2SO_3^- \qquad (2\text{-}72)$$
$$\quad\ 22 \qquad\qquad\qquad\qquad\ \ 23$$

via

$$SO_3^- \cdot + RCH{=}CH_2 \;\rightarrow\; R\dot{C}HCH_2SO_3^- \qquad (2\text{-}73)$$

$$R\dot{C}HCH_2SO_3^- + HSO_3^- \;\rightarrow\; RCH_2CH_2SO_3^- + SO_3^- \cdot \qquad (2\text{-}74)$$

additions. When ethylene is used as the alkene, telomers result—an observation that supports the proposed free-radical chain sequence for this reaction.

Phosphorus compounds having at least one hydrogen bonded to phosphorus participate in free-radical addition reactions. Diethyl phosphite (24) adds to alkenes in peroxide-induced reactions yielding a diethyl alkanephosphonate 25 as the reaction product. As in the reactions of mercaptans, the hydrogen bonded to the heteroatom is the most reactive toward radical displacement.

$$\overset{\displaystyle O}{\underset{\displaystyle }{\|}}\qquad\qquad\qquad\qquad\overset{\displaystyle O}{\underset{\displaystyle }{\|}}$$
$$HP(OC_2H_5)_2 + RCH{=}CH_2 \;\rightarrow\; RCH_2CH_2P(OC_2H_5)_2 \qquad (2\text{-}75)$$
$$\qquad 24 \qquad\qquad\qquad\qquad\qquad 25$$

via

$$\overset{\displaystyle O}{\underset{\displaystyle }{\|}}\qquad\qquad\qquad\qquad\overset{\displaystyle O}{\underset{\displaystyle }{\|}}$$
$$\cdot P(OC_2H_5)_2 + RCH{=}CH_2 \;\rightarrow\; R\dot{C}HCH_2P(OC_2H_5)_2 \qquad (2\text{-}76)$$

$$\overset{\displaystyle O}{\underset{\displaystyle }{\|}}\qquad\quad\overset{\displaystyle O}{\underset{\displaystyle }{\|}}$$
$$R\dot{C}HCH_2P(OC_2H_5)_2 + HP(OC_2H_5)_2 \;\rightarrow\;$$

$$\qquad\qquad\qquad\qquad\overset{\displaystyle O}{\underset{\displaystyle }{\|}}\qquad\qquad\overset{\displaystyle O}{\underset{\displaystyle }{\|}}$$
$$RCH_2CH_2P(OC_2H_5)_2 + \cdot P(OC_2H_5)_2 \quad (2\text{-}77)$$

Other addition reactions will be discussed in more detail in Chapters 6 and 7. The preceding examples suffice to illustrate both the versatility of the addition reaction and that a more complete appreciation of the reactivities of free radicals and substrates is required.

C. ELIMINATION AND DISPLACEMENT

The products of some free-radical reactions can be explained in terms of a chain sequence consisting of a displacement reaction and a β-elimination reaction. Reactions that will be described in this section are those in which the fragmenting radical is not readily formed in an addition reaction but must be generated in some other manner, namely a suitable displacement reaction.

The β-elimination reaction is a product-forming chain propagating reaction in that an unsaturated compound results as a product in the reaction. The best examples of chain reactions composed of only an elimination reaction and a displacement reaction are those involving alkoxy (26) and α-alkoxyalkyl (27) free radicals. Both classes of radicals undergo β-elimination yielding a compound that has a carbonyl function and a chain-carrying alkyl free radical.

$$
\begin{array}{c}
R_3CO \cdot \\
\textbf{26}
\end{array}
\searrow
\quad R_2C{=}O + R \cdot
$$

$$
\begin{array}{c}
R_2\overset{\cdot}{C}OR \\
\textbf{27}
\end{array}
\nearrow
\tag{2-78}
$$

The alkoxy free radical can be formed by addition of an alkyl radical to a carbonyl in such a manner that the alkyl radical bonds with the carbon of this function. Such reactions are rare but do occur (see Chapter 8). An α-alkoxyalkyl radical results if the alkyl radical adds to the carbonyl with bonding at the oxygen. Again, such reactions rarely occur, largely because of the low reactivity of the carbonyl function toward addition. Both radicals do fragment, however, when they are generated by some other means than additions to the carbonyl function.

The light-induced reaction of t-alkyl hypochlorites yields ketones and alkyl chlorides. The reactions likely proceed by a chain process in which an alkoxy radical undergoes β-elimination yielding the ketone and an alkyl radical. A subsequent displacement reaction of the alkyl radical with the parent hypochlorite yields the alkyl chloride and the chain-carrying alkoxy radical. The decomposition of t-amyl hypochlorite (28) yielding acetone and ethyl chloride is illustrative of this reaction.

$$
\underset{\substack{|\\ CH_3}}{\overset{\substack{CH_3\\ |}}{C_2H_5COCl}} \xrightarrow{\ h\nu\ } C_2H_5Cl + CH_3\overset{\overset{\textstyle O}{\textstyle \|}}{C}CH_3
\tag{2-79}
$$

via

$$\underset{\overset{|}{CH_3}}{\overset{\overset{CH_3}{|}}{C_2H_5CO\cdot}} \longrightarrow C_2H_5\cdot + CH_3\overset{O}{\overset{||}{C}}CH_3 \qquad (2\text{-}80)$$

$$C_2H_5\cdot + \underset{\overset{|}{CH_3}}{\overset{\overset{CH_3}{|}}{C_2H_5COCl}} \longrightarrow C_2H_5Cl + \underset{\overset{|}{CH_3}}{\overset{\overset{CH_3}{|}}{C_2H_5CO\cdot}} \qquad (2\text{-}81)$$

The peroxide-induced reactions of acetals involve a chain sequence in which the displacement occurs on a carbon-bonded hydrogen yielding an α,α-dialkoxyalkyl radical which undergoes β-elimination of an alkyl free radical with concurrent formation of an ester. Diethyl butyral (29) can be converted to, among some other products, ethyl butyrate 30 and ethane. In a similar chain

$$\underset{\overset{|}{OC_2H_5}}{\overset{\overset{OC_2H_5}{|}}{n\text{-}C_3H_7CH}} \xrightarrow[125°]{Per.} n\text{-}C_3H_7\overset{O}{\overset{||}{C}}-OC_2H_5 + C_2H_6 \qquad (2\text{-}82)$$
$$\mathbf{29} \qquad\qquad\qquad \mathbf{30}$$

via

$$\underset{\overset{|}{OC_2H_5}}{\overset{\overset{OC_2H_5}{|}}{n\text{-}C_3H_7C\cdot}} \longrightarrow n\text{-}C_3H_7\overset{O}{\overset{||}{C}}OC_2H_5 + C_2H_5\cdot \qquad (2\text{-}83)$$

$$C_2H_5\cdot + \underset{\overset{|}{OC_2H_5}}{\overset{\overset{OC_2H_5}{|}}{n\text{-}C_3H_7CH}} \longrightarrow C_2H_6 + \underset{\overset{|}{OC_2H_5}}{\overset{\overset{OC_2H_5}{|}}{n\text{-}C_3H_7C\cdot}} \qquad (2\text{-}84)$$

sequence, cyclic acetals rearrange to linear esters. 2-Phenyl dioxalane (31) is converted to ethyl benzoate (32) in a peroxide-induced reaction involving β-elimination of the alkyl moiety of the intermediate radical 33.

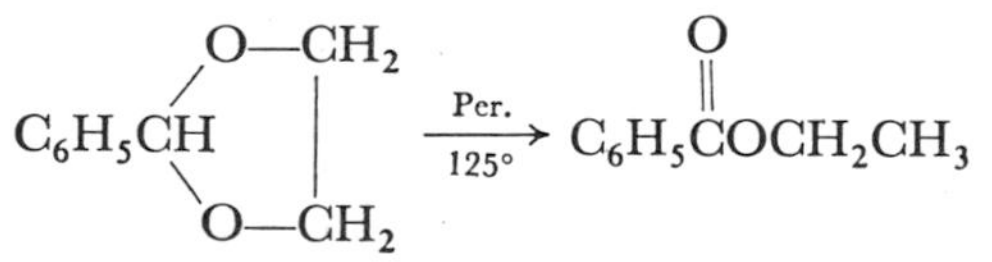

$$C_6H_5CH\overset{O-CH_2}{\underset{O-CH_2}{\big<}} \xrightarrow[125°]{Per.} C_6H_5\overset{O}{\overset{||}{C}}OCH_2CH_3 \qquad (2\text{-}85)$$
$$\mathbf{31} \qquad\qquad\qquad \mathbf{32}$$

via

$$
\underset{\textbf{33}}{C_6H_5\overset{\displaystyle O-CH_2}{\overset{\diagup}{\underset{\diagdown}{C}}}\;\cdot\;\underset{O-CH_2}{\big|}} \;\rightarrow\; C_6H_5\overset{\displaystyle O\,\cdot\,CH_2}{\overset{\diagup}{\underset{\diagdown}{C}}}\;\underset{O-CH_2}{\big|} \tag{2-86a}
$$

$$
C_6H_5\overset{\displaystyle O\,\cdot\,CH_2}{\overset{\diagup}{\underset{\diagdown}{C}}}\;\underset{O-CH_2}{\big|} \;+\; C_6H_5\overset{\displaystyle O-CH_2}{\overset{\diagup}{\underset{\diagdown}{\dot C H}}}\;\underset{O-CH_2}{\big|} \;\rightarrow \tag{2-86b}
$$

$$
C_6H_5\overset{O}{\overset{\|}{C}}-OCH_2CH_3 + C_6H_5\overset{\displaystyle O-CH_2}{\overset{\diagup}{\underset{\diagdown}{\dot C}}}\;\underset{O-CH_2}{\big|}
$$

In some cases displacement may produce a free radical that is the adduct radical of a reversible addition reaction. Such an adduct radical may be expected to fragment readily yielding an unsaturated product. The conversion of β-hydroxysulfides (**34**) to ketones and mercaptans is illustrative of such a reaction. Although thiyl radicals add readily to double bonds, the reaction is

$$
\underset{\underset{\textbf{34}}{\overset{\displaystyle |}{OH}}}{RSCH_2\overset{\displaystyle |}{C}HR'} \xrightarrow{\text{Per.}} RSH + CH_3\overset{O}{\overset{\|}{C}}R' \tag{2-87}
$$

via

$$
RS\cdot + RSCH_2\underset{\overset{\displaystyle |}{OH}}{CHR'} \longrightarrow RSH + RSCH_2\underset{\overset{\displaystyle |}{OH}}{\dot C R'} \tag{2-88}
$$

$$
RSCH_2\underset{\overset{\displaystyle |}{OH}}{\dot C R'} \longrightarrow CH_2{=}\underset{\overset{\displaystyle |}{OH}}{CR'} + RS\cdot \tag{2-89}
$$

$$
 \longrightarrow CH_3\overset{O}{\overset{\|}{C}}R'
$$

reversible, and a free radical that would have the same structure as the adduct radical produced in addition of a thiyl radical to an unsaturated linkage should be expected to undergo β-elimination. The reactions of β-hydroxysulfides are interesting in that the unsaturated species formed in the β-elimination, namely an enol, is converted to a product that does not react with the eliminated thiyl radical. Furthermore, displacement occurs quite specifically on the α-hydrogen of the parent β-hydroxysulfide yielding only the radical capable of undergoing the desired elimination reaction.

D. ADDITION AND ELIMINATION

Certain rearrangement reactions can be explained in terms of a free-radical chain sequence in which an addition reaction produces a free radical that undergoes β-elimination of a radical other than that which added to the unsaturated linkage. For example, equilibration of the butenyl bromides (**35** and **36**) in light and peroxide-induced reactions likely proceeds by such a chain sequence. The adduct radical produced in the addition reaction can eliminate either of the two bromine atoms. Similarly, the butenyl methyl sulfides can be isomerized to equilibrium mixtures.

$$\underset{\underset{\textbf{35}}{\overset{|}{\underset{Br}{}}}}{CH_2\text{=}CHCHCH_3} \overset{h\nu}{\rightleftharpoons} \underset{\underset{\textbf{36}}{(cis \text{ and } trans)}}{BrCH_2CH\text{=}CHCH_3} \qquad (2\text{-}90)$$

via

$$Br\cdot + \underset{\overset{|}{\underset{Br}{}}}{CH_2\text{=}CHCHCH_3} \rightleftharpoons \underset{\overset{|}{\underset{Br}{}}}{BrCH_2\dot{C}HCHCH_3}$$

$$\rightleftharpoons BrCH_2CH\text{=}CHCH_3 + Br\cdot \quad (2\text{-}91)$$

Other examples of rearrangements resulting from a chain sequence consisting of an addition-β-elimination sequence can be found in Chapter 8.

E. ADDITION, ELIMINATION, AND DISPLACEMENT

Kharasch additions to certain unsaturated compounds have been observed to yield products in which there has been rearrangement of the unsaturated moiety. Many of these reactions can be explained in terms of a chain sequence which includes a β-elimination reaction of the adduct radical. For example, addition of $BrCCl_3$ to allyl bromide (**37**) yields a reaction mixture that can be explained in terms of a mechanism that includes a β-elimination reaction (2-94) step in the chain sequence.

$$2BrCCl_3 + 2\underset{\textbf{37}}{CH_2\text{=}CHCH_2Br} \rightarrow$$

$$BrCH_2CHBrCH_2Br + Cl_3CCH_2CHBrCH_2CCl_3 \quad (2\text{-}92)$$

via

$$Cl_3C\cdot + CH_2\text{=}CHCH_2Br \rightarrow Cl_3CCH_2\dot{C}HCH_2Br \qquad (2\text{-}93)$$

$$Cl_3CCH_2\dot{C}HCH_2Br \rightarrow Cl_3CCH_2CH{=}CH_2 + Br\cdot \qquad (2\text{-}94)$$

$$Br\cdot + CH_2{=}CHCH_2Br \rightarrow BrCH_2\dot{C}HCH_2Br \qquad (2\text{-}95)$$

$$BrCH_2\dot{C}HCH_2Br + BrCCl_3 \rightarrow BrCH_2CHBrCH_2Br + Cl_3C\cdot \quad (2\text{-}96)$$

$$Cl_3C\cdot + Cl_3CCH_2CH{=}CH_2 \rightarrow Cl_3CCH_2\dot{C}HCH_2CCl_3 \qquad (2\text{-}97)$$

$$Cl_3CCH_2\dot{C}HCH_2CCl_3 + BrCCl_3 \rightarrow Cl_3CCH_2CHBrCH_2CCl_3 + Cl_3C\cdot$$

$$(2\text{-}98)$$

If the β-elimination reaction of the adduct radical involves the opening of a ring, a single addition product is formed. A classical example of such a reaction is the addition of carbon tetrachloride to β-pinene (**38**)—a reaction in which a cyclobutane ring system is lost as the result of a β-elimination reaction of the adduct radical.

(2-99)

38

via

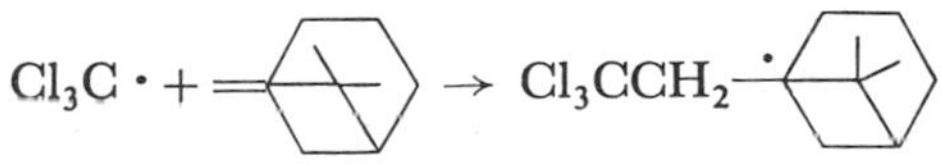

(2-100)

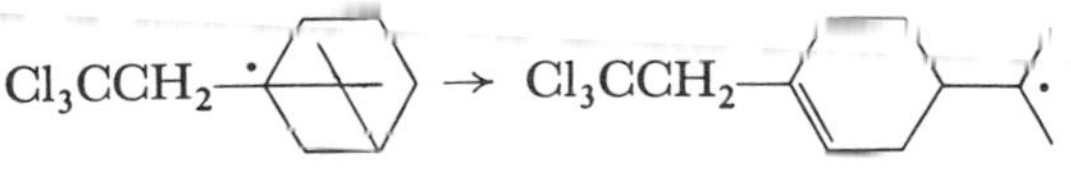

(2-101)

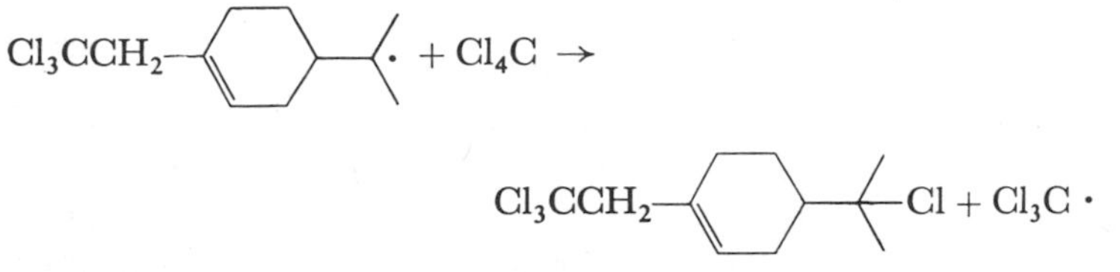

(2-102)

F. REARRANGEMENT AND DISPLACEMENT

Although rearranged products are obtained in some of the reactions involving β-elimination reactions, there are rearrangements in which a 1,2-shift is one of the radical-propagating steps. Since the 1,2-shift is not a product-forming reaction, a displacement reaction is very often another chain-propagating reaction in the overall mechanism. The chlorination of t-butyl benzene (**39**) yielding 1-phenyl-2-chloro-2-methyl-propane (**40**) involves such a chain sequence. As pointed out in Chapter 1 (see also Chapter 9), the groups that are

$$C_6H_5C(CH_3)_3 + Cl_2 \rightarrow C_6H_5CH_2CCl(CH_3)_2 \qquad (2\text{-}103)$$
$$\quad\;\; \mathbf{39} \qquad\qquad\qquad\qquad \mathbf{40}$$

via

$$Cl\cdot + C_6H_5C(CH_3)_3 \rightarrow HCl + \underset{\displaystyle \cdot CH_2}{C_6H_5C(CH_3)_2} \qquad (2\text{-}104)$$

$$\underset{\displaystyle \cdot CH_2}{C_6H_5C(CH_3)_2} \rightarrow \underset{\displaystyle C_6H_5CH_2}{\cdot C(CH_3)_2} \qquad (2\text{-}105)$$

$$C_6H_5CH_2\dot{C}(CH_3)_2 + Cl_2 \rightarrow C_6H_5CH_2CCl(CH_3)_2 \qquad (2\text{-}106)$$
$$+ Cl\cdot$$

capable of migration in 1,2-shifts of this sort are somewhat limited. Consequently, formation of rearranged products in halogenation reactions are rare and occur only when a chain-carrying free radical is less stable than the rearranged radical and a group capable of migration is positioned on an atom that is *beta* to the radical site.

G. ADDITION, REARRANGEMENT, AND DISPLACEMENT

A free radical capable of undergoing a 1,2-shift can be formed in an addition reaction. A displacement reaction by the rearranged radical on an adding reagent will then yield an addition product in which rearrangement is evident. The addition of $BrCCl_3$ to 3,3,3-trichloropropene (**41**) is illustrative of such a reaction. The displacement reaction (2-110) on the $BrCCl_3$, as evidenced by the structure of the addition product, is performed by a free radical that resulted from the 1,2-shift (2-109) of a chlorine atom in the initially formed adduct radical.

$$BrCCl_3 + CH_2{=}CHCCl_3 \rightarrow Cl_3CCH_2CHClCCl_2Br \qquad (2\text{-}107)$$
$$\mathbf{41}$$

via

$$Cl_3C\cdot + CH_2{=}CHCCl_3 \rightarrow Cl_3CCH_2\dot{C}HCCl_3 \qquad (2\text{-}108)$$

$$Cl_3CCH_2\dot{C}HCCl_3 \rightarrow \underset{\displaystyle Cl}{Cl_3CCH_2CH\dot{C}Cl_2} \qquad (2\text{-}109)$$

$$Cl_3CCH_2CHCl\dot{C}Cl_2 + BrCCl_3 \rightarrow Cl_3CCH_2CHClCCl_2Br + Cl_3C\cdot \quad (2\text{-}110)$$

Other examples of chain sequences that involve addition, rearrangement, and displacement reactions as the chain-propagating reactions can be found in Chapter 9.

H. HYDROGEN ATOM AND ELECTRON-TRANSFER REACTIONS

The oxidation of a secondary alcohol to a ketone by di-t-butyl peroxide is a free-radical chain reaction that involves a displacement reaction and a hydrogen atom transfer reaction from the alcohol derived α-hydroxyalkyl radical to the peroxide.

$$R_2CHOH + (CH_3)_3COOC(CH_3)_3 \rightarrow R_2C{=}O + 2(CH_3)_3COH \tag{2-111}$$

via

$$(CH_3)_3CO\cdot + R_2CHOH \rightarrow (CH_3)_3COH + R_2\dot{C}OH \tag{2-112}$$

$$R_2\dot{C}OH + (CH_3)_3COOC(CH_3)_3 \rightarrow R_2C{=}O + (CH_3)_3COH +$$
$$(CH_3)_3CO\cdot \tag{2-113}$$

Copper is a good example of an ion capable of an oxidation state change of one. It has been found to induce some interesting reactions which can be explained by a chain mechanism in which Cu^I and Cu^{II} are participants. The

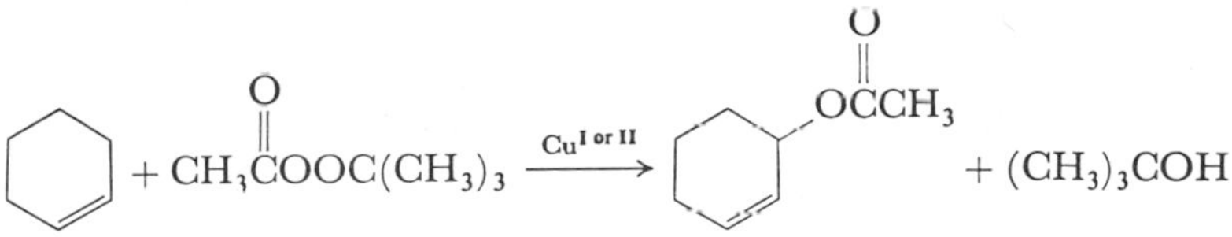

$$(2\text{-}114)$$

via

$$Cu^IBr + CH_3\overset{O}{\overset{\|}{C}}OOC(CH_3)_3 \rightarrow CH_3\overset{O}{\overset{\|}{C}}O^- + (Cu^{II}Br)^+ + (CH_3)_3CO\cdot \tag{2-115}$$

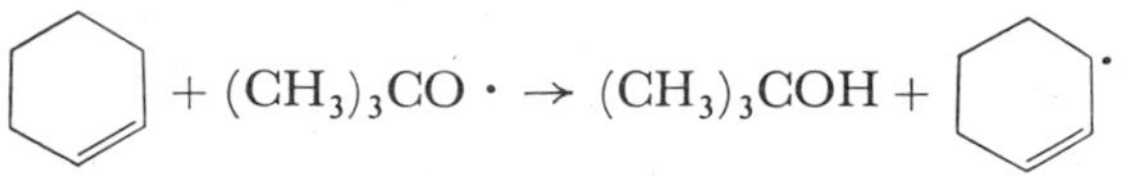

$$+ (CH_3)_3CO\cdot \rightarrow (CH_3)_3COH + \tag{2-116}$$

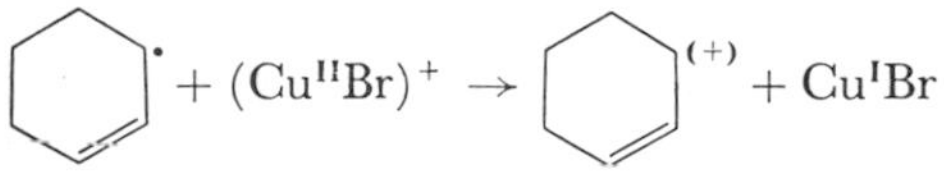

$$+ (Cu^{II}Br)^+ \rightarrow \tag{2-117}$$

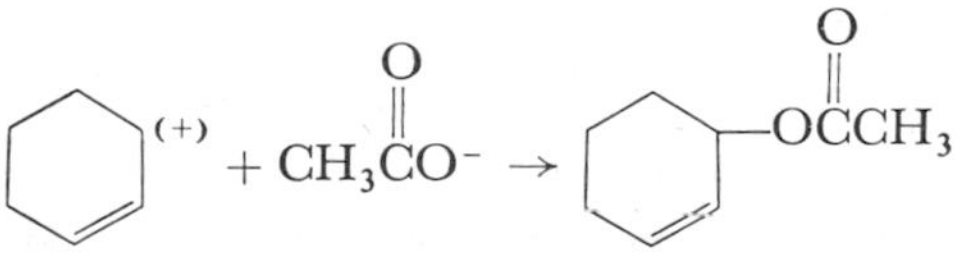

$$(2\text{-}118)$$

reaction of cyclohexene (**42**) with *t*-butyl peracetate (**43**) yielding the acetate ester **44** is an example of such a reaction. The reaction may well be somewhat more complicated than shown in the chain sequence in that the metal ion may be involved in various kinds of complexes with negative ions in the solution which impart redox characteristics to the ions causing the reaction to proceed smoothly. It is pertinent, however, that a reasonable chain sequence can be arrived at that does include the copper ions. The involvement of copper is necessary since the course of the reaction of cyclohexene with *t*-butyl peracetate in the absence of copper is different. Copper is only a catalyst in the reaction and need not be present in stoichiometric amounts. A more complete discussion of the redox reactions of free radicals can be found in Chapter 11.

II. Product Formation in Termination Reactions

In all of the reactions given in the previous sections, the reactants were converted to products in a sequence of chain-propagating reactions. Although each of the chain sequences must have involved some termination reaction, the amount of product produced in this process was negligible compared to the amounts of products formed in the chain sequence. If the chain sequence were long, the amount of initiation required would also be small. There are some free-radical reactions in which the product of the reaction is formed in either a radical coupling reaction or a disproportionation reaction. It follows, of course, that unless there may be some unusual feature, such as is the case in vinyl polymerization, involved in the formation of the radicals that participate in the termination process, large amounts of initiation are going to be required in order to achieve an appreciable amount of reaction. Consequently, we find that most of the free-radical reactions in which the product results from a termination process are those where introduction of free radicals in relatively large amounts can be accomplished effectively. In some cases this may be done by use of a chemical initiator while in others effective radical-producing photochemical processes are employed.

A. VINYL POLYMERIZATION

The conversion of many simple vinyl compounds to high molecular weight polymers is a reaction in which the product, namely the polymer molecule, is formed in a termination reaction. The reaction does, however, have all of the earmarks of a true free-radical chain process in that only very small amounts of initiation are required to cause conversion of a large amount of the reactant to product. The reason is that the chain-propagating reaction in the reaction, namely addition, is not a product-forming process. If only the reactions in 2-119 through 2-123 were involved (see Chapter 12), the actual number of molecules of polymer produced can be no greater than the number of initiator molecules that decompose to form radicals. The high conversion of the reactant

$$\text{Init.} \rightarrow 2R' \cdot \tag{2-119}$$

$$R' \cdot + CH_2 = CHX \rightarrow R'CH_2\dot{C}HX \tag{2-120}$$

$$R'CH_2\dot{C}HX + CH_2 = CHX \rightarrow R'CH_2CHCH_2\dot{C}HX \tag{2-121}$$
$$\overset{|}{X}$$

$$R'CH_2CHCH_2\dot{C}HX + nCH_2 = CHX \rightarrow R' \left(\begin{matrix} CH_2CH \\ | \\ X \end{matrix}\right)_{n+1} CH_2\dot{C}HX$$
$$\overset{|}{X}$$

$$\tag{2-122}$$

$$2R' \left(\begin{matrix} CH_2CH \\ | \\ X \end{matrix}\right)_{n+1} CH_2\dot{C}HX \rightarrow$$

$$R' \left(\begin{matrix} CH_2CH \\ | \\ X \end{matrix}\right)_{n+1} CH_2CHCHCH_2 \left(\begin{matrix} CHCH_2 \\ | \\ X \end{matrix}\right)_{n+1} -R' \tag{2-123}$$

to product results from the chain-propagating addition reaction 2-122 producing large free radicals that couple to form the polymer molecule. Some of the principles of free-radical vinyl polymerization kinetics as they pertain to free-radical chain reactions are expanded upon in Chapter 12.

B. OXIDATIVE DIMERIZATIONS

Reactions of certain alkyl aromatics with peroxides yield products that are produced in the termination reaction. For example, decomposition of diacetyl peroxide (45) in cumene (46) yields 2,3-dimethyl-2,3-diphenylbutane (47) as a reaction product. It follows that the number of moles of product formed will be no greater than the number of moles of peroxide that decompose. Although reaction of the methyl radical with cumene is a chain-propagating displacement

$$\underset{45}{CH_3\overset{O}{\overset{||}{C}}O\overset{O}{\overset{||}{O}}CCH_3} + 2\underset{\underset{46}{\overset{|}{CH_3}}}{\overset{\overset{|}{CH_3}}{C_6H_5CH}} \rightarrow 2CH_4 + 2CO_2 + \underset{\underset{47}{\overset{|}{CH_3}\ \overset{|}{CH_3}}}{\overset{\overset{|}{CH_3}\ \overset{|}{CH_3}}{C_6H_5C-CC_6H_5}} \tag{2-124}$$

via

$$CH_3\overset{O}{\overset{||}{C}}O\overset{O}{\overset{||}{O}}CCH_3 \rightarrow 2CH_3 \cdot + 2CO_2 \tag{2-125}$$

$$CH_3 \cdot + C_6H_5CH(CH_3)_2 \rightarrow CH_4 + C_6H_5\dot{C}(CH_3)_2 \tag{2-126}$$

$$2C_6H_5\dot{C}(CH_3)_3 \rightarrow \underset{\underset{CH_3}{|}}{\overset{\overset{CH_3}{|}}{C_6H_5C}}-\underset{\underset{CH_3}{|}}{\overset{\overset{CH_3}{|}}{CC_6H_5}} \qquad (2\text{-}127)$$

reaction, the cumyl radical that is produced does not react in a reaction to regenerate a free radical but only in the coupling reaction.

In a similar manner, derivatives of succinic acid can be prepared by reaction of either acids or esters with peroxides. The oxidative dimerization of isobutyric acid (**48**) for example, yields tetramethyl succinic acid (**49**) by a mechanism similar to that of the reaction of cumene.

$$2\underset{\textbf{48}}{(CH_3)_2CHCO_2H} + CH_3\overset{\overset{O}{||}}{C}O O\overset{\overset{O}{||}}{C}CH_3 \rightarrow \underset{\underset{\underset{\textbf{49}}{CO_2H}}{|}}{\overset{\overset{CO_2H}{|}}{(CH_3)_2CC(CH_3)_2}} + 2CO_2 + 2CH_4$$

$$(2\text{-}128)$$

via

$$CH_3\overset{\overset{O}{||}}{C}OO\overset{\overset{O}{||}}{C}CH_3 \rightarrow 2CH_3\cdot + 2CO_2 \qquad (2\text{-}129)$$

$$CH_3\cdot + (CH_3)_2CHCO_2H \rightarrow CH_4 + (CH_3)_2\dot{C}CO_2H \qquad (2\text{-}130)$$

$$2(CH_3)_2\dot{C}CO_2H \rightarrow \underset{\underset{CO_2H}{|}}{\overset{\overset{CO_2H}{|}}{(CH_3)_2CC(CH_3)_2}} \qquad (2\text{-}131)$$

These reactions are somewhat limited in their usefulness since they require stoichiometric amounts of initiator. There are occasions, however, when they may prove to be one of the better methods of preparing certain compounds. The formation of 2,3-dimethoxy-2,3-diphenylbutane (**51**) by the oxidative dimerization of α-methoxyethylbenzene (**50**) with di-*t*-butyl peroxide is one such case.

$$\underset{\textbf{50}}{2C_6H_5\overset{\overset{OCH_3}{|}}{C}HCH_3} + (CH_3)_3COOC(CH_3)_3 \rightarrow$$

$$2(CH_3)_3COH + \underset{\underset{CH_3O}{|}}{\overset{\overset{CH_3}{|}}{C_6H_5C}}-\underset{\underset{OCH_3}{|}}{\overset{\overset{CH_3}{|}}{CC_6H_5}} \qquad (2\text{-}132)$$

$$\textbf{51}$$

via

$$(CH_3)_3COOC(CH_3)_3 \rightarrow 2(CH_3)_3CO\cdot \qquad (2\text{-}133)$$

$$(CH_3)_3CO\cdot + C_6H_5\overset{|}{\underset{OCH_3}{C}}HCH_3 \rightarrow (CH_3)_3COH + C_6H_5\overset{|}{\underset{OCH_3}{\dot{C}}}CH_3 \quad (2\text{-}134)$$

$$2C_6H_5\overset{|}{\underset{OCH_3}{\dot{C}}}CH_3 \rightarrow C_6H_5\overset{CH_3}{\underset{CH_3O}{\overset{|}{C}}}\overset{CH_3}{\underset{OCH_3}{\overset{|}{C}}}C_6H_5 \qquad (2\text{-}135)$$

C. PHOTOCHEMICAL REDUCTION OF KETONES

Certain aryl ketones are reduced in photochemical reactions by alcohols. For example, illumination of a solution of benzophenone (52) in isopropyl alcohol yields benzpinacol (53) as a reaction product with concurrent oxidation on the alcohol to acetone. The mechanism of this reaction involves formation of triplet benzophenone (54), a species with radical qualities, from a photochemically excited state of benzophenone. A displacement reaction on the isopropyl alcohol by this triplet species yields a benzhydrol radical and an α-hydroxyalkyl radical which transfers a hydrogen atom to a molecule of benzophenone in the ground state, yielding acetone and another benzhydrol radical. Coupling of two benzhydrol radicals yields the benzpinacol. Although there are radical-propagating steps in this sequence of reactions, the reaction is not a chain process since no more than one molecule of benzpinacol can be formed from each diradical produced in the photochemical initiation step.

$$2(C_6H_5)_2C{=}O + (CH_3)_2CHOH \xrightarrow{h\nu}$$
$$\mathbf{52}$$

$$(C_6H_5)_2\overset{|}{\underset{HO}{C}}{-}\overset{|}{\underset{OH}{C}}(C_6H_5)_2 + (CH_3)_2C{=}O \quad (2\text{-}136)$$
$$\mathbf{53}$$

via

$$(C_6H_5)_2C{=}O \rightarrow [(C_6H_5)_2C{=}O]^* \rightarrow (C_6H_5)_2\dot{C}O\cdot \qquad (2\text{-}137)$$
$$\mathbf{54}$$

$$(C_6H_5)_2\dot{C}O\cdot + (CH_3)_2CHOH \rightarrow (C_6H_5)_2\dot{C}OH + (CH_3)_2\dot{C}OH \quad (2\text{-}138)$$

$$(CH_3)_2\dot{C}OH + (C_6H_5)_2C{=}O \rightarrow (CH_3)_2C{=}O + (C_6H_5)_2\dot{C}OH \quad (2\text{-}139)$$

$$2(C_6H_5)_2\dot{C}OH \rightarrow (C_6H_5)_2\overset{|}{\underset{HO}{C}}{-}\overset{|}{\underset{OH}{C}}(C_6H_5)_2 \qquad (2\text{-}140)$$

Chapter 3

Kinetic Aspects of Free-Radical Chain Reactions

I. Introduction

Most of the free-radical reactions discussed here are those in which the products are formed from the reactants in a chain sequence of free-radical propagating reactions. It was pointed out in Chapter 2 that initiation and termination reactions had to be taken into account as part of the overall reaction although these steps are relatively unimportant as product-forming reactions if the chain sequence is long. However, when the overall kinetics of the reaction are considered, initiation and termination play very significant roles. The kinetic aspects of free-radical chain reactions involve an interesting balance of the rates of the initiation and termination steps as well as of the individual radical-propagating reactions that comprise the chain sequence itself.

Although an understanding of their overall kinetics, which is covered in the first part of this chapter, is necessary to appreciate fully free-radical chain reactions, the usefulness of such kinetic studies is somewhat limited for reasons that will become obvious. The second part of this chapter is concerned with some kinetic methods that have proven valuable in learning about the reactivities of both radicals and of substrates with which they react in radical-propagating reactions.

II. Steady-State Assumptions

A. GENERAL CONSIDERATIONS

Consider a simple free-radical chain reaction such as that of A_2 with XY as shown in 3-1. This reaction can be proposed to have a mechanism where

$$A_2 + XY \rightarrow AX + AY \qquad (3-1)$$

initiation occurs by some unspecified process yielding two $A\cdot$ radicals, the chain sequence consists of two displacement reactions involving only $A\cdot$ and $Y\cdot$ as chain-carrying radicals, and termination is accomplished only by coupling of two of the chain carrying $Y\cdot$ radicals. (Other termination reactions are, of course, possible.) The stoichiometry of the reaction, if the chain sequence is

$$A_2 \xrightarrow{k_2} 2A\cdot \qquad (3-2)$$

$$A\cdot + XY \xrightarrow{k_3} AX + Y\cdot \qquad (3-3)$$

$$Y\cdot + A_2 \xrightarrow{k_4} AY + A\cdot \qquad (3-4)$$

$$2Y\cdot \xrightarrow{k_5} Y_2 \qquad (3-5)$$

long, is determined by reactions 3-3 and 3-4 and the amount of Y_2 formed in the termination step 3-5 is negligible. The rate of the reaction, however, is determined not only by the chain sequence but also by the initiation reaction (3-2) and the termination reaction (3-5). At first glance it may appear to be a somewhat complex matter to arrive at a rate law that will take all of these factors into consideration. It turns out, however, to be a relatively simple problem because of certain kinetic situations that exist if the reaction proceeds by the mechanism shown. Radical-propagating reactions are always first order in free radicals and, if either a displacement, an addition, or a redox reaction is involved, are also first order in the substrate. The latter can be present in comparatively large concentrations. Even with rate constants about 10^4–10^6 times smaller than those for the termination reactions, the propagating reaction can proceed at rates faster than the termination reaction. (See Table 3-1 for representative rate constants for both radical propagating and destroying reactions.)

The kinetic chain length can be defined as the rate of one of the propagating reactions (R_p) divided by the rate of termination reaction (R_t). It is evident

$$\text{Kinetic chain length} = \frac{R_p}{R_t} = \frac{k_4[\text{Y} \cdot][\text{A}_2]}{k_5[\text{Y} \cdot]^2} = \frac{k_4[\text{A}_2]}{k_5[\text{Y} \cdot]} \qquad (3\text{-}6)$$

from equation 3-6 that the kinetic chain length is inversely proportional to the concentration of the chain-carrying radical. Consider a reaction in which the rate constant for the propagating reaction is 10^4 moles liter^{-1} sec^{-1} and that for the termination reaction 10^7. If the concentration of the substrate A_2 is 10 moles per liter, the concentrations of the chain-carrying radical $\text{Y} \cdot$ would be 10^{-5} moles per liter in order to have a kinetic chain length of 1000 for the reaction.

The kinetic situation presented in the hypothetical reaction 3-2 through 3-5 (slow introduction of free radicals in the initiation step, rapid chain-propagating reactions, and a very fast termination reaction) dictate that the radical concentrations be small if a long kinetic chain length is to be expected, and this concentration does not change appreciably. At the outset of the reaction, no free radicals are present. Decomposition of A_2 produces free radicals that participate in either radical-propagating processes in which there will be no change in the overall radical concentration, or the termination reaction in which there is a change in radical concentration. When decomposition of A_2 first starts, the radical concentration is extremely low, so low that the rate of removal of the radicals by the termination reaction is slower than the rate of their introduction by the initiation process. Thus, at the very beginning of the reaction, the radical concentration increases until it becomes high enough for the termination process to remove them. In time the concentration is such that they are removed in the termination reaction at the same rate at which they are being produced in the initiation reaction. When this point is reached, the radical concentration remains essentially constant even though radicals are being produced in the initiation reaction, are reacting in propagating reactions and are being consumed in the termination process. This concentration is called the steady-state concentration and generally is attained within a few seconds in most free-radical chain reactions.

An important generalization can be reached concerning free-radical chain reactions based on the assumption that a steady-state concentration of radicals is maintained in free-radical chain reactions. *The rate of initiation must be equal to the rate of termination.* Much use is made of this generalization in deriving the kinetic rate laws for free-radical chain reactions.

TABLE 3-1 Absolute Rate Constants for Reactions of Free Radicals

Reaction	Temperature (°C)	k ($l\,mole^{-1}\,sec^{-1}$)	Reference
Radical-Propagating Reactions			
$(CH_3)_3CO\cdot + C_6H_5CH_3 \rightarrow (CH_3)_3COH + C_6H_5CH_2\cdot$	24	$(1.2 \pm 0.4) \times 10^3$	1
$(CH_3)_3CO\cdot + C_6H_5CH_3 \rightarrow (CH_3)_3COH + C_6H_5CH_2\cdot$	30	7.2×10^4	2
$(CH_3)_3CO\cdot + C_6H_{12}{}^a \rightarrow (CH_3)_3COH + C_6H_{11}\cdot$	24	$(26 \pm 4) \times 10^3$	1
$(CH_3)_3CO\cdot + HCCl_3 \rightarrow (CH_3)_3COH + Cl_3C\cdot$	24	3×10^3	1
$Cl_3C\cdot + (CH_3)_3COCl \rightarrow Cl_4C + (CH_3)_3CO\cdot$	24	$(1.2 \pm 0.4) \times 10^3$	1
$(C_6H_5)_2\dot{C}O\cdot + C_6H_5CH_3 \rightarrow (C_6H_5)_2\dot{C}OH + C_6H_5CH_2\cdot$	—	2×10^4	3
$(C_6H_5)_2\dot{C}O\cdot + (C_6H_5)_2CHOH \rightarrow (C_6H_5)_2\dot{C}OH + (C_6H_5)_2\dot{C}OH$	—	5×10^6	3
$\sim CH_2\overset{\displaystyle C_6H_5}{\underset{\displaystyle C_6H_5}{CH\cdot}} + CH_2{=}\overset{\displaystyle C_6H_5}{CH} \rightarrow \sim CH_2\overset{\displaystyle C_6H_5}{CH}CH_2\overset{\displaystyle C_6H_5}{CH}\cdot$	30	5.5×10^1	4
$\sim CH_2\overset{\displaystyle OAc}{CH\cdot} + CH_2{=}\overset{\displaystyle OAc}{CH} \rightarrow \sim CH_2\overset{\displaystyle OAc}{CH}CH_2\overset{\displaystyle OAc}{CH}\cdot$	25	1.01×10^3	5
$(CH_3)_3CO\cdot \rightarrow (CH_3)_2C{=}O + CH_3\cdot$	51	$1 \times 10^{4\,b}$	6

Radical-Destroying Reactions

Reaction			
$2Cl_3C\cdot \rightarrow C_2Cl_6$	24	5×10^7	1
$2(CH_3)_3CO\cdot \rightarrow (CH_3)_3COOC(CH_3)_3$	24	$(7 \pm 3) \times 10^7$	1
$2I\cdot \rightarrow I_2$	25	8.4×10^9	7
$2\sim CH_2CH\cdot \rightarrow \sim CH_2CH-CHCH_2\sim$ $\quad\mid \qquad\qquad\qquad \mid$ $\quad C_6H_5 \qquad\quad C_6H_5\ C_6H_5$	30	2.5×10^7	4
$2\sim CH_2CH\cdot \rightarrow \sim CH_2CH-CHCH_2\sim$ $\quad\mid \qquad\qquad\qquad \mid$ $\quad OAc \qquad\qquad OAc\ OAc$	25	3.06×10^7	5

[a] Cyclohexane.
[b] sec^{-1}.

Not only does the overall concentration of free radicals reach a steady state, but each of the chain-carrying radicals has its own steady-state concentration. This conclusion is reached if the assumption is made that the rate of termination cannot be erratic but must proceed at a constant rate if it is to proceed at the same rate as the unimolecular initiation reaction. This means that if only one of the chain-carrying radicals is involved in termination, its concentration cannot change. Consequently, if there is an overall steady-state concentration of free radicals, the other chain-carrying free radical must also have a constant concentration.

The assumption that each chain-carrying free radical has its own steady-state concentration leads to another important generalization concerning free-radical chain reactions. *Each propagating reaction in the chain sequence must have the same rate.* The validity of this generalization is arrived at in the following manner. If a long kinetic chain length is maintained in the reaction, the best opportunity for change in either chain-carrying radical arises in the chain-propagating reactions since each propagating reaction may occur several hundred times before the radical becomes involved in termination. Consider for example Y·, a chain-carrying radical produced in the propagating reaction 3-3 but consumed in 3-4. If $[Y\cdot]$ is to remain constant in order to participate in the termination reaction in such a way that termination proceeds at a constant rate, the rate at which it is produced in 3-3 must equal the rate at which it is consumed in 3-4.

B. DERIVATION OF KINETIC RATE LAWS

With the information from the preceding section concerning the concentration and rate relationships that exist in free-radical chain reactions, it becomes a relatively simple matter to derive a rate equation for such reactions. Applying these concepts to the hypothetical reaction 3-1 and assuming the mechanism to be the sequence of reactions 3-2 through 3-5, the rate of the overall reaction

$$\text{Rate} = \frac{-d[XY]}{dt} = \frac{-d[A_2]}{dt} = \frac{d[AX]}{dt} = \frac{d[AY]}{dt} \tag{3-7}$$

can be expressed as the rate of either of the two chain-propagating reactions 3-3 or 3-4. Neither of these expressions is particularly useful since both include

$$\text{Rate} = \frac{-d[XY]}{dt} = \frac{d[AX]}{dt} = k_3[A\cdot][XY] \tag{3-8}$$

or

$$\text{Rate} = \frac{-d[A_2]}{dt} = \frac{d[AY]}{dt} = k_4[Y\cdot][A_2] \tag{3-9}$$

the concentration of a chain-carrying radical. Although some things are known about these concentrations, namely that they are small and are steady-state concentrations, it is not a simple matter to express them in any definite numerical concentration units and thereby obtain a rate. However, a rather useful relationship based on the steady-state assumption exists, and the concentrations of chain-carrying radicals can be found in terms of the reactants. Assuming a steady-state concentration of all the radicals, the rates of initiation and termination are equal and from this a value for $[Y\cdot]$ can be obtained in terms

$$k_2[A_2] = 2k_5[Y\cdot]^2 \qquad \text{or} \qquad [Y\cdot] = \left(\frac{k_2}{2k_5}\right)^{1/2} [A_2]^{1/2} \qquad (3\text{-}10)$$

of $[A_2]$. Substitution of this value for $[Y\cdot]$ in equation 3-9 results in the rate law 3-11, a rate law for reaction 3-1 in terms of one of the reactants and the

$$\text{Rate} = k_4 \left(\frac{k_2}{2k_5}\right)^{1/2} [A_2]^{3/2} \qquad (3\text{-}11)$$

reaction-rate constants of some of the individual steps in the overall mechanism.

Not all free-radical chain reactions are quite as simple as the example just shown. First, the termination reaction need not be restricted to coupling of just one of the chain-carrying radicals. We shall see that the rate expression depends very much on the particular termination reaction involved. Secondly, the nature of the initiation reaction 3-2 was not specified. It will be of interest to note how the initiation process, whether it is a chemical initiator or a light-induced reaction, enters into the rate expression. Finally, it is of value to derive rate laws for reactions in which the kinetic chain length is short and the initiator is one of the reactants. The rate laws for most free-radical reactions are overall three-halves order in the reactants. This results from the fact that initiation is generally a unimolecular reaction yielding two radicals, the rate-limiting step (or steps) in the chain sequence are generally bimolecular (unimolecular in chain-carrying radicals and unimolecular in substrates), and termination is a bimolecular reaction involving chain-carrying radicals. If initiation is a bimolecular process, the rate law becomes overall second order. If a uni-molecular rearrangement or β-elimination becomes a rate-limiting factor in the chain sequence, the rate law may be something less than three-halves order overall. The derivations of kinetic rate laws for various kinds of reactions are shown in the next section. Rate laws for free-radical chain reactions are of limited value in determining the general mechanism of the reaction. It is satisfying, however, to find that the derived rate laws are the same as the experimentally observed rate laws in many cases, lending support to the assumptions made concerning the nature of these reactions.

C. EFFECT OF TERMINATION ON DERIVED RATE LAWS

1. Unimolecular Initiation, Bimolecular Chain-Propagation and Termination. Consider the hypothetical reaction 3-12, the mechanism of

$$AB + XY \xrightarrow{\text{Bz}_2\text{O}_2} AX + BY \qquad (3\text{-}12)$$

which is one in which the initiation is a unimolecular reaction and both the chain propagation and termination reactions are bimolecular. The chain

$$Bz_2O_2 \xrightarrow{k_{13}} 2BzO\cdot \qquad (3\text{-}13)$$

$$BzO\cdot + XY \xrightarrow{k_{14}} BzOX + Y\cdot \qquad (3\text{-}14)$$

$$Y\cdot + AB \xrightarrow{k_{15}} YB + A\cdot \qquad (3\text{-}15)$$

$$A\cdot + XY \xrightarrow{k_{16}} AX + Y\cdot \qquad (3\text{-}16)$$

$$2A\cdot \xrightarrow{k_{17}} A_2 \qquad (3\text{-}17)$$

$$2Y\cdot \xrightarrow{k_{18}} Y_2 \qquad (3\text{-}18)$$

$$A\cdot + Y\cdot \xrightarrow{k_{19}} AY \qquad (3\text{-}19)$$

sequence 3-15 and 3-16 is assumed to be very long; therefore, only a small amount of benzoyl peroxide is required for initiation and consequently little BzOX is produced in reaction 3-14. Furthermore, very little termination product, whether it is A_2, Y_2, or AY, is formed.

The rate of reaction 3-12 can be given by either of the chain-propagating steps, and the choice of which of these two rate expressions would serve us best

$$\text{Rate} = k_{15}[Y\cdot][AB] = k_{16}[A\cdot][XY] \qquad (3\text{-}20)$$

In deriving a kinetic rate law depends on the termination step that is operative. Consider first termination by reaction 3-17, the coupling of two A· radicals. In this case, expressing the rate as that of reaction 3-16, as shown in equation 3-21,

$$\text{Rate} = -\frac{d[XY]}{dt} = \frac{d[AX]}{dt} = k_{16}[A\cdot][XY] \qquad (3\text{-}21)$$

is advantageous since the rate of initiation must equal the rate of termination,

$$k_{13}[Bz_2O_2] = 2k_{17}[A\cdot]^2. \qquad (3\text{-}22)$$

Solving 3-22 for [A·], and substituting this value (3-23) in 3-21 we obtain a

$$[A\cdot] = \left(\frac{k_{13}}{2k_{17}}\right)^{1/2}[Bz_2O_2]^{1/2} \qquad (3\text{-}23)$$

kinetic rate expression for the reaction in terms of the reactants.

$$\text{Rate} = k_{16}\left(\frac{k_{13}}{2k_{17}}\right)^{1/2}[Bz_2O_2]^{1/2}[XY] \qquad (3\text{-}24)$$

If reaction 3-18 is the termination reaction, it is better to use reaction 3-15 to express the rate.

$$\text{Rate} = -\frac{d[AB]}{dt} = \frac{d[BY]}{dt} = k_{15}[Y\cdot][AB] \qquad (3\text{-}25)$$

From the steady-state assumption

$$k_{13}[Bz_2O_2] = 2k_{18}[Y\cdot]^2 \qquad (3\text{-}26)$$

which gives a value for $[Y\cdot]$, namely

$$[Y\cdot] = \left(\frac{k_{13}}{2k_{18}}\right)^{1/2} [Bz_2O_2]^{1/2}. \qquad (3\text{-}27)$$

Substitution of this value for $[Y\cdot]$ in 3-25 results in the kinetic rate law 3-28 for the reaction when terminated by coupling of $Y\cdot$ radicals, a rate law quite different from that derived for the reaction when terminated by coupling of $A\cdot$ radicals.

$$\text{Rate} = k_{15}\left(\frac{k_{13}}{2k_{18}}\right)^{1/2} [Bz_2O_2]^{1/2}[AB] \qquad (3\text{-}28)$$

In order to derive a kinetic rate law for the reaction when the cross-termination reaction 3-19 is operative, the concentration of one of the chain-carrying free-radicals must be determined in terms of the other. The steady-state assumption leads to 3-29 from which one of the radical concentrations can be

$$k_{13}[Bz_2O_2] = k_{19}[A\cdot][Y\cdot] \qquad (3\text{-}29)$$

found in terms of the other. The choice will depend on which of the two chain-propagating reactions is used to express the rate. For example, using reaction 3-15, it would be necessary to find a suitable value for $[Y\cdot]$. Keeping in mind that the rates of each of the propagating reactions are the same (3-20), solving for $[A\cdot]$ gives

$$[A\cdot] = \frac{k_{15}[Y\cdot][AB]}{k_{16}[XY]} \qquad (3\text{-}30)$$

which, when substituted in 3-29, gives

$$k_{13}[Bz_2O_2] = \frac{k_{19}k_{15}[Y\cdot]^2[AB]}{k_{16}[XY]} \qquad (3\text{-}31)$$

from which $[Y\cdot]$ can be found. Substitution in 3-25 results in the rate law for a chain reaction in which the cross-termination process 3-19 occurs. The

$$\text{Rate} = \left(\frac{k_{13}k_{15}k_{16}}{k_{19}}\right)^{1/2} [Bz_2O_2]^{1/2}[XY]^{1/2}[AB]^{1/2} \qquad (3\text{-}32)$$

reader should satisfy himself that the same expression results using equation 3-16 to express the rate and getting the termination reaction rate in terms of $[A\cdot]$.

Although all different, the three kinetic rate laws have some important similarities. Each reaction has a rate law that is overall three-halves order in terms of all reactants, always half order in terms of the initiator. Furthermore, a ratio of the reaction rate constant of the initiation reaction to the half power over that of the particular termination reaction to the half power is present in each rate expression. The reaction rate constants for the propagating reactions always appear in a total first power term.

The question naturally arises as to which termination reaction is actually operative in a reaction or do they all possibly occur simultaneously. Some situations do exist where only one may be operative. A reaction in which $Y \cdot$ is very reactive in its propagating reaction 3-15 but in which $A \cdot$ is not reactive presents such a situation. If AB and XY are present in comparable concentrations, this would mean that $k_{15} \gg k_{16}$, and since reactions 3-15 and 3-16 have the same rate, the steady-state concentration of $[A \cdot]$ would have to be very much larger than that of $[Y \cdot]$. The most likely termination reaction in such a case would be the coupling of two $A \cdot$ radicals and the kinetic rate expression would be equation 3-24. Examination of 3-24 reveals that it contains the reaction rate constant and the concentration term of the reactant in reaction 3-16, the reaction that involves the less reactive of the two chain carrying radicals. Since both 3-15 and 3-16 must have the same rate, 3-16 cannot be referred to as a slow and, therefore, rate-determining reaction. Rather, reaction 3-16 can be considered to be the rate-limiting step in the chain sequence.

If the converse were true, namely $A \cdot$ is very reactive in its chain-propagating reaction and $Y \cdot$ is much less reactive, termination would most likely proceed by coupling of two $Y \cdot$ radicals, and the kinetic rate law would be 3-28. In this case, reaction 3-15 is the rate-limiting step and it should be noted that the reaction rate constant, k_{15}, as well as the concentration of the reactant AB, appear in this rate expression.

If cross-termination is to occur, the steady-state concentrations of each chain-carrying radical must be about the same. This will be true if both chain-propagating reactions have about the same reaction rate constants and neither could be expected to be the rate-limiting step. The derived kinetic rate law in this case has the reaction rate constants and concentration terms of both chain-propagating reactions.

2. Bimolecular Initiation, Chain-Propagation, and Termination Reactions.

Free-radical chain reactions that are initiated in some bimolecular process that gives rise to two free radicals (either chain-carrying radicals or radicals that could start two chains) will have somewhat different rate laws from those just derived. Reaction 3-12 could be initiated by a bimolecular process of AB and XY yielding the chain-carrying radicals $A \cdot$ and $Y \cdot$ (and the anomalous product BX which will be present in only small amounts if the

$$AB + XY \xrightarrow{\ k_{33}\ } BX + A \cdot + Y \cdot \qquad (3\text{-}33)$$

kinetic chain length of the reaction is long, see below). Invoking the steady-state assumption (3-34) and substituting the value for $[A \cdot]$ found in 3-30 into 3-34

$$k_{33}[AB][XY] = k_{19}[A \cdot][Y \cdot] \qquad (3\text{-}34)$$

gives 3-35, an expression for $[Y \cdot]$ in terms of the reactants. Substituting this

$$[Y \cdot] = \left(\frac{k_{33} k_{16}}{k_{15} k_{19}}\right)^{1/2} [XY] \tag{3-35}$$

value in 3-25 results in the rate law 3-36.

$$\text{Rate} = \left(\frac{k_{33} k_{15} k_{16}}{k_{19}}\right)^{1/2} [XY][AB] \tag{3-36}$$

Cross-termination is most likely to occur in bimolecularly initiated reactions because of the nature of the species involved. Generally one reactant is an oxidizing agent and the other a reducing agent. Both chain-propagating steps will be rapid (see Chapter 4) and, consequently, the steady-state concentrations of the chain-carrying radicals will be comparable. Furthermore, the polar nature (again see Chapter 4) of each chain-carrying radical also dictates preferential cross-termination.

Derivations of the rate laws for reactions initiated by 3-33 but terminated by the coupling reactions 3-17 or 3-18 result in the rate expressions 3-37 and 3-38, respectively Such rate laws for bimolecularly initiated reactions may be

$$\text{Rate} = k_{16} \left(\frac{k_{33}}{2 k_{17}}\right)^{1/2} [XY]^{3/2} [AB]^{1/2} \tag{3-37}$$

$$\text{Rate} = k_{15} \left(\frac{k_{33}}{2 k_{18}}\right)^{1/2} [XY]^{1/2} [AB]^{3/2} \tag{3-38}$$

observed if the substrate of one chain-propagating reaction is present in only small concentrations. For example, if the ratio $[XY]:[AB]$ concentration were very low, the steady-state concentration of $[A \cdot]$ would necessarily be larger than that of $[Y \cdot]$ and termination by 3-17 would occur making the rate law 3-37 operative.

3. Unimolecular Initiation and Chain-Propagation, and Bimolecular Termination.
Radical rearrangements and β-eliminations are generally rapid processes if the chain reactions have a comparatively long kinetic chain length. Although these unimolecular chain-propagating reactions most often are not rate-limiting steps in the chain sequence, it is of interest to examine the consequences if such reactions do become rate limiting. Consider the reaction 3-39

$$CB + XY \rightarrow AX + BY \tag{3-39}$$

in which the chain-carrying radical $C \cdot$ rearranges to $A \cdot$ as shown. A system having three chain-carrying free radicals has six different termination paths

$$Bz_2O_2 \rightarrow 2BzO \cdot \tag{3-13}$$

$$BzO \cdot + XY \rightarrow BzOX + Y \cdot \tag{3-14}$$

$$Y\cdot + CB \rightarrow BY + C\cdot \qquad (3\text{-}40)$$

$$C\cdot \rightarrow A\cdot \qquad (3\text{-}41)$$

$$A\cdot + XY \rightarrow AX + Y\cdot \qquad (3\text{-}16)$$

available. The rate laws resulting from terminations that involve only $Y\cdot$ and $A\cdot$, radicals that participate only in bimolecular radical propagating reactions, will be similar to those derived earlier. If the chain-carrying radical $C\cdot$ becomes involved in termination, the rate law for the reaction is quite different. Consider, for example, the rate law if termination occurred only by the coupling reaction 3-42, which would be the case if the rearrangement reaction 3-41 were the

$$2C\cdot \xrightarrow{\;k_{42}\;} C_2 \qquad (3\text{-}42)$$

rate-limiting reaction in the chain sequence. The rate could be expressed by 3-43, and invoking the steady-state assumption, namely that the rates of the in-

$$\text{Rate} = k_{41}[C\cdot] \qquad (3\text{-}43)$$

itiation reaction 3-13 and termination reaction 3-42 are the same, $[C\cdot]$ is found in terms of the initiator concentration. Substituting this value for $[C\cdot]$ in 3-41

$$[C\cdot] = \left(\frac{k_{13}}{2k_{42}}\right)^{1/2} [Bz_2O_2]^{1/2} \qquad (3\text{-}44)$$

results in 3-45, a rate law that is overall half order in reactants. This rate law

$$\text{Rate} = k_{41}\left(\frac{k_{13}}{2k_{42}}\right)^{1/2} [Bz_2]^{1/2} \qquad (3\text{-}45)$$

is reasonable for the proposed mechanism since neither reactant is involved in the rate-limiting step of the chain reaction.

If termination were to occur by a reaction involving $C\cdot$ and, for example, $Y\cdot$, both 3-40 and 3-41 would have to be regarded as rate-limiting steps. The

$$C\cdot + Y\cdot \xrightarrow{\;k_{46}\;} CY \qquad (3\text{-}46)$$

rate law derived on the basis of the steady state-assumption is

$$\text{Rate} = \left(\frac{k_{13}k_{40}k_{41}}{k_{46}}\right)^{1/2} [Bz_2O_2]^{1/2}[CB]^{1/2}. \qquad (3\text{-}47)$$

It should be noted that this rate law is overall first order, the consequence of having a unimolecular chain-propagating reaction as a rate-limiting step in the chain sequence. If chain termination occurs by 3-48, the rate law shown in 3-49 results.

$$C\cdot + A\cdot \xrightarrow{\;k_{48}\;} CA \qquad (3\text{-}48)$$

$$\text{Rate} = \left(\frac{k_{13}k_{16}k_{41}}{k_{48}}\right)^{1/2} [Bz_2O_2]^{1/2}[XY]^{1/2} \qquad (3\text{-}49)$$

The identical rate laws result if the nature of reaction 3-50 is such that the radical C· undergoes β-elimination yielding A· as a chain-carrying radical

$$C· \rightarrow A· + D \tag{3-50}$$

along with some unsaturated product D. Although it is a product-forming reaction, the fragmentation is a unimolecular process.

D. LIGHT-INDUCED REACTIONS

The reaction of AB and XY could be initiated by photolysis of AB.

$$AB \xrightarrow[h\nu]{k_{51}} A· + B· \tag{3-51}$$

$$B· + XY \xrightarrow{k_{52}} BX + Y· \tag{3-52}$$

$$Y· + AB \xrightarrow{k_{15}} BY + A· \tag{3-15}$$

$$A· + XY \xrightarrow{k_{16}} AX + Y· \tag{3-16}$$

$$2A· \xrightarrow{k_{17}} A_2 \tag{3-17}$$

$$2Y· \xrightarrow{k_{18}} Y_2 \tag{3-18}$$

$$A· + Y· \xrightarrow{k_{19}} AY \tag{3-19}$$

The derivation of rate laws for light-induced reactions is similar to those outlined previously for unimolecularly initiated chain reactions. The only difference is that the rate of initiation is no longer a thermal decomposition of initiator, a rate that is determined by the concentration of the initiator and the rate constant for its decomposition reaction. The rate of initiation is determined by the concentration of the compound undergoing photolysis (see below), its efficiency in absorbing light and by the intensity of the incident light. Thus, the rate of initiation (R_i) is given by

$$R_i = k_{51} I[AB] \tag{3-53}$$

where k_{51} is a constant related to the efficiency displayed by the light in causing photolysis AB, and I is the intensity of the incident light. If termination were to occur by coupling of two A· radicals, the steady-state concentration of A· would be given by 3-54 and 3-55 is the derived rate law for the light-induced

$$[A·] = \left(\frac{k_{51} I[AB]}{2k_{17}}\right)^{1/2} \tag{3-54}$$

reaction of AB and XY having this particular termination process.

$$Rate = k_{16} \left(\frac{k_{51} I}{2k_{17}}\right)^{1/2} [XY][AB]^{1/2} \tag{3-55}$$

In a similar manner, 3-56 can be derived as the rate law if coupling of two

$$\text{Rate} = k_{15}\left(\frac{k_{51}I}{2k_{18}}\right)^{1/2}[\text{AB}]^{3/2} \qquad (3\text{-}56)$$

Y· radicals (reaction 3-18) is the termination reaction. The rate expression for the reaction when cross-coupling (reaction 3-19) is operative is given by 3-57.

$$\text{Rate} = \left(\frac{k_{15}k_{16}k_{51}I}{k_{19}}\right)^{1/2}[\text{XY}]^{1/2}[\text{AB}] \qquad (3\text{-}57)$$

The only significant difference between the rate laws for the light-induced reactions and those for the corresponding peroxide-induced reactions is that the reaction rate constant for the initiator decomposition reaction, namely k_{13}, is replaced by $k_{51}I$ and the concentration term of the initiator, $[\text{Bz}_2\text{O}_2]^{1/2}$, is replaced by $[\text{AB}]^{1/2}$. It should be noted that the rate of the reaction is dependent on the square root of the intensity of the illumination as well.

The above kinetic rate laws apply for photochemically induced reactions only if part of the light is absorbed. If the concentration of the reagent undergoing photolysis is very large or the intensity of the radiation is low, the rate of photolysis will depend only on the latter and is independent of the concentration of the reagent. Thus

$$R_i = k_{51}I \qquad (3\text{-}58)$$

and the rate expressions 3-55, 3-56, and 3-57 become 3-59, 3-60, and 3-61, respectively.

$$\text{Rate} = k_{16}\left(\frac{k_{51}I}{2k_{17}}\right)^{1/2}[\text{XY}] \qquad (3\text{-}59)$$

$$\text{Rate} = k_{15}\left(\frac{k_{51}I}{2k_{18}}\right)^{1/2}[\text{AB}] \qquad (3\text{-}60)$$

$$\text{Rate} = \left(\frac{k_{15}k_{16}k_{51}I}{k_{19}}\right)^{1/2}[\text{XY}]^{1/2}[\text{AB}]^{1/2} \qquad (3\text{-}61)$$

E. SHORT CHAIN REACTIONS

There are situations in which kinetic analysis of reactions with a short kinetic chain length may be of value. A reaction in which only chain-carrying radicals are produced in the initiation process and the termination products are the same as those produced in the chain sequence presents such a case. The stoichiometry of the overall reaction is the same as the stoichiometry of the chain sequence. The hypothetical reaction of A_2 with XY will serve to point out the kinetic analysis of such a process.

$$\text{A}_2 + \text{XY} \longrightarrow \text{AX} + \text{AY} \qquad (3\text{-}62)$$

via

$$A_2 \xrightarrow{k_{63}} 2A\cdot \tag{3-63}$$

$$A\cdot + XY \xrightarrow{k_{64}} AX + Y\cdot \tag{3-64}$$

$$Y\cdot + A_2 \xrightarrow{k_{65}} AY + A\cdot \tag{3-65}$$

$$A\cdot + Y\cdot \xrightarrow{k_{66}} AY \tag{3-66}$$

Derivation of a rate law for this reaction must take into account that appreciable amounts of A_2 are consumed in both the chain-propagating reaction 3-65 and the initiation reaction 3-63 and its rate of reaction is given by

$$\frac{-d[A_2]}{dt} = k_{63}[A_2] + k_{64}[A\cdot][XY] \tag{3-67}$$

or

$$\frac{-d[A_2]}{dt} = k_{63}[A_2] + k_{65}[Y\cdot][A_2]. \tag{3-68}$$

Finding a suitable value for $[Y\cdot]$ in terms of the reactants would give a rate law for the reaction. In order to do so, a steady-state concentration for each radical is assumed. Thus, if

$$\frac{d[A\cdot]}{dt} = 0$$

then

$$k_{63}[A_2] + k_{65}[Y\cdot][A_2] = k_{64}[A\cdot][XY] + k_{66}[A\cdot][Y\cdot]. \tag{3-69}$$

Also, if

$$\frac{d[Y\cdot]}{dt} = 0$$

then

$$k_{64}[A\cdot][XY] = k_{65}[Y\cdot][A_2] + k_{66}[A\cdot][Y\cdot]. \tag{3-70}$$

Equations 3-69 and 3-70, two simultaneous equations with the two unknowns $[A\cdot]$ and $[Y\cdot]$, can be solved for either $[A\cdot]$ or $[Y\cdot]$. Adding 3-69 and 3-70 results in 3-71

$$k_{63}[A_2] = 2k_{66}[A\cdot][Y\cdot] \tag{3-71}$$

which gives the value of $[A\cdot]$ shown in 3-72.

$$[A\cdot] = \frac{k_{63}[A_2]}{2k_{66}[Y\cdot]} \tag{3-72}$$

Substitution of this value for $[A\cdot]$ in 3-70 gives 3-73

$$\frac{k_{63}k_{64}[A_2][XY]}{2k_{66}[Y\cdot]} = k_{65}[Y\cdot][A_2] + \frac{k_{63}}{2}[A_2] \tag{3-73}$$

which can be simplified to the quadratic equation 3-74.

$$2k_{65}k_{66}[A_2][Y\cdot]^2 + k_{63}k_{66}[A_2][Y\cdot] - k_{63}k_{64}[A_2][XY] = 0 \tag{3-74}$$

Solving 3-74 for $[Y\cdot]$ (3-75)

$$[Y\cdot] = \frac{-k_{63}}{4k_{65}} \pm \frac{k_{63}}{2k_{65}} \left(\frac{1}{4} + 2\frac{k_{64}k_{65}}{k_{63}k_{66}}[XY]\right)^{1/2} \tag{3-75}$$

and substituting this value in 3-68 results in the rate law 3-76.

$$\text{Rate} = \left[\frac{3k_{63}}{4} \pm \frac{k_{63}}{2}\left(\frac{1}{4} + 2\frac{k_{64}k_{65}}{k_{63}k_{66}}[XY]\right)^{1/2}\right][A_2] \tag{3-76}$$

Equation 3-71 can, of course, be solved for $[Y\cdot]$ which can be substituted in either 3-69 or 3-70 giving a quadratic equation that can be solved for $[A\cdot]$. Substitution of this value for $[A\cdot]$ in 3-67 also results in the rate equation 3-76.

This rate law, although somewhat more complex, has some of the same features as the rate law for a reaction having a long kinetic chain length. It is overall three-halves order in the reactants, half order in the reactant XY and first order in A_2. All of the reaction rate constants are found in the rate law since each step plays a significant role in determining the course of the reaction.

III. Kinetic Analyses of Some Chain Reactions

A. AUTOXIDATION OF HYDROCARBONS

Although the number is smaller than we might expect, some free-radical chain reactions have been subjected to careful kinetic analysis The results are interesting in that they do support many of the assumptions made in deriving kinetic rate laws for chain reactions on the basis of the steady-state assumption. One of the better examples is found in the kinetic analysis of the reactions of various reagents (e.g. alkanes, alkenes, aldehydes, ethers, etc) with molecular oxygen (see Chapter 11). The initiating and chain-propagating steps of the chemically initiated autoxidations of such compounds, reactions that yield a

$$RH + O_2 \xrightarrow{\text{Init.}} ROOH \tag{3-77}$$

hydroperoxide as the product, is given by the sequence 3-78 through 3-85 including the three possible termination reactions 3-83 through 3-85. The chain sequence 3-81 and 3-82 accounts for the stoichiometry of the reaction 3-77

$$\text{Initiator} \xrightarrow{k_{78}} 2R'\cdot \tag{3-78}$$

$$R'\cdot + O_2 \xrightarrow{k_{79}} R'OO\cdot \tag{3-79}$$

$$R'OO\cdot + RH \xrightarrow{k_{80}} R'OOH + R\cdot \tag{3-80}$$

$$R\cdot + O_2 \xrightarrow{k_{81}} ROO\cdot \tag{3-81}$$

$$ROO\cdot + RH \xrightarrow{k_{82}} ROOH + R\cdot \tag{3-82}$$

$$2R \cdot \xrightarrow{k_{83}} R_2 \tag{3-83}$$

$$R \cdot + ROO \cdot \xrightarrow{k_{84}} ROOR \tag{3-84}$$

$$2ROO \cdot \xrightarrow{k_{85}} ROOR + O_2 \tag{3-85}$$

and, if the kinetic chain length is long, the amounts of R'OOH (an initiator fragment product) and any of the termination products ROOR or R_2 should be negligible. Examination of the chain sequence shows that the chain-propagating reaction, 3-81, is different from any of those encountered in Chapter 2 in that it is a coupling reaction. However, since molecular oxygen is a diradical, only two of the three radical sites in the reactants are destroyed in the coupling process and a free radical is produced that can propagate the chain sequence. As might be expected, the chain-propagating coupling reaction 3-81 has a very high reaction rate constant. On the other hand, the displacement reaction 3-82 has a rate constant more like those of other chain-propagating reactions that are unimolecular in free radicals. Thus, if $k_{81} \gg k_{82}$ and a steady-state concentration exists for each of the chain-carrying radicals, then

$$\frac{d[ROO \cdot]}{dt} = 0 = \frac{d[R \cdot]}{dt} \tag{3-86}$$

meaning that since

$$k_{81}[R \cdot][O_2] = k_{82}[ROO \cdot][RH], \tag{3-87}$$

$$[ROO \cdot] \gg [R \cdot] \tag{3-88}$$

providing, of course, a suitably high concentration of oxygen is always present. Of the three possible termination processes, the dimerization reaction of two ROO· radicals, namely 3-85, would seem the most probable. Using 3-85 as the termination reaction and invoking the steady-state assumption for the chain-carrying radicals, the derived rate law for autoxidation is given by 3-89. This is the experimentally observed kinetic rate law that most autoxidations

$$Rate = k_{82} \left(\frac{k_{78}}{2k_{85}} \right)^{1/2} [I]^{1/2} [RH] \tag{3-89}$$

follow if a sufficient concentration of oxygen is maintained.[8] If the concentration of oxygen is decreased, termination by 3-84 and ultimately by 3-83 will occur (see Chapter 11) and other rate laws are observed.

B. HALOGENATIONS WITH t-BUTYL HYPOCHLORITE

Another reaction that has been subjected to careful kinetic analysis is the chlorination of various compounds with t-butyl hypochlorite. The mechanism

$$(CH_3)_3COCl + RH \rightarrow (CH_3)_3COH + RCl \tag{3-90}$$

of these light-induced reactions with all possible termination reactions is given by 3-91 through 3-97. Reactions 3-93 and 3-94 describe the chain sequence

$$(CH_3)_3COCl \xrightarrow[(h\nu)]{k_{91}} (CH_3)_3CO\cdot + Cl\cdot \qquad (3\text{-}91)$$

$$Cl\cdot + RH \longrightarrow HCl + R\cdot \qquad (3\text{-}92)$$

$$(CH_3)_3CO\cdot + RH \xrightarrow{k_{93}} (CH_3)_3COH + R\cdot \qquad (3\text{-}93)$$

$$R\cdot + (CH_3)_3COCl \xrightarrow{k_{94}} RCl + (CH_3)_3CO\cdot \qquad (3\text{-}94)$$

$$2R\cdot \xrightarrow{k_{95}} R_2 \qquad (3\text{-}95)$$

$$2(CH_3)_3CO\cdot \xrightarrow{k_{96}} (CH_3)_3COOC(CH_3)_3 \qquad (3\text{-}96)$$

$$(CH_3)_3CO\cdot + R\cdot \xrightarrow{k_{97}} (CH_3)_3COR \qquad (3\text{-}97)$$

that accounts for the stoichiometry of the halogenation reaction. The reaction of t-butyl hypochlorite and toluene ($R = C_6H_5CH_2$—) in carbon tetrachloride, using low toluene:t-butyl hypochlorite ratios, followed the rate given in 3-98

$$Rate = K'[RH]^{0.92}[t\text{-BuOCl}]^{0.65}I^{0.35} \qquad (3\text{-}98)$$

where I is the intensity of the illumination and K' is a term that includes all of the reaction rate constants as well as the efficiency of the photolytic process in initiating this reaction.[1] This rate law suggests that termination occurs almost entirely by the coupling of two t-butoxy radicals (reaction 3-96) since derivation of the kinetic rate law based on this assumption results in 3-99.

$$Rate = k_{93}\left[\frac{k_{91}I}{2k_{96}}\right]^{1/2}[RH][t\text{-BuOCl}]^{1/2} \qquad (3\text{-}99)$$

Interestingly, the chlorination of chloroform ($R = Cl_3C$—) with t-butyl hypochloride follows a different kinetic rate law, namely 3-100.[1] The derived

$$Rate = K'[t\text{-BuOCl}]^{1.30}I^{0.5} \qquad (3\text{-}100)$$

rate law based on coupling of two chain-carrying $Cl_3C\cdot$ radicals (3-95) as the termination step is 3-101. Apparently trichloromethyl radicals are not as reactive

$$Rate = k_{94}\left[\frac{k_{91}I}{2k_{95}}\right]^{1/2}[t\text{-BuOCl}]^{3/2} \qquad (3\text{-}101)$$

as benzyl radicals in displacement reactions with t-butyl hypochlorite and in the chlorination of chloroform, reaction 3-94 becomes the rate-limiting step in the chain sequence. (See Chapter 5 for more details concerning the kinetics of this reaction).

C. BROMINATION WITH BROMOTRICHLOROMETHANE

If a reaction is terminated by the coupling reaction of only one of the chain-carrying radicals, this must mean that one of the reactions in the chain sequence is exerting a limiting effect on not only the overall rate but also on the kinetic chain length of the reaction. In some cases this effect may be so severe that it causes the reaction to have a short kinetic chain length and termination products may be detected. The light-induced reaction of bromotrichloromethane with toluene is such a reaction.[9] In a photochemically induced reaction,

$$BrCCl_3 + C_6H_5CH_3 \xrightarrow{h\nu} HCCl_3 + C_6H_5CH_2Br \qquad (3\text{-}102)$$

the kinetic chain length (determined by comparing the amounts of product formed relative to hydrogen bromide) was found to be about 20 and hexachloroethane was detected as a reaction product. The mechanism in 3-103 through 3-107 is suggested for this reaction based on these observations. The

$$BrCCl_3 \xrightarrow[h\nu]{k_{103}} Br\cdot + Cl_3C\cdot \qquad (3\text{-}103)$$

$$Br\cdot + RH \longrightarrow HBr + R\cdot \qquad (3\text{-}104)$$

$$Cl_3C\cdot + RH \xrightarrow{k_{105}} HCCl_3 + R\cdot \qquad (3\text{-}105)$$

$$R\cdot + BrCCl_3 \xrightarrow{k_{106}} RBr + Cl_3C\cdot \qquad (3\text{-}106)$$

$$2Cl_3C\cdot \xrightarrow{k_{107}} C_2Cl_6 \qquad (3\text{-}107)$$

derived kinetic rate law for this reaction is 3-108. Making use of the principles

$$\text{Rate} = k_{105} \left[\frac{k_{103} I}{2k_{107}} \right]^{1/2} [RH][BrCCl_3]^{1/2} \qquad (3\text{-}108)$$

developed earlier, the kinetic rate expression for the benzoyl peroxide-induced reaction of toluene, a reaction that should involve the same chain-propagating reactions and the same termination process, would be given by 3-109 where k_i

$$\text{Rate} = k_{105} \left[\frac{k_i}{2k_{107}} \right]^{1/2} [RH][Bz_2O_2]^{1/2} \qquad (3\text{-}109)$$

is the reaction rate constant for the unimolecular decomposition of benzoyl peroxide. The kinetics for this reaction were found experimentally to follow the

$$\text{Rate} = k'[RH]^{0.98} [Bz_2O_2]^{0.58} [BrCCl_3]^{0.13} \qquad (3\text{-}110)$$

rate law given in 3-110[10] which parallels closely the derived rate law.

D. OXIDATION OF SECONDARY ALCOHOLS WITH t-BUTYL PEROXIDE

The oxidation of secondary alcohols with t-butyl peroxide is a reaction in which one of the reactants, namely the peroxide, is involved in both the initiation reaction and the chain sequence. The mechanism for the reaction in which the alcohol is oxidized to the ketone with concurrent reduction of the peroxide to t-butyl alcohol is given by 3-112 through 3-115. Support for the cross-termination reaction 3-115 comes from the kinetic analysis of the reaction. In this process,

$$R_2CHOH + (CH_3)_3COOC(CH_3)_3 \longrightarrow R_2C{=}O + 2(CH_3)_3COH \quad (3\text{-}111)$$

via

$$(CH_3)_3COOC(CH_3)_3 \xrightarrow{k_{112}} 2(CH_3)_3CO \cdot \quad (3\text{-}112)$$

$$(CH_3)_3CO \cdot + R_2CHOH \xrightarrow{k_{113}} (CH_3)_3COH + R_2\dot{C}OH \quad (3\text{-}113)$$

$$R_2\dot{C}OH + (CH_3)_3COOC(CH_3)_3 \xrightarrow{k_{114}}$$

$$R_2C{=}O + (CH_3)_3COH + (CH_3)_3CO \cdot \quad (3\text{-}114)$$

$$R_2\dot{C}OH + (CH_3)_3CO \cdot \xrightarrow{k_{115}} R_2C{=}O + (CH_3)_3COH \quad (3\text{-}115)$$

the kinetic chain length is very likely not long since the induced decomposition of the peroxide by the α-hydroxyalkyl radical (see Chapter 10) accounts for only about 60–70% of the reaction of the peroxide. This estimation is made from the observation that the rate of decomposition in the secondary alcohol is only about 2 to 3 times faster than the unimolecular decomposition reaction 3-112. The rate of decomposition of the peroxide becomes

$$\text{Rate} = k_{112}[\text{Per.}] + k_{114}[R_2\dot{C}OH][\text{Per.}] \quad (3\text{-}116)$$

and invoking the steady-state assumption, the derived rate law in terms of the reactants would be

$$\text{Rate} = \left[\frac{3}{4}k_{112} \pm \frac{k_{112}}{2}\left(\frac{1}{4} + 2\frac{k_{113}k_{114}}{k_{112}k_{115}}[R_2CHOH]\right)^{1/2} \right][\text{Per.}] \quad (3\text{-}117)$$

The rate studies of the reaction of t-butyl peroxide in several secondary alcohols show that the reaction is first order in peroxide during the first two or three half-lives of the peroxide, when the reactants are present in about comparable amounts (5–10:1 alcohol:peroxide).[11] After much of the peroxide has been consumed, the reaction is no longer first order in peroxide indicating that the cross-termination process 3-115 is operative only if the chain-carrying radicals are present in about comparable amounts. As the concentration of peroxide is decreased during the course of the reaction, the ratio of the steady-state concentrations of the chain-carrying species increases in favor of the α-hydroxyalkyl radicals, and termination by the coupling of two α-hydroxyalkyl radicals occurs. The occurrence of part of the termination by 3-118 effects the overall rate law

$$2R_2\dot{C}OH \xrightarrow{k_{118}} \underset{\underset{\displaystyle OH \quad OH}{|\quad\quad|}}{R_2C-CR_2} \quad (3\text{-}118)$$

for the reaction since, if only termination by 3-118 occurs, the rate law would be

$$\text{Rate} = k_{112}[\text{Per.}] + k_{114}\left(\frac{k_{112}}{2k_{118}}\right)^{1/2}[\text{Per.}]^{3/2}. \tag{3-119}$$

Examination of the products obtained in reactions of secondary alcohols with t-butyl peroxide in which all of the peroxide has been consumed indicates the presence of coupling products of alcohol-derived radicals. Reactions of secondary alcohols in which the t-butyl peroxide is present in only small concentrations are reported to yield largely glycols,[12] an observation consistent with the kinetic analysis of the reactions at comparatively high and low peroxide concentration (see Chapter 10).

IV. Competition Studies

Although the kinetic analyses of free-radical chain reactions can be informative concerning certain details of the overall mechanism, such kinetic studies are of limited value in obtaining information about the reactivities of free radicals and substrates in chain-propagating reactions. Another kinetic approach to free-radical chemistry is available that is not only more informative but experimentally easier to handle. This approach is to use competition reactions, that is, reactions in which a given free radical has a choice of two or more chain-propagating reactions that it might follow.

A. SINGLE SUBSTRATE WITH TWO OR MORE REACTION SITES

Probably the simplest of all competition studies is that in which a single reactant participates in two or more chain-propagating reactions. Consider, for example, the chlorination of propane. Examination of the products of this reaction shows that both n-propyl chloride and isopropyl chloride are formed. These products suggest that the chlorine atom is capable of reacting either with the primary or secondary hydrogens of propane in the displacement reaction.

$$\text{Cl}\cdot + \text{CH}_3\text{CH}_2\text{CH}_3 \xrightarrow{k_{pri}} \text{HCl} + \text{CH}_3\text{CH}_2\dot{\text{C}}\text{H}_2 \tag{3-120}$$

$$\text{CH}_3\text{CH}_2\dot{\text{C}}\text{H}_2 + \text{Cl}_2 \longrightarrow \text{CH}_3\text{CH}_2\text{CH}_2\text{Cl} + \text{Cl}\cdot \tag{3-121}$$

$$\text{Cl}\cdot + \text{CH}_3\text{CH}_2\text{CH}_3 \xrightarrow{k_{sec}} \text{HCl} + \text{CH}_3\dot{\text{C}}\text{HCH}_3 \tag{3-122}$$

$$\text{CH}_3\dot{\text{C}}\text{HCH}_3 + \text{Cl}_2 \longrightarrow \underset{\underset{\text{Cl}}{|}}{\text{CH}_3\text{CHCH}_3} + \text{Cl}\cdot \tag{3-123}$$

The ratio of the reaction rate constants, k_{sec}/k_{pri}, is a measure of the reactivity of secondary hydrogens with respect to primary hydrogens of propane toward attack by a chlorine atom. In such a competition study, the ratio of reaction rate constants is equal to the product ratio multiplied by a statistical factor (3 in this case) equating the number of different hydrogens. Since

$$\frac{d[CH_3CHClCH_3]}{dt} = k_{sec}[Cl\cdot][C_3H_8] \tag{3-124}$$

and

$$\frac{d[CH_3CH_2CH_2Cl]}{dt} = k_{pri}[Cl\cdot][C_3H_8], \tag{3-125}$$

after dividing 3-124 by 3-125 and integrating from time zero to some finite time t, 3-126 results. In order to obtain k_{sec}/k_{pri}, only the relative amounts and

$$\frac{k_{sec}}{k_{pri}} = \frac{(CH_3CHClCH_3)_t}{(CH_3CH_2CH_2Cl)_t} \times 3 \tag{3-126}$$

not the actual concentrations of the products are required. The ratio of the reaction rate constants, since they are both those of second-order reactions, becomes unitless.

The relative rates of elimination of alkyl radicals from an alkoxy radical is an example of a unimolecular reaction of a single free radical that may follow more than a single path. The relative rates of β-elimination of alkyl radicals from alkoxy radicals can be measured by determination of the amounts of the ketones (or alkyl chlorides in the case of decompositions of t-alkyl hypochlorites)

$$R\cdot + R'COR'' \tag{3-127}$$

$$\begin{array}{c} R \\ | \\ R'-CO\cdot \xrightarrow{k_{128}} R'\cdot + RCOR'' \\ | \\ R'' \end{array} \tag{3-128}$$

$$R''\cdot + RCOR' \tag{3-129}$$

$$R\cdot(or\ R'\cdot\ or\ R''\cdot) + RR'R''COCl \rightarrow$$

$$RR'R''CO\cdot + RCl\ (or\ R'Cl\ or\ R''Cl) \tag{3-130}$$

produced in the reaction. Thus, for example,

$$\frac{k_{127}}{k_{128}} = \frac{(R'COR'')}{(RCOR'')} = \frac{(RCl)}{(R'Cl)} \tag{3-131}$$

where the amounts of the products formed are the terms in parentheses.

Obtaining kinetic data by product analysis where only one substrate is involved in the reactions with a particular radical can be used quite generally if appropriate analysis of the products is available. The only limitations are that the products examined are produced only as a result of the particular chain-propagating reactions under examination and that the kinetic order of the chain-propagating reactions is the same. An example of a reaction in which the latter limitation becomes important is the determination of the rate of fragmentation of a t-butoxy radical relative to its rate of hydrogen abstraction. Consider the following chain reaction, one in which t-butyl hypochlorite reacts with some substrate RH yielding RCl, t-butyl alcohol, acetone, and methane. Determina-

$$R \cdot + (CH_3)_3COCl \longrightarrow RCl + (CH_3)_3CO \cdot \qquad (3\text{-}132)$$

$$(CH_3)_3CO \cdot + RH \xrightarrow{k_{133}} (CH_3)_3COH + R \cdot \qquad (3\text{-}133)$$

$$(CH_3)_3CO \cdot \xrightarrow{k_{134}} CH_3COCH_3 + CH_3 \cdot \qquad (3\text{-}134)$$

$$CH_3 \cdot + (CH_3)_3COCl \longrightarrow CH_3Cl + (CH_3)_3CO \cdot \qquad (3\text{-}135)$$

tion of the amount of t-butyl alcohol relative to that acetone (or methane) is not necessarily a measure of the ease of fragmentation of the t-butoxy radicals. The ratio k_{133}/k_{134} is dependent on the nature of RH since the rate constant for this reaction is that for the reaction of the t-butoxy radical with RH. Although product analysis would give some information if different RHs were used as far as the reactivities of the RHs toward hydrogen abstraction by t-butoxy radicals are concerned, care must be taken since the rate of formation of t-butyl alcohol depends on both the concentration and structure of RH whereas the rate of formation of acetone (or methane) does not (precluding any unusual solvent effects that may influence the rate of the unimolecular reaction, see Chapter 5). The ratio k_{133}/k_{134} has the units of liters mole^{-1} in this case. A meaningful relationship concerning the reactivities of various species toward reaction with the t-butoxy radical relative to the fragmentation reaction as measured by the relative amounts of t-butyl alcohol and acetone formed can be ascertained if the concentrations of the various species [RH] are always the same or appropriate corrections made to account for differences in [RH].

$$\frac{d[t\text{-BuOH}]}{d[CH_3COCH_3]} = \frac{k_{133}[RH]}{k_{134}} \qquad (3\text{-}136)$$

B. TWO SUBSTRATES IN DIRECT COMPETITION

1. Nonreversible Systems. Reaction of two different reactants with a free radical will obviously result in more rapid consumption of the more reactive of the two. Consider the reactions of A and B, both of which can react (in either a displacement or an addition reaction) with some radical $X \cdot$. The latter results from interaction of XY with the radical produced in the reactions of A and B with $X \cdot$. If A is more reactive than B, A will be consumed faster than B and

$$X \cdot + A \xrightarrow{k_A} A \cdot \xrightarrow{XY} AY + X \cdot \qquad (3\text{-}137)$$

$$X \cdot + B \xrightarrow{k_B} B \cdot \xrightarrow{XY} BY + X \cdot \qquad (3\text{-}138)$$

AY will be produced faster than BY. A direct measure of the relative reactivity of A with respect to B is the ratio of reaction rate constants, k_A/k_B. Determination of this reactivity ratio is accomplished by measuring how much of A and B have reacted (or AY and BY have been formed) at time t in a competition reaction in which known amounts of A and B are present at the beginning of the reaction. The rates at which A and B react (or AY and BY are formed) are given by 3-139 and 3-140, respectively.

$$\frac{-d[A]}{dt} = \frac{d[AY]}{dt} = k_A[X\cdot][A] \tag{3-139}$$

$$\frac{-d[B]}{dt} = \frac{d[BY]}{dt} = k_B[X\cdot][B] \tag{3-140}$$

Dividing 3-139 by 3-140 and rearranging gives 3-141.

$$\frac{d[A]/[A]}{d[B]/[B]} = \frac{k_A}{k_B} \tag{3-141}$$

Integration of this equation over the limits of the initial concentration to the concentration at time t results in

$$\frac{\log(A_o/A_f)}{\log(B_o/B)_f} = \frac{k_A}{k_B} \tag{3-142}$$

where the subscripts "o" and "f" refer to the amounts of A and B present at the beginning of the competition reaction and the amounts remaining at time t, respectively.

2. **Reversible Systems.** Experimentally competition studies such as those outlined above generally are easily performed. The initial amounts of the reactants are known and it is necessary only to find a suitable analytical method of determining the amounts of each reactant remaining after the reaction has proceeded for a period of time. This experimentally determined relative reactivity ratio, which shall be referred to for the rest of this chapter as $\bar{P}$ where

$$\bar{P} = \frac{\log(A_o/A_f)}{\log(B_o/B_f)} \tag{3-143}$$

must be evaluated with some care. In the case of a competition study such as outlined in reactions 3-137 and 3-138, $\bar{P}$ is truly a measure of the relative reactivity ratio of the two substrates A and B toward reaction with $X\cdot$. This need not always be the case and the meaning of $\bar{P}$ can become more complex if either one or both of the reactions being compared involve a reversible reaction of the substrate with the reacting free radical. Consider the following situation:

$$X\cdot + A \underset{k_{-A}}{\overset{k_A}{\rightleftharpoons}} A\cdot \xrightarrow[k_{AY}]{XY} AY + X\cdot \tag{3-144}$$

$$X\cdot + B \underset{k_{-B}}{\overset{k_B}{\rightleftharpoons}} B\cdot \xrightarrow[k_{BY}]{XY} BY + X\cdot \tag{3-145}$$

If the equilibria between $A \cdot$ and A and $X \cdot$ and between $B \cdot$ and B and $X \cdot$ are established much faster than either $A \cdot$ or $B \cdot$ can react with XY, the rates at which A and B are consumed become dependent on the rates of the second step of the reaction, namely the reactions of $A \cdot$ and $B \cdot$ with XY. These rates depend on the equilibria concentrations of $A \cdot$ and $B \cdot$, on the respective reaction rate constants of $A \cdot$ and $B \cdot$ with XY and on the concentration of XY. The steady-state concentrations of $[A \cdot]$ and $[B \cdot]$ are given by 3-147 and 3-149, respectively.

$$k_A[A][X \cdot] = k_{-A}[A \cdot] + k_{AY}[A \cdot][XY] \tag{3-146}$$

and

$$[A \cdot] = \frac{k_A[X \cdot][A]}{k_{-A} + k_{AY}[XY]} \tag{3-147}$$

$$k_B[B][X \cdot] = k_{-B}[B \cdot] + k_{BY}[B \cdot][XY] \tag{3-148}$$

and

$$[B \cdot] = \frac{k_B[X \cdot][B]}{k_{-B} + k_{BY}[XY]} \tag{3-149}$$

The rate of reaction of A relative to B (or rate of formation of AY relative to BY) is given by 3-150.

$$\frac{d[A]}{d[B]} = \frac{k_A k_{AY}[A](k_{-B} + k_{BY}[XY])}{k_B k_{BY}[B](k_{-A} + k_{AY}[XY])} \tag{3-150}$$

After rearrangement and integration (assuming $[XY]$ is large and undergoes essentially no change), equation 3-151 is obtained. It is obvious that the experimentally determined relative reactivity ratio $\bar{P}$ measures a more complex

$$\bar{P} = \frac{\log (A_o/A_f)}{\log (B_o/B_f)} = \frac{k_A k_{AY}(k_{-B} + k_{BY}[XY])}{k_B k_{BY}(k_{-A} + k_{AY}[XY])} \tag{3-151}$$

quantity in this case, one that involves all of the reaction rate constants as well as the concentration of XY.

Examination of equation 3-151 shows that if $k_{BY}[XY] \gg k_{-B}$ and $k_{AY}[XY] \gg k_{-A}$, which would be the case if the first step is not reversible, the equation reduces to 3-142. On the other hand, if $k_{-A} \gg k_{AY}[XY]$ and $k_{-B} \gg k_{BY}[XY]$, the measured relative reactivity ratio $\bar{P}$ becomes

$$\bar{P} = \frac{(k_A/k_{-A}) k_{AY}}{(k_B/k_{-B}) k_{BY}} \tag{3-152}$$

and the equilibrium constants for the reversible reactions are largely responsible for the observed relative reactivity ratios. Although reaction rate constants for the reactions of $A \cdot$ and $B \cdot$ with XY are involved, the concentration of XY no longer plays a role in determining $\bar{P}$.

An interesting situation arises if one reaction is reversible and the other is not as shown in 3-153 and 3-154. The rate equation for the relative change in

$$X \cdot + A \xrightarrow{k_A} A \cdot \xrightarrow[k_{AY}]{XY} AY + X \cdot \tag{3-153}$$

$$X \cdot + B \underset{k_{-B}}{\overset{k_B}{\rightleftarrows}} B \cdot \xrightarrow{\overset{XY}{}}_{k_{BY}} BY + X \cdot \tag{3-154}$$

the concentrations of A and B based on steady-state assumptions becomes

$$\frac{d[A]}{d[B]} = \frac{k_A[A][X \cdot]}{k_B k_{BY}[XY][B][X \cdot]/k_{-B} + k_{BY}[XY]} \tag{3-155}$$

or

$$\frac{d[A]/[A]}{d[B]/[B]} = \frac{k_A(k_{-B} + k_{BY}[XY])}{k_B k_{BY}[XY]} \tag{3-156}$$

Further simplification and integration gives

$$\bar{P} = \frac{k_A k_{-B}}{k_B k_{BY}[XY]} + \frac{k_A}{k_B} \tag{3-157}$$

which shows that if the reversibility of one of the reactions is an important factor, the concentration of the reactant XY will play a role in determining $\bar{P}$. The rate of reaction of B, the component involved in the reversible reaction, depends on [XY]. If $k_{BY}[XY] \gg k_{-B}$, the reaction is no longer reversible and the first term of the equation becomes zero and $\bar{P}$ is again simply a measure of k_A/k_B, the relative reactivity of A with respect to B toward reaction with $X \cdot$.

Experimentally $\bar{P}$ should be inversely proportional to [XY] for a system such as given in 3-153 and 3-154. A plot of $\bar{P}$ against [XY] would have a slope of $k_A k_{-B}/k_B k_{BY}$ and the intercept would be k_A/k_B. This intercept would be at a point where the first term of the equation is zero, a situation that would result only if the concentration of XY is infinitely high. If this were the case, a reversible reaction of $X \cdot$ and B would not exist since it would be expected to react infinitely fast with XY.

Various competition studies will also appear throughout the remainder of this text. Of particular interest are the competition reactions encountered in vinyl polymerization reactions. The details of these particular competition reactions are discussed in those sections where the need for the data is required.

References

1. J. A. Howard and K. U. Ingold, *J. Amer. Chem. Soc.*, **88**, 4725, 4726 (1966).
2. C. Walling and V. P. Kurkov, *ibid.*, **88**, 4727 (1966); **89**, 4895 (1967).
3. G. S. Hammond, W. P. Baker, and W. M. Moore, *ibid.*, **83**, 2795 (1961).
4. M. S. Matheson, E. E. Aver, E. B. Bevilacqua, and E. J. Hart, *ibid.*, **73**, 1700 (1951).
5. H. Kwart, H. S. Broadbent, and P. D. Bartlett, *ibid.*, **72**, 1060 (1950).
6. G. R. McMillan, *ibid.*, **82**, 2422 (1960).
7. F. W. Lampe and R. M. Moyer, *ibid.*, **76**, 2140 (1954).

8. L. Bateman, *Quart. Revs.*, **8**, 147 (1954).

9. E. S. Huyser, *J. Amer. Chem. Soc.*, **82**, 391 (1960).

10. E. S. Huyser and K. E. Schmude, unpublished results.

11. E. S. Huyser and C. J. Bredeweg, *J. Amer. Chem. Soc.*, **86**, 2401 (1964).

12. K. Schwetlick, W. Geyer, and H. Hartmann, *Angew. Chem.*, **72**, 779 (1960).

Chapter 4

Structure and Reactivity Factors in Reactions of Free Radicals

I. General Considerations

The significant aspects of the kinetic analyses of free-radical chain reactions discussed in the previous chapter can be readily summarized as follows: chain-propagating reactions must proceed at rates that are faster than the termination reactions if a long kinetic chain length for the chain sequence is to be expected. In this chapter, some of the factors that affect the rates of the chain-propagating reactions of free radicals will be examined. An appreciation of these factors will make it possible to understand why some reactions have a long kinetic chain length whereas others, which might be expected to be good chain reactions, have either a short kinetic chain length or are not chain reactions at all.

The fact that termination reactions are bimolecular in free radicals whereas the propagating reactions are unimolecular is a most significant factor in free-radical chain reactions. This necessitates small radical concentrations if long kinetic chain lengths are to be expected. This concentration factor alone is not enough, however, to assure that the radical-propagating reactions will be fast enough to compete with the termination reactions. As a generalization, it can be expected that termination reactions have low activation energy requirements whereas the activation energy requirements of the propagating reactions are not only higher but may well cover a larger range. Even if radical concentrations are kept low, the propagating reactions will not compete with termination if the activation energy of the propagating reaction is too high. The discussions that follow, therefore, will be centered on the chain-propagating reactions because these reactions are generally the limiting factor in the overall free-radical reaction. An appreciation of the factors that affect the activation energy requirements for radical-propagating reaction require an understanding of the resonance, polar, and steric aspects of these reactions.

II. Resonance Factors

The role of resonance in affecting the rates of free-radical reactions is important but can be easily overemphasized. While it is essential, however, to have an understanding of the resonance factors encountered in radical reactions, it is equally important to know the limitations that must be put on the use of resonance concepts in evaluating the factors that determine the rate of a radical reaction.

A. BOND DISSOCIATION ENERGIES

A bond dissociation energy is defined as the energy required to break a chemical bond homolytically. The result of such a cleavage would be the formation of two species, each with an unpaired electron. Bond dissociation energies differ from bond energies. This can best be illustrated by comparing the bond dissociation energy of a carbon-hydrogen bond in methane with the bond energies of the carbon-hydrogen bonds of the same hydrocarbon. The bond dissociation energy is that energy required to break a single bond and in this case would yield a hydrogen atom and a methyl radical. The bond energy of

$$CH_4 \rightarrow CH_3 \cdot + H \cdot + 104 \text{ kcal/mole} \tag{4-1}$$

the carbon-hydrogen bond is the average energy required to break all of the covalent bonds divided by four, the number of bonds involved. The bond

$$CH_4 \rightarrow C + 4H \cdot + 398 \text{ kcal/mole} \tag{4-2}$$

energy in methane (398 kcal/4) is very nearly the same as the bond dissociation energy although two very different processes are involved. The applicability of bond dissociation energies to free-radical chemistry should be obvious since the species formed are of the type encountered in free-radical reactions. Bond energies, on the other hand, offer no significant information directly pertinent to free-radical processes.

B. RESONANCE ENERGIES OF RADICALS

The bond dissociation energy of a carbon-hydrogen bond in methane is peculiar to methane and is the consequence of the species produced, namely a hydrogen atom and a methyl radical. The bond dissociation energy of a carbon-

$$CH_3CH_2—H \rightarrow CH_3CH_2 \cdot + H \cdot + 98 \text{ kcal/mole} \qquad (4\text{-}3)$$

hydrogen bond of ethane is different because the homolytic cleavage of a carbon-hydrogen bond in ethane produces an ethyl radical rather than a methyl radical. The lower energy requirement for the homolytic cleavage of a carbon-hydrogen bond in ethane is due to the greater resonance stabilization of the ethyl compared to the methyl radical. The stabilities of the alkyl radicals increase with the extent of hyperconjugative delocalizing possible as evidenced by the bond dissociation energies of carbon-hydrogen bonds leading to primary, secondary, and tertiary radicals. Assigning a resonance energy of 0 kcal/mole to the methyl radical, the stabilization of the ethyl, isopropyl, and t-butyl radicals would be 6.0, 9.5, and 13.0 kcal/mole, respectively (see Table 4-1).

TABLE 4-1   Carbon-Hydrogen Bond-Dissociation Energies $(\Delta H_d)^a$

Bond	Radical Formed	ΔH_d (kcal/mole)	Resonance Energy of Radical[b]
$CH_3—H$	$CH_3 \cdot$	104.0 ± 1	0.0
$C_2H_5—H$	$C_2H_5 \cdot$	98.0 ± 1	6.0 ± 2
$(CH_3)_2CH—H$	$(CH_3)_2\dot{C}H$	94.5 ± 1	9.5 ± 2
$(CH_3)_3C—H$	$(CH_3)_3C \cdot$	91.0 ± 1	13.0 ± 2
$CH_2{=}CHCH_2—H$	$CH_2{=}CH\dot{C}H_2$	85 ± 1	19 ± 2
$C_6H_5CH_2—H$	$C_6H_5\dot{C}H_2$	85 ± 1	19 ± 2
$CH_3COCH_2—H$	$CH_3CO\dot{C}H_2$	92 ± 3	12 ± 4
$NCCH_2—H$	$NC\dot{C}H_2$	86 ± 3	18 ± 4
$CF_3—H$	$F_3C \cdot$	106 ± 1	—
$Cl_3C—H$	$Cl_3C \cdot$	95.7 ± 1	8.3 ± 2
$CH_2{=}CH—H$	$CH_2{=}\dot{C}H$	104 ± 2	0 ± 2
$C_6H_5—H$	$C_6H_5 \cdot$	104 ± 2	0 ± 2

[a] Bond dissociation energies in Tables 4-1 and 4-2 are taken from compilations given in J. A. Kerr, *Chem. Rev.*, **66**, 465 (1966), and S. W. Benson, *J. Chem. Ed.*, **42**, 502 (1965).

[b] Relative to CH_3.

The bond dissociation energy of an allylic hydrogen of propylene indicates that the allylic radical is significantly stabilized by delocalization. In this case, two equivalent structures contribute to the resonance-stabilized hybrid radical.

$$CH_2\!\!=\!\!CHCH_2\cdot \;\leftrightarrow\; \cdot CH_2CH\!\!=\!\!CH_2$$

Similarly, carbon radicals having α-phenyl, α-carbonyl, and α-cyano groups

are stabilized to a significant degree because of delocalization of the unpaired electron and this is reflected in the bond dissociation energies of carbon-hydrogen bonds that lead to these species.

Comparison of the bond dissociation energies of the carbon-hydrogen bonds in chloroform and fluoroform suggests that the chlorine atom has a greater stabilizing effect than fluorine possibly because of delocalization of the unpaired

electron into the empty d-orbitals of chlorine. Fluorine has no empty orbitals of low enough energy to allow for any significant delocalization of the unpaired electron on the halogen atoms of the trihalomethyl radical. Bromine should be expected to display a stabilizing effect similar to that of chlorine. Both phenyl and vinyl radicals are relatively less stable than most alkyl radicals.

Table 4-2 lists other bond dissociation energies useful in the study of free-radical reactions. Before applying these bond dissociation energies to some further aspects of free-radical reactions, we shall discuss some of these energies in terms of the radicals produced.

The dissociation energies of the carbon-carbon bonds are considerably smaller than those of most carbon-hydrogen bonds. As will be pointed out later in this chapter, cleavage of carbon-carbon bonds by a displacement is not as common in free-radical reactions as might be expected. However, the formation of a carbon-carbon bond is a very important process in free-radical addition reactions. It is worth noting that the bond dissociation energies of the different

TABLE 4-2   Bond Dissociation Energies^a

Bond	ΔH_d (kcal/mole)		Bond	ΔH_d (kcal/mole)
O‖ CH$_3$C—H	88	± 2	F—F	38
			Cl—Cl	58
HOCH$_2$—H	92	± 2	Br—Br	46
HOCH(CH$_3$)—H	90	± 2	I—I	36.1
ĊH$_2$—CH$_2$—H	40	± 2	CH$_3$—F	108	± 3
CH$_3$O—H	102	± 2	Cl$_3$C—F	106	± 3
(CH$_3$)$_3$CO—H	103	± 2	CH$_3$—Cl	84	± 2
CH$_3$S—H	88	$\pm ?$	C$_2$H$_5$—Cl	81	± 2
C$_6$H$_5$S—H	75	$\pm ?$	(CH$_3$)$_2$CH—Cl	81	± 2
CH$_3$—CH$_3$	88	± 2	(CH$_3$)$_3$C—Cl	79	± 2
C$_6$H$_5$CH$_2$—CH$_3$	70	± 2	C$_6$H$_5$CH$_2$—Cl	68	± 2
CH$_3$—OH	91	± 2	Cl$_3$C—Cl	73	± 2
CH$_3$O—CH$_3$	80	± 5	CH$_3$—Br	70	± 2
HO—OH	51	± 1	C$_2$H$_5$—Br	69	± 2
(CH$_3$)$_3$CO—OH	44	± 3	(CH$_3$)$_2$CH—Br	68	± 2
(CH$_3$)$_3$CO—OC(CH$_3$)$_3$	37.4	± 1	(CH$_3$)$_3$C—Br	63	± 2
			C$_6$H$_5$CH$_2$—Br	51	± 1
			Cl$_3$C—Br	54	± 2
O‖ O‖ CH$_3$CO—OCCH$_3$	30	± 2	CH$_3$—I	56.3	± 1
			C$_6$H$_5$CH$_2$—I	40	± 1
H$_2$N—NH$_2$	56	± 2	F$_3$C—I	54	± 2
(CH$_3$)$_2$N—N(CH$_3$)$_2$	42	± 5	H—F	135.8
HO—Cl	60	± 3	H—Cl	103.0
HO—Br	56	± 3	H—Br	87.5
			H—I	71.3

^{a} See footnote ^{a} of Table 4.1.

carbon-carbon bonds parallel the stabilities of the radicals produced by the homolytic cleavage.

The dissociation energy of a β-hydrogen from an ethyl radical leads to a value for a π-bond. Examination of the consequence of this homolytic cleavage

$$\cdot CH_2CH_2—H \rightarrow CH_2{=}CH_2 + H\cdot + 40 \text{ kcal/mole} \qquad (4\text{-}4)$$

shows that the olefinic bond of ethylene will be produced. The energy required to break a carbon-hydrogen bond of the ethyl radical is 40 kcal/mole whereas breaking a similar carbon-hydrogen bond in ethane is 98 kcal/mole. The conclusion is that the bond dissociation energy of the π-electron system in ethylene is 58 kcal/mole.

The remaining bond dissociation energies in Table 4-2 will be encountered in the discussion of other reactions.

C. ENTHALPIES OF RADICAL REACTIONS

Along with giving some concept of the relative stabilities of free radicals, the bond dissociation energies allow us to determine the enthalpies (*vide infra*) of many radical-propagating reactions that comprise a chain sequence and, consequently, the enthalpies of the overall reaction. Although informative, other information is also often required to predict either the feasibility or the course of a free-radical reaction.

1. Determining Reaction Enthalpies. The same amount of energy required for the homolytic cleavage of a given bond is released when the radicals produced are allowed to recombine to form the same chemical bond. Bond dissociation energies make it possible to determine the energies (enthalpies) of free-radical reactions in which bonds are made and broken. The following examples illustrate how this is done and also point out certain limitations in using reaction enthalpies for determining the course of a free-radical reaction.

Consider the reaction of a chlorine atom with methane, a radical-propagating

$$Cl \cdot + CH_3—H \rightarrow HCl + CH_3 \cdot \qquad (4-5)$$

reaction in which a carbon-hydrogen bond is broken—a process requiring energy—and a hydrogen-chlorine bond is formed—a process releasing energy.

Break	$CH_3—H$	+104 kcal/mole
Make	H—Cl	−103 kcal/mole
	$\Delta H = +$	1 kcal/mole

The reaction is possibly endothermic to the extent of about 1 kcal/mole. Reaction of a methyl radical with molecular chlorine is, however, a comparatively exothermic reaction. The reaction of chlorine and methane to form

$$CH_3 \cdot + Cl_2 \rightarrow CH_3Cl + Cl \cdot \qquad (4-6)$$

Break	Cl—Cl	+58 kcal/mole
Make	$CH_3—Cl$	−84 kcal/mole
	$\Delta H = −26$	kcal/mole

hydrogen chloride and methyl chloride is an exothermic reaction and the enthalpy of the chain reaction is the sum of the two chain-propagating reactions.

Since chlorination of methane is a reaction having a long kinetic chain length,

$$Cl\cdot + CH_4 \rightarrow HCl + CH_3\cdot \qquad \Delta H = +\ 1 \text{ kcal/mole}$$

$$CH_3\cdot + Cl_2 \rightarrow CH_3Cl + Cl\cdot \qquad \Delta H = -26 \text{ kcal/mole}$$

$$\overline{CH_4 + Cl_2 \rightarrow CH_3Cl + HCl \qquad \Delta H = -25 \text{ kcal/mole}}$$

$$(4\text{-}7)$$

most of the reactants are converted to products in the chain sequence, and the enthalpy of the chain sequence is essentially the enthalpy of this particular chlorination reaction. Contributions to the overall energy of the reaction by the energy released in the termination process (a coupling reaction that releases much energy) or by the initiation reaction (which requires energy) are negligible due to the comparatively small number of bonds made or broken in these processes if the reaction has a long kinetic chain length.

Another reaction that illustrates how bond dissociation energies can be used to determine the enthalpy of the overall reaction is the free-radical polymerization of an unsaturated compound such as ethylene. The mechanism of the overall reaction involves initiation, chain propagation, and termination. The polymer molecules actually obtained as the product are produced in the termination reaction. However, if a high molecular weight polymer is formed, the kinetic chain length of the chain-propagating reaction 4-10 must be very

$$\text{Init.} \rightarrow 2R\cdot \qquad (4\text{-}8)$$

$$R\cdot + CH_2{=}CH_2 \rightarrow RCH_2\dot{C}H_2 \qquad (4\text{-}9)$$

$$RCH_2CH_2\cdot + nCH_2{=}CH_2 \rightarrow R(CH_2CH_2)_nCH_2\dot{C}H_2 \qquad (4\text{-}10)$$

$$2R(CH_2CH_2)_nCH_2CH_2\cdot \rightarrow R(CH_2CH_2)_{2n+2}R \qquad (4\text{-}11)$$

long. This means that the energetics are determined almost entirely in the chain-propagating reaction, namely addition of a primary alkyl radical to the unsaturated linkage of ethylene. The contribution to the enthalpy of this reaction to the enthalpy of the overall reaction would be large, whereas that of the termination reaction would, in comparison, be very small as would be the energy required to decompose the initiator into radical fragments. In the propagating reaction, a carbon-carbon bond is formed and the π-electron system of ethylene is broken. The dissociation energies of the bonds involved in

Make	$\sim CH_2CH_2{-}CH_2CH_2\cdot$	-80 kcal/mole
Break	$CH_2 {\cdots} CH_2$	$+58$ kcal/mole
		$\overline{\Delta H = -22 \text{ kcal/mole}}$

this reaction predict that the heat of polymerization of ethylene should be approximately -22 kcal/mole, a value very close to that observed experimentally.

The following examples point out some calculations that may be useful in terms of gaining insight into radical reactions. Consider the bromination of toluene with molecular bromine. In this reaction, both chain-propagating

$$Br \cdot + C_6H_5CH_3 \rightarrow HBr + C_6H_5CH_2 \cdot \qquad \Delta H = -2.5 \text{ kcal/mole} \quad (4\text{-}12)$$

$$C_6H_5CH_2 \cdot + Br_2 \rightarrow C_6H_5CH_2Br + Br \cdot \qquad \Delta H = -5 \quad \text{kcal/mole} \quad (4\text{-}13)$$

$$C_6H_5CH_3 + Br_2 \rightarrow C_6H_5CH_2Br + HBr \qquad \Delta H = -7.5 \text{ kcal/mole} \quad (4\text{-}14)$$

reactions are exothermic and, consequently, the overall reaction is exothermic. However, the bromination of methane presents a somewhat different problem.

$$Br \cdot + CH_4 \rightarrow HBr + CH_3 \cdot \qquad \Delta H = +16.5 \text{ kcal/mole} \quad (4\text{-}15)$$

$$CH_3 \cdot + Br_2 \rightarrow CH_3Br + Br \cdot \qquad \Delta H = -24 \quad \text{kcal/mole} \quad (4\text{-}16)$$

$$CH_4 + Br_2 \rightarrow CH_3Br + HBr \qquad \Delta H = - 7.5 \text{ kcal/mole}$$

Again the overall reaction is exothermic but one of the reactions in the chain sequence is endothermic. Bromination of methane has, at best, a comparatively short kinetic chain length because, even though the thermochemistry predicts that the overall reaction should be favorable, one step in the chain sequence is not. This particular example illustrates the limitation of simply determining the enthalpy of the overall reaction and the necessity of knowing the enthalpies of the individual steps in the chain sequence to determine the feasibility of certain radical reactions.

2. Limitations of Use of Enthalpies. Some generalizations can be made concerning resonance. A resonance-stabilized radical is generally less reactive than one that has little or no resonance stabilization. When several reactions that have *very similar character* are compared (see Figure 4.1), the more exothermic reactions (or least endothermic) are faster. For example, the order of reactivity of the hydrogens in an alkane toward displacement by a radical $X \cdot$ would be tertiary > secondary > primary, provided only tertiary, secondary, and primary alkyl radicals are produced in the reaction. The exceptions to these generalizations, due to polar and steric factors, are of particular interest.

The energy profiles for some hypothetical reactions (Figures 4.2 and 4.3) illustrate the limitations of the thermochemistry in determining the feasibility of a given reaction. Paths A and D of Figures 4.2 and 4.3, respectively, are those expected for the indicated reactions. It is possible, for polar or steric reasons, that an exothermic reaction may have a high activation energy requirement

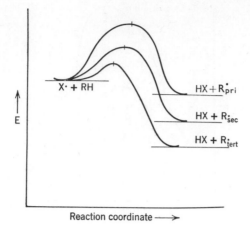

Fig. 4.1.   Energy profiles for similar reactions.

(B in Figure 4.2) or an endothermic reaction, provided it is not too endothermic, may have a fairly low activation energy requirement (C in Figure 4.3). Examination of the enthalpies of the individual steps in a chain sequence is, therefore, not sufficient to allow for predictions concerning the feasibility of a chain sequence.

3. The Hammond Postulate. A useful kinetic concept in free-radical chemistry is the "Hammond postulate." In general, an exothermic reaction has a low activation energy requirement and, as illustrated in Figure 4.4, a transition state that resembles the reactants of the reaction more than it resembles the

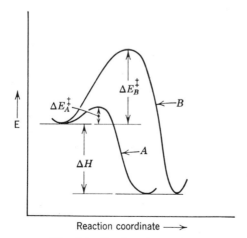

Fig. 4.2.   Exothermic reaction.

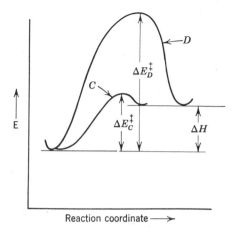

Fig. 4.3.   Endothermic reaction.

products. An endothermic reaction has a higher activation energy and a transition state that resembles the products more than the reactants (Figure 4.5). The possible importance of resonance stabilization of the free radicals produced in a chain-propagating reaction will most likely appear in reactions in which there is considerable productlike character in the transition state. The course of certain radical-propagating reactions, because they may involve a rapid chain-propagating step with a low activation energy requirement, is often dictated by factors other than the resonance energy of the radical produced in the reaction.

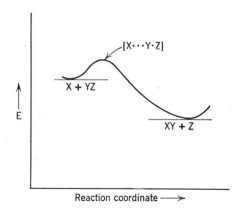

Fig. 4.4.   Low activation energy reaction.

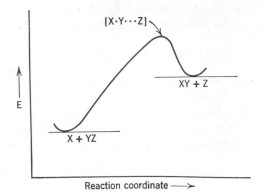

Fig. 4.5.   High activation energy reaction.

III. Polar Effects

A. TYPES OF POLAR EFFECTS

Two kinds of polar effects have been observed in free-radical chain-propagating reactions. One of these, an inductive effect, is found in the reactions of reactive electrophylic radicals such as the chlorine atom. The other, which can be referred to as the donor-acceptor polar effect, is somewhat more subtle but more common in radical reactions. Our discussion of polar effects in this chapter deals mainly with the donor-acceptor polar effect. The inductive effects found in chlorine atom reactions are discussed in detail in Chapter 5.

B. DONOR-ACCEPTOR CONCEPT

At the outset, we would not expect reactions that involve electrically neutral free radicals and substrates that produce neutral products to be influenced by polar effects. It appears, however, that the transition states of many such reactions have polar character. It is not always possible to state precisely the nature of this polar character, but enough is known that some predictions can be made concerning its importance to the transition state of the reaction.

The donor-acceptor concept can be illustrated by a hypothetical free-radical chain-propagating displacement reaction between a radical $R\cdot$ and a substrate AB yielding RA and $B\cdot$, a reaction that has a transition state in which there is

$$R\cdot + AB \rightarrow [R\cdot\cdot A\cdot\cdot B] \rightarrow RA + B\cdot \qquad (4\text{-}17)$$

some bond-making between $R\cdot$ and A and some bond-breaking between A and B. The resonance stabilization of radicals $R\cdot$ and $B\cdot$ probably plays a role in determining the energy of the transition state of this reaction. The energy of the transition state could be lowered by allowing a partial separation of charges to appear in the transition state of the reaction. The reaction of $R\cdot$

$$R\cdot + AB \rightarrow \left[R^{\delta+}\cdots A \cdots \overset{\delta-}{B}\right] \rightarrow RA + B\cdot \qquad (4\text{-}18)$$

and AB, for example, could involve a transition state in which the radical $R\cdot$ assumes some positive (or cationic) character and the B moiety of AB assumes negative (or anionic) character. This transition state can be regarded as a resonance hybrid of the two extremes, namely the nonpolar transition state and that in which there is complete transfer of an electron from $R\cdot$ to B. The

$$[R\cdots A\cdots B] \leftrightarrow \left[R^+\cdots A\cdots \overset{-}{B}\right]$$

energy of the hybrid transition state would be lower than that of a reaction in which there is no delocalization of the electrons (see Figure 4.6). The extent of the delocalization of the charges depends on the importance of the cannonical structure having complete separation of charges as a contributor to the hybrid and is dictated by the nature of $R\cdot$ and B.

A polar transition state might be expected to be solvated to some extent by a polar solvent thereby lowering the activation energy requirement for the reaction relative to that found if the reaction were carried out in a less polar medium. Such solvent effects generally are not observed in chain-propagating reactions possibly because the enthalpy change is offset by the entropy change in the temperature range of most reactions. Solvent effects that are observed in radical-propagating reactions (Chapter 5) often involve extensive interaction of the solvent with a chain-carrying radical.

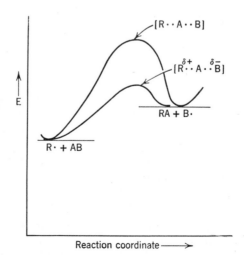

Fig. 4.6. Effect of polar contribution to transition state energy requirement.

Free radicals that assume cationic character in transition states are called donor radicals. In order to function as a donor radical, an acceptor species must also be present. In the case cited above, the substrate with which the donor radical reacted served as the acceptor. The donor-acceptor process could have been reversed; the attacking radical could have been the acceptor and the substrate the donor.

$$\text{R} \cdot + \text{AB} \rightarrow \left[\text{R}^{\delta-} \cdot \cdot \text{A} \cdot \cdot \overset{\delta+}{\text{B}} \right] \rightarrow \text{RA} + \text{B} \cdot \qquad (4\text{-}19)$$

A donor radical can be defined as one that would yield a more stable cation if it lost an electron than anion if it gained an electron.

$$\underset{\text{Less stable than } \text{R}^{(+)}}{\text{R}^{(-)}} \xleftarrow{+e^{(-)}} \underset{\text{Donor radical}}{\text{R} \cdot} \xrightarrow{-e^{(-)}} \underset{\text{More stable than } \text{R}^{(-)}}{\text{R}^{(+)}}$$

$$(4\text{-}20)$$

A donor substrate is any species that will yield a donor-free radical in a chain-propagating reaction. An acceptor radical is defined as one that gives a more stable anion on gaining an electron than cation by losing an electron. An acceptor substrate is one that would yield an acceptor radical in a chain-

$$\underset{\text{More stable than } \text{R}^{(+)}}{\text{R}^{(-)}} \xleftarrow{+e^{(-)}} \underset{\text{Acceptor radical}}{\text{R} \cdot} \xrightarrow{-e^{(-)}} \underset{\text{Less stable than } \text{R}^{(-)}}{\text{R}^{(+)}}$$

$$(4\text{-}21)$$

propagating reaction. These definitions will become clearer when illustrated with reactions of donor and acceptor radicals with acceptor and donor substrates, respectively.

C. DONOR RADICALS AND SUBSTRATES

The definition of a donor radical in the previous section suggests that alkyl radicals would be good donor radicals. A species such as a t-butyl radical, for example, on loss of an electron yields the comparatively stable t-butyl carbonium ion. Furthermore, a t-butyl radical is a better donor radical than an ethyl radical since tertiary carbonium ions are more stable than primary carbonium ions. The stabilities of the allylic and benzylic carbonium ions suggest that allyl and benzyl radicals should be good donor radicals.

Donor substrates are defined as species that yield donor radicals in radical-propagating reactions. If we consider hydrogen abstraction (displacement on hydrogen) as a means of converting a substrate to a free radical, isobutane, propylene, and toluene would be donor substrates since they produce donor radicals. Isobutylene, butadiene, and styrene are donor substrates since addition

$$\textit{Donor substrates} \qquad \textit{Donor radicals}$$

$$R \cdot + (CH_3)_3CH \rightarrow RH + (CH_3)_3C \cdot \tag{4-22}$$

$$R \cdot + CH_2{=}CHCH_3 \rightarrow RH + CH_2{=}CHCH_2 \cdot \tag{4-23}$$

$$R \cdot + C_6H_5CH_3 \rightarrow RH + C_6H_5CH_2 \cdot \tag{4-24}$$

yields a radical in each case with essentially the same donor qualities of the t-butyl, allyl, and benzyl radicals.

$$\textit{Donor substrates} \qquad \textit{Donor radicals}$$

$$R \cdot + CH_2{=}C(CH_3)_2 \rightarrow RCH_2\dot{C}(CH_3)_2 \tag{4-25}$$

$$R \cdot + CH_2{=}CHCH{=}CH_2 \rightarrow RCH_2\dot{C}HCH{=}CH_2 \tag{4-26}$$

$$R \cdot + CH_2{=}CHC_6H_5 \rightarrow RCH_2\dot{C}HC_6H_5 \tag{4-27}$$

Table 4-3 lists some of the more commonly encountered donor radicals and the general structure of the substrates from which they can be derived either in displacement or addition reactions. As shown, the addition would occur with bonding of the adding radical to the β-carbon relative to the functional group.

Substituents in the *meta* and *para* positions of toluenes that affect the stability of a benzylic carbonium ion will affect the behavior of substituted toluenes as donors. For example, p-nitrotoluene is a comparatively poor donor relative to toluene in benzylic hydrogen abstractions whereas p-methoxytoluene is a better donor than toluene in similar reactions.

The ability of oxygen to stabilize a positive charge on an adjacent carbon by delocalization of the charge makes α-hydroxy- and α-alkoxyalkyl radicals

$$|\overset{\frown}{\overset{\downarrow}{O}}{-}C^{(+)}\stackrel{\nearrow}{\searrow} \leftrightarrow |\overset{(+)}{O}{=}C\stackrel{\nearrow}{\searrow}$$

excellent donor radicals. Many free-radical chain reactions involving alcohols and ethers proceed readily because of their donor qualities. Similarly, α-amino alkyl radicals are good donor radicals and many amines show donor charac-

$$-\overset{\frown}{\overset{\downarrow}{N}}{-}C^{(+)}\stackrel{\nearrow}{\searrow} \leftrightarrow -\overset{(+)}{N}{=}C\stackrel{\nearrow}{\searrow}$$

teristics in their free-radical reactions. The stability of the acylium ion due to delocalization causes the acyl radicals to display definite donor qualities in their

$$\stackrel{\nearrow}{\searrow}\overset{\frown}{O}{\overset{(+)}{=}}C- \leftrightarrow \overset{+}{\underline{O}}{\equiv}C-$$

reactions and aldehydes to act as donor substrates.

TABLE 4-3   Donor Radicals and Substrates

Name	General Structure	Substrate Displacement	Substrate Addition
Primary alkyl	$R\dot{C}H_2$	RCH_2—H	CH_2=CH_2
Secondary alkyl	$R_2\dot{C}H$	R_2CH—H	RCH=CH_2
Tertiary alkyl	$R_3C\cdot$	R_3C—H	R_2C=CH_2
Allylic	$\!>\!\!C\!=\!C\!-\!\overset{\shortmid}{C}\!\cdot$	$\!>\!\!C\!=\!\overset{\shortmid}{C}\!\overset{\shortmid}{C}\!-\!H$	$\!>\!\!C\!=\!\overset{\shortmid}{C}\!-\!\overset{\shortmid}{C}\!=\!CH_2$
Benzylic	$Ar\overset{\shortmid}{\underset{\shortmid}{C}}\!\cdot$	$Ar\overset{\shortmid}{\underset{\shortmid}{C}}\!-\!H$	$Ar\overset{\shortmid}{C}\!=\!CH_2$
α-Hydroxyalkyl	$HO\overset{\shortmid}{\underset{\shortmid}{C}}\!\cdot$	$HO\overset{\shortmid}{\underset{\shortmid}{C}}\!-\!H$	$HO\overset{\shortmid}{C}\!=\!CH_2$
α-Alkoxyalkyl	$RO\overset{\shortmid}{\underset{\shortmid}{C}}\!\cdot$	$RO\overset{\shortmid}{\underset{\shortmid}{C}}\!-\!H$	$RO\overset{\shortmid}{C}\!=\!CH_2$
α-Aminoalkyl	$\!>\!\!N\!-\!\overset{\shortmid}{\underset{\shortmid}{C}}\!\cdot$	$\!>\!\!N\!-\!\overset{\shortmid}{\underset{\shortmid}{C}}\!-\!H$	$\!>\!\!N\!-\!\overset{\shortmid}{C}\!=\!CH_2$
Acyl	$-\overset{\overset{\textstyle O}{\|}}{C}\!\cdot$	$-\overset{\overset{\textstyle O}{\|}}{C}\!-\!H$	O=C=CH_2

D. ACCEPTOR RADICALS AND SUBSTRATES

Halogen atoms, as expected on the basis of the definition presented previously, are acceptor radicals. The bromine atom shows a stronger tendency to accept an electron becoming a bromide ion than to lose an electron becoming a

$$Br^{(+)} \xleftarrow{-e^{(+)}} Br\cdot \xrightarrow{+e^{(-)}} Br^{(-)} \tag{4-28}$$

bromonium ion and, therefore, bromine atoms act as acceptor radicals in most of their reactions. Likewise, chlorine atoms, although prone to show another type of polar character (see Chapter 5) also have acceptor characteristics as do iodine atoms, alkoxy radicals, and thiyl radicals. In each case, the anion that would result upon gaining an electron by the radical is considerably more stable than the cation formed on loss of an electron. Some of the more commonly encountered acceptor radicals along with acceptor substrates which could yield such radicals are listed in Table 4-4. Certain atoms and functionalities bonded to a carbon atom with an unpaired electron render acceptor qualities to that radical. The trihalomethyl radicals, for example, function as acceptor radicals

TABLE 4-4   Acceptor Radicals and Substrates

		Substrates	
Name	General Structure	Displacement	Addition
Chlorine atom	Cl·	Cl—Cl, HCl	—
Bromine atom	Br·	Br—Br, HBr	—
Alkoxy	RO·	ROX, ROOR	—
Thiyl	RS·	RSH, RSSR	—
Trihalomethyl	X_3C·	X_3C—X, X_3C—H	—
α-Haloalkyl	$\overset{\displaystyle\mid}{\underset{\displaystyle\mid}{X—C}}$·	$\overset{\displaystyle\mid}{XC}$—X, $\overset{\displaystyle\mid}{XC}$—H	$\overset{\displaystyle}{\underset{\displaystyle\mid}{XC}}$=$CH_2$
α-Cyanoalkyl	$N{\equiv}C\overset{\displaystyle\mid}{\underset{\displaystyle\mid}{C}}$·	$N{\equiv}C\overset{\displaystyle\mid}{C}$—H	$N{\equiv}CCH$=CH_2
α-Carbonylalkyl	$\overset{\displaystyle O}{\overset{\displaystyle \|}{—C}}\overset{\displaystyle\mid}{\underset{\displaystyle\mid}{C}}$·	$\overset{\displaystyle O}{\overset{\displaystyle \|}{—C}}\overset{\displaystyle}{\underset{\displaystyle\mid}{C}}$—H	$\overset{\displaystyle O}{\overset{\displaystyle \|}{C}}$—CH=$CH_2$

in both displacement and addition reactions. The carbanion species produced from the trichloromethyl radical by gaining an electron is stabilized by delocalization of the charge into the *d*-orbitals of the halogen atoms. Both bromine and chlorine are able to stabilize a charge in this fashion as evidenced by the

$$Cl—C^{(-)}\begin{smallmatrix}Cl\\[-2pt]\\Cl\end{smallmatrix} \leftrightarrow \begin{smallmatrix}Cl\\[-2pt]\\Cl\end{smallmatrix}C\begin{smallmatrix}Cl^-\\[-2pt]\\Cl\end{smallmatrix}$$

greater acidities of both bromoform and chloroform relative to unsubstituted alkanes. Fluorine, on the other hand, does not have this ability. Acceptor qualities for the trifluoromethyl radical might, however, be expected owing to the electronegativity of the fluorine atoms. The reactions of the trifluoromethyl radical are, for the most part, very rapid with most substrates because the radical possesses very little resonance stabilization (see Table 4-1).

The dihaloalkyl radicals display acceptor qualities for very much the same reason as the trihalomethyl radicals. We might expect that a single halogen such as chlorine could display the same effect only to a smaller extent. On the other hand, it could stabilize a positive charge in the same manner described earlier for oxygen functions. If this were the case, a single α-halogen might render donor qualities to the radical. By similar reasoning, the di- and trihalo-

$$|\bar{\text{Cl}}-\text{C} \cdot \leftrightarrow |\overset{+}{\underline{\text{Cl}}}=\text{C}$$

alkyl radicals might be expected to be donor radicals. Apparently the delocalization of the negative charge is more effective in these cases than is the stabilization of the positive charge.

α-Cyanoalkyl radicals are good acceptors because the negative charge can be delocalized into the cyano group. Similarly, the α-carbonyl groups in

$$\text{N}{\equiv}\text{C}{-}\overset{\cdot}{\text{C}}{-} \leftrightarrow \bar{\text{N}}{=}\text{C}{=}\text{C}$$

ketone- and ester-derived radicals render acceptor qualities to these radicals.

$$\overset{\cdot}{\text{O}}{=}\text{C}{-}\overset{\cdot}{\text{C}}{-} \leftrightarrow \bar{\text{O}}{-}\text{C}{=}\text{C}$$

The radicals derived from malonic ester and acetoacetic ester show strong acceptor characteristics because delocalization of the negative charge is more extensive than when a single functional group is present. Malonic ester and

$$R = OC_2H_5 \text{ or } CH_3$$

acetoacetic ester show strong acceptor character in reactions that involve abstraction of an active methylene hydrogen.

E. CRITERION FOR DONOR-ACCEPTOR POLAR EFFECTS

A donor radical can display its donor characteristics only if it has an acceptor substrate with which it can react. Similarly, an acceptor radical requires a donor substrate in order for the radical to display its polar qualities. An interesting consequence of such requirements is that chain-propagating reactions involving donor radicals must produce acceptor radicals and vice versa. The following examples illustrate this point.

The chlorination of an alkane with carbon tetrachloride proceeds by the two step sequence shown, the first step of which involves abstraction of a hydrogen from the donor alkane by the acceptor trichloromethyl radical. This reaction yields a donor alkyl radical that reacts as shown with the polyhalomethane regenerating the trichloromethyl radical.

$$Cl_3C \cdot + HR \rightarrow \left[\overset{\delta-}{Cl_3C} \cdot \cdot H \cdot \cdot \overset{\delta+}{R} \right] \rightarrow Cl_3CH + R \cdot \qquad (4\text{-}29)$$

$$R \cdot + ClCCl_3 \rightarrow \left[\overset{\delta+}{R} \cdot \cdot Cl \cdot \cdot \overset{\delta-}{CCl_3} \right] \rightarrow RCl + Cl_3C \cdot \qquad (4\text{-}30)$$

The peroxide-induced alkylation of malonic ester with 1-octene, a reaction that proceeds by the following sequence of chain-propagating reactions (see Chapter 6), also illustrates this principle. The addition reaction of the malonic

$$(C_2H_5OCO)_2\dot{C}H + CH_2{=}CHC_6H_{13}\text{-}n \rightarrow$$

$$\left[(C_2H_5OCO)_2\overset{\delta-}{\dot{C}H} \cdot \cdot CH_2 \overset{\delta+}{\cdots} \overset{}{\dot{C}}HC_6H_{13}\text{-}n \right] \rightarrow$$

$$\begin{matrix} (C_2H_5OCO)_2CHCH_2\dot{C}H \\ | \\ C_6H_{13}\text{-}n \end{matrix} \qquad (4\text{-}31)$$

$$(C_2H_5OCO)_2CHCH_2\dot{C}H + H_2C(CO_2C_2H_5)_2 \rightarrow$$
$$\begin{matrix} | \\ C_6H_{13}\text{-}n \end{matrix}$$

$$\left[\begin{matrix} (C_2H_5OCO)_2CHCH_2\overset{\delta+}{\dot{C}H} \cdot \cdot H \cdot \cdot \overset{\delta-}{CH}(CO_2C_2H_5)_2 \\ | \\ C_6H_{13}\text{-}n \end{matrix} \right] \rightarrow$$

$$(C_2H_5OCO)_2CHC_8H_{17}\text{-}n + (C_2H_5OCO)_2\dot{C}H \qquad (4\text{-}32)$$

ester-derived radical, an acceptor radical, to 1-octene yields an alkyl-derived donor radical. In the hydrogen atom abstraction reaction, the donor alkyl-derived radical yields the acceptor ester-derived radical. In both reactions, transition states having separation of charges owing to the donor and acceptor qualities of the species involved would be expected.

Some unimolecular chain-propagating reactions also show definite polar effects. Fragmentations of alkoxy radicals involve transition states in which there is separation of charge if the eliminated group is an alkyl radical (see

$$R-\underset{|}{\overset{|}{C}}-O^{-} \;\rightarrow\; \left[\overset{\delta+}{R}\cdots\overset{|}{\underset{|}{C}}\overset{\delta-}{\cdots O}\right] \;\rightarrow\; R\cdot + \;\overset{}{\underset{}{>}}C{=}O \qquad (4\text{-}33)$$

Chapter 8). In this case an acceptor radical yields a donor radical in a chain-propagating reaction.

IV. Steric Effects

A. DISPLACEMENT REACTIONS

The fact that almost all displacement reactions occur on univalent hydrogen or halogen atoms is a reflection of the importance of steric effects in these reactions. The enthalpic requirements for displacements on carbon are often not appreciably different from those on hydrogen or halogen. Compare the displacement on a benzylic hydrogen of toluene by a chlorine atom with the hypothetical reaction of chlorine with ethyl benzene in which the displacement

$$Cl\cdot + C_6H_5CH_3 \;\rightarrow\; HCl + C_6H_5\dot{C}H_2 \qquad (4\text{-}34)$$

Make	HCl	-103 kcal/mole
Break	$C_6H_5CH_2{-}H$	$+\,85$ kcal/mole

$$\Delta H = -\;18 \text{ kcal/mole}$$

occurs on the β-carbon of ethyl benzene. The two reactions are similar in many regards in that they both are exothermic, both involve the same attacking

$$Cl\cdot + C_6H_5CH_2CH_3 \;\rightarrow\; CH_3Cl + C_6H_5CH_2\cdot \qquad (4\text{-}35)$$

Make	CH_3Cl	-84 kcal/mole
Break	$C_6H_5CH_2{-}CH_3$	$+70$ kcal/mole

$$\Delta H = -14 \text{ kcal/mole}$$

radical (the chlorine atom), and both produce the same product radical (the benzyl radical). Whereas the displacement on hydrogen is a commonly encountered reaction of the chlorine atom with alkanes and alkyl aromatics, similar displacements on carbon are rare, possible exceptions are those involving the cyclopropane ring system. The reaction most likely does not occur

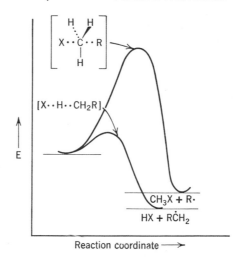

Fig, 4.7. Energy profiles for displacement reactions on hydrogen and on carbon.

readily because of a high activation energy requirement for the displacement on carbon caused by the sterically hindered approach of a free radical to an atom coordinated with two or more other atoms or groups (Fig. 4.7).

Displacements on bivalent oxygen in the oxygen-oxygen linkage of certain peroxides are known. Formation of acylals in the reactions of an α-alkoxyalkyl radical with diacyl peroxides constitutes such a displacement. These displace-

$$R_2C\cdot + O-O \rightarrow \left[R_2\overset{\delta+}{C}\cdot\cdot O\cdot\cdot\overset{\delta-}{O} \right] \rightarrow R_2COAc + \cdot OAc \quad (4-36)$$
$$\quad | \quad\quad | \quad | \quad\quad\quad | \quad | \quad | \quad\quad\quad\quad |$$
$$\quad OR \quad Ac \; Ac \quad\quad\quad OR \; Ac \; Ac \quad\quad\quad\quad OR$$

ments are also subject to steric factors because a similar reaction does not appear to occur with an α-alkoxyalkyl radical and t-butyl peroxide. The oxygens in this peroxide occupy a position similar to that of the primary carbons in the neopentyl halides which are unreactive in S_N2 reactions because of the steric effects exerted by the β-methyl groups.

B. ADDITION REACTIONS

The reactivities of unsaturated linkages toward addition by free radicals are influenced by steric factors. In general, 1,1-disubstituted ethylenes are more reactive toward radical addition than terminal alkenes but 1,2-disubstituted alkenes are far less reactive. The lower reactivity of nonterminally unsaturated compounds toward addition of free radicals generally is ascribed to a steric effect.

The direction of addition of a free radical to a terminal unsaturation is generally such that the adding radical bonds to the least substituted carbon. This direction of addition can be accounted for in terms of only resonance and polar factors in most cases. The adduct radical formed by bonding to the terminal carbon would not only be the more stable of the two but any polar factors that might be operative could exert their influence in the transition state of the reaction. However, steric factors must also play a significant role in determining this particular orientation since addition to a symmetrically disubstituted ethylene yields an adduct radical with essentially the same energy. Apparently formation of the σ-bond between the adding radical and the substituted carbon of an alkene linkage is sterically hindered by the groups bonded to the carbon.

Not all additions occur with bonding of the adding free radical to the least substituted atom of the unsaturated linkage. In certain cases the resonance and polar effects may outweigh the steric factors and an adduct radical is formed in which bonding occurs on the more substituted of the two sites in the unsaturated link. An example is the addition of an acyl radical to mesityl oxide.

$$
\underset{\substack{\|\\ O}}{R C} \cdot + \underset{\substack{| \\ CH_3}}{\overset{CH_3}{C}} = \underset{\substack{| \\ COCH_3}}{\overset{H}{C}} \rightarrow \left[\underset{\substack{\| \\ O}}{\overset{\delta+}{R C}} \cdot \cdot \underset{\substack{| \\ CH_3}}{\overset{CH_3}{C}} \vcentcolon\vcentcolon \underset{\substack{\\ COCH_3}}{\overset{H}{C^{\delta-}}} \right] \rightarrow
$$

$$ (4\text{-}37) $$

$$
\underset{\substack{\| | \\ OCH_3}}{R C \overset{CH_3}{C}} {-}\dot{C}HCOCH_3 \xrightarrow{\text{RCHO}} \underset{\substack{\| | \\ OCH_3}}{R C \overset{CH_3}{C}}{-}CH_2COCH_3 + \underset{\substack{\| \\ O}}{R C} \cdot
$$

C. TERMINATION REACTIONS

The "galvinoxyl" radical owes its stability, in part, to steric effects. Although the radical couples (or disproportionates) readily with other free radicals, it does not do so with itself. While the radical is no doubt stabilized to a significant extent by resonance, it is also true that the radical sites, namely the oxygen atoms, are quite possibly sterically hindered from attaining the proximity necessary for either bonding in a coupling reaction or transferring an electron in a redox reaction. The approach of smaller, and possibly more reactive, radicals to the oxygens apparently is not sterically hindered to the same extent and reaction can occur. It is difficult in this case to separate completely the resonance and steric contributions that give the radical its particular properties

The latter must, however, play a role of some significance since without the
t-butyl groups flanking the oxygens, a radical is obtained that shows no particularly high degree of stability.

"galvinoxyl"

Chapter 5

Halogenation Reactions

I. Introduction

Many hydrogen containing compounds participate in chain reactions with molecular halogens (principally bromine or chlorine), reactions in which an exchange of a hydrogen atom for a halogen atom occurs. Since halogenation

$$X_2 + RH \rightarrow HX + RX \qquad (5\text{-}1)$$

via

$$X\cdot + RH \rightarrow HX + R\cdot \qquad (5\text{-}2)$$

$$R\cdot + X_2 \rightarrow RX + X\cdot \qquad (5\text{-}3)$$

reactions provide one of the better means of introducing a functionality into an alkane, there is much to be gained in determining the specificity (or lack of it) in exchanging the hydrogens with halogens in compounds that have several different kinds of hydrogens.

Hydrogen-halogen exchange reactions also occur between hydrogen-containing compounds and halogen-containing compounds such as XY where X is a halogen. The mechanisms for such reactions are quite similar to those of the molecular halogens except that a radical $Y\cdot$ (e.g., $Cl_3C\cdot$, $RO\cdot$) acts as the hydrogen abstracting radical.

$$XY + RH \rightarrow HY + RX \qquad (5\text{-}4)$$

via

$$Y\cdot + RH \rightarrow HY + R\cdot \qquad (5\text{-}5)$$

$$R\cdot + XY \rightarrow RX + Y\cdot \qquad (5\text{-}6)$$

In view of the importance of halogenation reactions, much work has been reported in this area. This chapter is concerned mainly with the aspects of these reactions that are illustrative of the reactivities of free radicals in chain-propagating reactions. More comprehensive surveys of halogenation reactions can be found in the general references listed at the end of this chapter.

II. Brominations with Molecular Bromine

A. THE CHAIN SEQUENCE

The energetics of the radical-propagating and overall bromination reactions

$$Br\cdot + RH \rightarrow HBr + R\cdot \qquad (5\text{-}7)$$

$$R\cdot + Br_2 \rightarrow RBr + Br\cdot \qquad (5\text{-}8)$$

of several hydrocarbons are shown in Table 5-1. Whereas bromination of toluene proceeds readily and reactions having long kinetic chain lengths are observed, methane is brominated less readily and the kinetic chain length of the reaction is comparatively short. The ease of brominating the simple alkanes is methane < ethane < propane (yielding isopropyl bromide) < isobutane (yielding t-butyl bromide). These observations indicate that the hydrogen abstraction reaction 5-7 is the limiting factor in the chain sequence.

The kinetics of the photo-induced bromination of methane in the gas phase support a mechanism in which hydrogen abstraction from methane by a bromine

$$Br_2 \xrightarrow{k_9} 2Br\cdot \qquad (5\text{-}9)$$

$$Br\cdot + RH \underset{k_{-10}}{\overset{k_{10}}{\rightleftarrows}} HBr + R\cdot \qquad (5\text{-}10)$$

$$R\cdot + Br_2 \xrightarrow{\ k_{11}\ } RBr + Br\cdot \tag{5-11}$$

$$2Br\cdot + M \xrightarrow{\ k_{12}\ } Br_2 + M \tag{5-12}$$

TABLE 5-1   Enthalpies of Bromination Reactions (Calculated from Bond-Dissociation Energies in Tables 4-1 and 4-2)

RH	ΔH for $Br\cdot + RH \rightarrow$ $HBr + R\cdot$ (kcal/mole)	ΔH for $R\cdot + Br_2 \rightarrow$ $RBr + Br\cdot$ (kcal/mole)	ΔH for $RH + Br_2 \rightarrow$ $HBr + RBr$ (kcal/mole)
CH_3—H	+16.5	−24	−7.5
CH_3CH_2—H	+10.5	−23	−12.5
$(CH_3)_2CH$—H	+7.0	−22	−15.0
$(CH_3)_3C$—H	+3.5	−17	−13.5
$C_6H_5CH_2$—H	−2.5	−5	−7.5

atom is the rate-limiting step in the chain sequence. It should be noted that the hydrogen abstraction reaction is a reversible process that becomes important if the concentration of hydrogen bromide is high enough to allow hydrogen bromide to compete with bromine for the methyl radicals. Another important factor is that coupling of two bromine atoms in the gas phase is an exothermic process ($\Delta H = -46$ kcal/mole) and generally occurs only in a termolecular reaction with a third body (M) which will help absorb this energy. The third body could be any of the species present in the system (Br_2, CH_4, CH_3Br or HBr). Consequently $[M]$, the concentration of the third body, is proportional to the pressure of the system. Furthermore, if the incident light is intense and not completely absorbed by bromine, the rate of initiation will be dependent on $[Br_2]$ and dependent of the light intensity I_0. If all of the light is absorbed, the rate of initiation depends only on I_0 and is independent of $[Br_2]$.

With these points in mind and employing the principles of steady-state rate law derivations developed in Chapter 3, the following rate laws can be derived

$$\frac{-d[Br_2]}{dt} = k_{10}\left(\frac{k_9 I_0}{2k_{12}}\right)^{1/2} \frac{[CH_4][Br_2]^{1/2}}{[M]^{1/2}(1 + (k_{-10}[HBr])/(k_{11}[Br_2]))} \tag{5-13a}$$

for this reaction. The rate law given in 5-13a assumes that the radiation available, is only partly absorbed by the bromine whereas in 5-13b the light is completely absorbed and the rate of initiation does not depend on $[Br_2]$. The experimentally

$$\frac{-d[Br_2]}{dt} = k_{10}\left(\frac{k_9 I_0}{2k_{12}}\right)^{1/2} \frac{[CH_4]}{[M]^{1/2}(1 + (k_{-10}[HBr])/(k_{11}[Br_2]))} \tag{5-13b}$$

observed rate law for a reaction involving intense illumination is 5-14 if the intensity of the illumination is high. This rate law applies only at the outset

$$\text{Rate} = k[Br_2]^{1/2}[CH_4]p^{-1/2} \tag{5-14}$$

of the reaction before the hydrogen bromide concentration becomes appreciable. The observation that the rate is inversely proportional to the square root of the pressure (p) is consistent with $[M]^{1/2}$ appearing in the denominator of both 5-13a and 5-13b. As long as the concentration of hydrogen bromide is negligible, the second term of the binomial in the denominator of either 5-13a or 5-13b is zero. It is significant that the presence of HBr in appreciable concentrations retards the reaction rate, a fact that is consistent with the derived rate law and is indicative of involvement of the methyl radicals with hydrogen bromide.

The derived rate laws would be different if termination reactions involved either coupling of two methyl radicals or cross-coupling of a bromine atom and a methyl radical. The observed rate law implicates coupling of two bromine atoms in the termination reaction. The comparatively difficult hydrogen atom abstraction from methane by bromine atoms causes the steady-state concentration ratio of the chain-carrying radicals to favor the bromine atoms. This will be necessary if this reaction is to proceed at the same rate as the relatively more facile reaction of the methyl radicals with molecular bromine.

The kinetic rate law is more complex if more than one termination reaction is operative. This is probably the case in the bromination of isobutane with molecular bromine. Both hydrogen abstraction from isobutane by the bromine atom and reaction of the t-butyl radical with bromine are facile reactions and, consequently, the steady-state concentrations of both chain-carrying radicals are comparable. Termination by reactions 5-18 and 5-19 as well as 4-17 becomes probable. Data for the gas phase reactions indicate that the ratio of the termina-

$$Br\cdot + (CH_3)_3CH \rightarrow HBr + (CH_3)_3C\cdot \tag{5-15}$$

$$(CH_3)_3C\cdot + Br_2 \rightarrow (CH_3)_3CBr + Br \tag{5-16}$$

$$2Br\cdot + M \rightarrow Br_2 + M \tag{5-17}$$

$$Br\cdot + (CH_3)_3C\cdot \rightarrow (CH_3)_3CBr \tag{5-18}$$

$$2(CH_3)_3C\cdot \rightarrow (CH_3)_3CC(CH_3)_3 \tag{5-19}$$

tion reactions 5-17, 5-18, and 5-19 is $1:0.38:0.20$.[2] The kinetic chain length for the bromination of isobutane at $127°C$ in the vapor phase was reported to be 4×10^9 compared to a kinetic chain length of 100 for the bromination of methane at $297°C$.

B. SPECIFICITY

1. Alkanes. The hydrogen abstraction reaction is the product-determining reaction of the chain sequence of most bromination reactions. If more than one

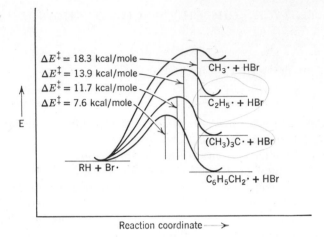

$\Delta E^{\ddagger} = 18.3$ kcal/mole
$\Delta E^{\ddagger} = 13.9$ kcal/mole
$\Delta E^{\ddagger} = 11.7$ kcal/mole
$\Delta E^{\ddagger} = 7.6$ kcal/mole

$CH_3\cdot + HBr$
$C_2H_5\cdot + HBr$
$(CH_3)_3C\cdot + HBr$
$RH + Br\cdot$
$C_6H_5CH_2\cdot + HBr$

E

Reaction coordinate $\longrightarrow$

Fig. 5.1.   Energy profiles for hydrogen atom abstractions by bromine atoms.

hydrogen is available for reaction with the bromine atom, more than a single type of radical may be produced by the hydrogen abstraction and, consequently, more than a single bromination product may result. A high degree of selectivity is displayed by the bromine atom as a hydrogen abstractor. Some insight into this selectivity is gained by examining the products obtained in the reactions of various alkanes, alkyl aromatics, and haloalkanes with molecular bromine.

The energy profiles for the reaction of bromine atoms with some simple alkanes are shown in Figure 5.1 along with the estimated activation energies[1, 2] for these hydrogen atom abstraction reactions.

The reactivity of alkyl hydrogens toward reaction with bromine atoms is in the order tertiary > secondary ≫ primary. Consequently, bromination of an

$$Br\cdot + CH_3\underset{\underset{CH_3}{|}}{\overset{\overset{CH_3}{|}}{C}}H \overset{k_{20}}{\underset{-HBr}{\nearrow}} CH_3\underset{\underset{CH_3}{|}}{\overset{\overset{CH_3}{|}}{C}}\cdot \xrightarrow{Br_2} CH_3\underset{\underset{CH_3}{|}}{\overset{\overset{CH_3}{|}}{C}}Br + Br\cdot \qquad (5\text{-}20)$$

$$\overset{k_{21}}{\underset{-HBr}{\searrow}} \cdot CH_2\underset{\underset{CH_3}{|}}{\overset{\overset{CH_3}{|}}{C}}H \xrightarrow{Br_2} BrCH_2\underset{\underset{CH_3}{|}}{\overset{\overset{CH_3}{|}}{C}}H + Br\cdot \qquad (5\text{-}21)$$

alkane such as isobutane yields essentially only *t*-butyl bromide with no appreciable amounts of isobutyl bromide. The relative reactivity ratio k_{20}/k_{21} is very large and has been estimated to be about 2000 at 146°C.[3] Similarly

$$\text{Br} \cdot + (CH_3)_2CHC(CH_3)_3 \rightarrow (CH_3)_2\overset{\cdot}{C}C(CH_3)_3 + HBr$$

$$\mathbf{1}$$

$$\text{Br} \cdot + (CH_3)_2CBrC(CH_3)_3 \xleftarrow{\;Br_2\;}$$

$$\mathbf{2}$$

(5-22)

bromination of 2,2,3-trimethylbutane (**1**) yields only the tertiary bromide, 2-bromo-2,2,3-trimethylbutane (**2**).

Brominations of alkanes with tertiary hydrogens such as 2,3-dimethylbutane (**3**) are complicated by formation of considerable amounts of dibromide.[4] These products likely result from the dehydrohalogenation of the tertiary bromide formed as the product of the free-radical chain reaction, followed by addition of molecular bromine to the unsaturated linkage. The dehydrohalogenation reaction occurs in the dark probably by an ionic mechanism.

$$\begin{array}{ccc}
CH_3\ CH_3 & & CH_3\ CH_3 \\
|\quad\ | & & |\quad\ | \\
CH_3C\!\!-\!\!CCH_3 + 2Br_2 & \xrightarrow[89\%]{55°} & CH_3C\!\!-\!\!CCH_3 + 2HBr \\
|\quad\ | & & |\quad\ | \\
H\ \ H & & Br\ \ Br \\
\mathbf{3} & & \mathbf{4}
\end{array}$$

(5-23)

via

$$\begin{array}{cc}
CH_3\ CH_3 & CH_3\ CH_3 \\
|\quad\ | & |\quad\ | \\
CH_3CHCHCH_3 \xrightarrow{Br_2} & CH_3CBrCHCH_3 \ \ (\text{Free-radical}) \\
\mathbf{3} &
\end{array}$$

(5-24)

$$\begin{array}{cc}
CH_3\ CH_3 & CH_3\ CH_3 \\
|\quad\ | & |\quad\ | \\
CH_3CBrCHBr \xrightarrow{-HBr} & CH_3C\!\!=\!\!CCH_3 \ \ (\text{Ionic})
\end{array}$$

$$\begin{array}{c}
CH_3\ CH_3 \\
|\quad\ | \\
\xrightarrow{Br_2} CH_3CBrCBrCH_3 \\
\mathbf{4}
\end{array}$$

(5-25)

Formation of dibromides is a complicating factor in the bromination of other compounds that have tertiary hydrogens on a carbon adjacent to a carbon having either secondary or tertiary hydrogens. Thus, the following cycloalkanes are brominated preferentially at the tertiary position in the free-radical process but considerable amounts of dibromides are found as reaction products.[5]

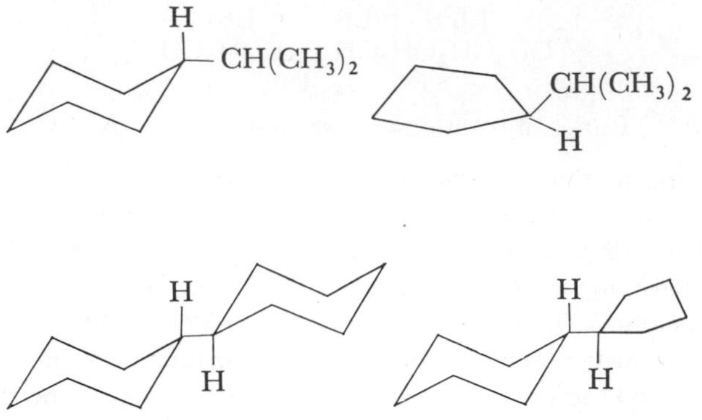

Norbornane (**5**) undergoes specific bromination yielding only bromination

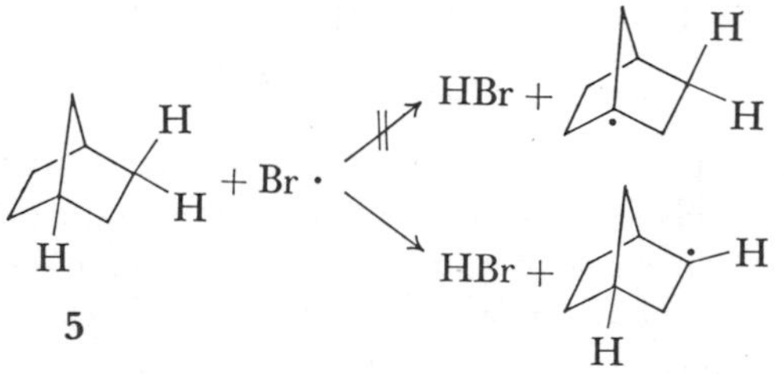

(5-26)

(5-27)

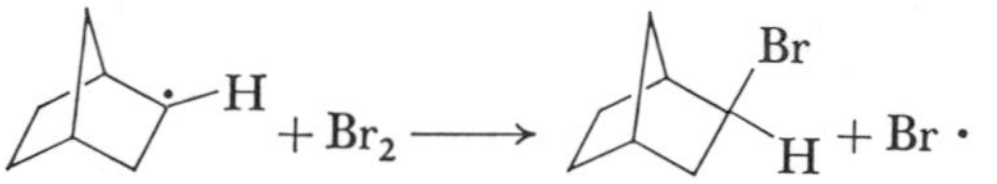

5

(5-28)

products that result from abstraction of a secondary hydrogen from the 2-position.[6] In this case, a bridgehead radical would be produced if a tertiary hydrogen were abstracted. The structural features of the bicyclic system are such that the internal strain introduced in the system by forming the planar sp^2 hybridized carbon at the bridgehead, would make the intermediate radical unstable. The low reactivity of the secondary hydrogens on the 7-carbon must reflect the low stability of the 7-norbornyl radical relative to the 2-norbornyl radical. The preferential formation of the *exo*-isomer over the *endo*-isomer is probably due to steric effects encountered by the bromine when it approaches from the *endo*-side of the radical rather than from the *exo*-side.

Bromination of *n*-heptane shows a preferential attack at the 2-position relative to the 3- or 4-position[7] in ratios of approximately 1.8:1:1. The preferred attack at the 2-position is possibly due to greater resonance stabilization of the radical or a polar transition state (*vide infra*) caused by the hyperconjugative effect of five hydrogens relative to four.

$$
\overset{1.8}{\underset{1.0}{CH_3}}\overset{}{CH_2}\overset{1.0}{CH_2}\overset{1.0}{CH_2}\overset{1.0}{CH_2}CH_2CH_3
$$

1.8   1.0   1.8

CH$_3$CH$_2$CH$_2$CH$_2$CH$_2$CH$_2$CH$_3$

1.0   1.0

Ratios of reactivities of secondary hydrogens

It would appear that the selectivities displayed by the bromine atom as a hydrogen abstractor could be explained solely on the basis of the resonance stabilities of the alkyl radicals formed in the displacement reaction. However, polar factors are important in these reactions. In the reactions of the alkanes with bromine atoms, the contribution of the acceptor qualities of the bromine atom to the transition state become more important with increasing donor ability of the alkyl radical produced. Since donor qualities of the alkyl radicals are determined by their respective carbonium ion stabilities, the relative activation energy requirements for the following hydrogen abstractions would be

$$
\begin{bmatrix} \overset{\delta-}{Br}\cdot\cdot H\cdot\cdot \overset{\delta+}{CH_3} \end{bmatrix} \gg \begin{bmatrix} \overset{\delta-}{Br}\cdot\cdot H\cdot\cdot \overset{\delta+}{CH_2R} \end{bmatrix} \gg \begin{bmatrix} \overset{\delta-}{Br}\cdot\cdot H\cdot\cdot \overset{\delta+}{CHR_2} \end{bmatrix} > \begin{bmatrix} \overset{\delta-}{Br}\cdot\cdot H\cdot\cdot \overset{\delta+}{CR_3} \end{bmatrix}.
$$

R = alkyl

Examination of the reactivities of various hydrogens toward abstraction by bromine atoms parallels this order of the contribution of the polar effect toward lowering the activation energy requirement more than the resonance stabilization of the radicals produced. Thus at 100°C, the relative reactivities of methane, ethane, propane, and isobutane toward hydrogen abstraction yielding methyl, ethyl, isopropyl, and t-butyl radicals are approximately $1 : 500 : 26,000 : 380,000$.[3] The difference between the ease of formation of the t-butyl radical and isopropyl radical is only a factor of about 12 to 15 whereas the difference between ease of formation of a secondary isopropyl radical and a primary ethyl radical is approximately 50. The methyl radical is most difficult to form by abstraction of a hydrogen from methane by a bromine atom.

2. **Alkyl Aromatics.** Toluene is readily converted to benzyl bromide in a light-induced reaction with molecular bromine. The kinetic chain length of this reaction is longer than that of the bromination of most alkanes with the possible exception of those that have tertiary hydrogens. The energetics of the reaction are such that both steps in the chain sequence are exothermic (see Table 5-1) and, in view of the long kinetic chain lengths, must both be rapid reactions.

$$Br_2 + C_6H_5CH_3 \rightarrow HBr + C_6H_5CH_2Br \tag{5-29}$$

via

$$Br\cdot + C_6H_5CH_3 \rightarrow HBr + C_6H_5\dot{C}H_2 \tag{5-30}$$

$$C_6H_5\dot{C}H_2 + Br_2 \rightarrow C_6H_5CH_2Br + Br\cdot \tag{5-31}$$

The reactivity of the benzylic hydrogens toward abstraction by bromine is reflected in the specificity observed in the bromination reactions of other alkyl aromatics. Bromination of ethylbenzene (6) yields only α-bromoethylbenzene (7) and cumene (8) yields only cumyl bromide (9). Free-radical bromination

$$C_6H_5CH_2CH_3 + Br_2 \rightarrow C_6H_5CHBrCH_3 + HBr \qquad (5\text{-}32)$$
$$\;\;67$$

$$C_6H_5CH(CH_3)_2 + Br_2 \rightarrow C_6H_5CBr(CH_3)_2 + HBr \qquad (5\text{-}33)$$
$$89$$

of t-butyl benzene (10) is not only slow but is accompanied by products resulting from electrophylic bromination.[8]

$$C_6H_5C(CH_3)_3 + Br_2 \rightarrow C_6H_5C(CH_3)_3CH_2Br \qquad (5\text{-}34)$$
$$10$$

The reactivities of benzylic hydrogens toward abstraction by bromine atoms depend on the extent and nature of substitution on the benzylic carbon atom. Table 5-2 lists the reactivities of several hydrocarbons relative to ethane toward hydrogen abstraction by bromine atoms reported by Russell and DeBoer.[9] In each case, the greater reactivity of benzylic hydrogens relative to the primary alkyl hydrogen of ethane is probably due to resonance stabilization of the benzylic radical produced in the reaction.

Substituents in the *meta* and *para* positions of the aromatic ring also exercise an effect on the reactivities of benzylic hydrogens. The nature of this effect is

TABLE 5-2 Relative Reactivities of Benzylic Hydrogens Toward Abstraction by Bromine Atoms[9] (R—H + Br· → R· + HBr)

RH	Reactivity
CH_3CH_2—H	1.00
$C_6H_5CH_2$—H	6.4×10^4
$C_6H_5CH(CH_3)$—H	1.0×10^6
$(C_6H_5)_2CH$—H	6.2×10^5
$C_6H_5C(CH_3)_2$—H	2.33×10^6
$(C_6H_5)_2C(CH_3)$—H	2.70×10^6
$(C_6H_5)_3C$—H	1.14×10^6

best appreciated in terms of a Hammett correlation[10] where k/k_0 is the reactivity ratio of the *meta* or *para* substituted benzene derivative relative to the unsubstituted derivative, ρ is a reaction parameter, and σ is the substituent parameter.[10]

$$\log \frac{k}{k_0} = \rho\sigma \tag{5-35}$$

A relatively poor correlation is observed when the data obtained from competitive brominations of substituted toluenes are subjected to treatment by the relationship given in 5-35. A lack of linear correlation between the σ-constant for the substituent and the $\log k/k_0$, in this case, cannot be attributed

$$\text{(5-36)}$$

$$\text{(5-37)}$$

to the fact that radical reactions are involved. The relationship also breaks down in many ionic reactions. Particularly noteworthy are the solvolysis reactions of substituted benzyl halides. Another set of substituent parameters, namely σ^+, was determined from the solvolysis rates of substituted cumyl chlorides (**11**) by Brown and Okamoto[11] and serve primarily as a measure of the effect the substituent displays in stabilizing cationic character in the benzylic position in the transition state of the reaction as shown in 5-38.

11

$$\text{(5-38)}$$

A surprisingly good correlation is observed between $\log k/k_0$ for the bromination of substituted toluenes when plotted against the σ^+ values for the substituents[12]. The value of ρ for the reactions appears to be dependent on the ratio of the bromine concentration to hydrogen bromide present in the system. A higher value for ρ (-1.36 at $80°C$) is observed when a large excess of bromine is present but is considerably smaller (-1.07 at $80°C$) when the concentration of bromine is low. This effect of bromine concentration is explained as resulting from reaction of the benzyl radicals with hydrogen bromine which develops a concentration comparable to that of bromine when the concentration of the latter is low.

$$X{-}\overset{\cdot}{C}H_2 + HBr \rightarrow \quad X{-}CH_3 + Br\cdot \qquad (5\text{-}39)$$

The transition state of the abstraction of benzylic hydrogens from toluene by bromine must resemble that of the solvolysis of the cumyl chlorides.[13] In terms

$$\left[X{-}\overset{\delta+}{C}H_2\cdot\cdot H\cdot\cdot\overset{\delta-}{Br} \right]$$

of the donor-acceptor concept, the bromine atom would be an acceptor radical and the toluene a donor substrate. The extent to which the polar character of the transition state contributes to lowering the activation energy requirement depends on the influence the substituent has on rendering donor qualities to the benzyl moiety which should parallel directly the effect that the substituent has on lowering the activation energy requirement for the solvolysis reactions of the benzyl halides.

3. Haloalkanes. The reactivity of the hydrogens on a carbon in a β-position relative to another bromine toward abstraction by bromine atoms is greater than might be expected. For example, the reactivities of the hydrogens on the 2-carbon of 1-bromobutane and on the 3-carbon of 2-bromobutane are decidedly more reactive toward abstraction by bromine atoms than are those of the corresponding chlorinated compounds (see Table 5-3).[14] The phenomenon appears to be peculiar to hydrogen abstraction by bromine atoms and is not observed in hydrogen abstractions by chlorine atoms. (The effects of electronegative substituents on hydrogen abstraction by chlorine atoms is discussed later in this chapter).

An explanation that best accommodates these observations is that the transition state of the hydrogen abstraction by bromine atoms is anchimerically assisted by the bromine bonded in the β-position. A transition state for the reaction such as shown in 5-40 involving a bridged species would be more likely in the case of the more polarizable bromine in the β-position than in the case of

TABLE 5-3   Reactivities of Hydrogens of Haloalkanes toward Abstraction by Halogen Atoms (60°C in Liquid Phase)[14]

Haloalkane	Abstraction by Halogen Atom	Site of Attack			
		1-carbon	2-carbon	3-carbon	4-carbon
$ClCH_2CH_2CH_2CH_3$	Cl·	0.158	0.478	1.00	0.397
$ClCH_2CH_2CH_2CH_3$	Br·	0.439	0.488	1.00	—
$BrCH_2CH_2CH_2CH_3$	Cl·	0.093	0.434	1.00	0.455
$BrCH_2CH_2CH_2CH_3$	Br·	0.062	5.78	1.00	—
$CH_3CHClCH_2CH_3$	Br·	—	1.00	0.085	—
$CH_3CHBrCH_2CH_3$	Br·	—	1.00	5.13	—

chlorine. The very reactive chlorine atom apparently does not require any neighboring group participation to proceed rapidly, whereas the less reactive bromine atom apparently makes use of any assistance available in attaining a transition state of low energy.

$$Br \cdot + H{-}C{-} \quad {-}C{-}Br \;\rightarrow\; \left[Br \cdots H \cdots C \quad Br \right] \;\rightarrow\; HBr \;+\; C \quad Br \quad (5\text{-}40)$$

The radicals produced in hydrogen abstractions such as these appear to have characteristics of a bridged species.[15] Bromination of *cis*-4-*t*-butylcyclohexyl bromide (**12**), a species that has its more stable conformation with the bromine in the axial position, yields a single dibromination product, namely *axial, axial* 1,2-dibromo-4-*t*-butylcyclohexane (**13**). Bromination of *trans*-4-*t*-butylcyclohexyl bromide (**14**), on the other hand, yields a mixture of isomeric dibromides. Also the bromination of the *cis*-isomer, which could involve the anchimeric assistance of the *axial*-bromine in the hydrogen abstraction without loss of the chair conformation of cyclohexane, was 15 times faster than the bromination

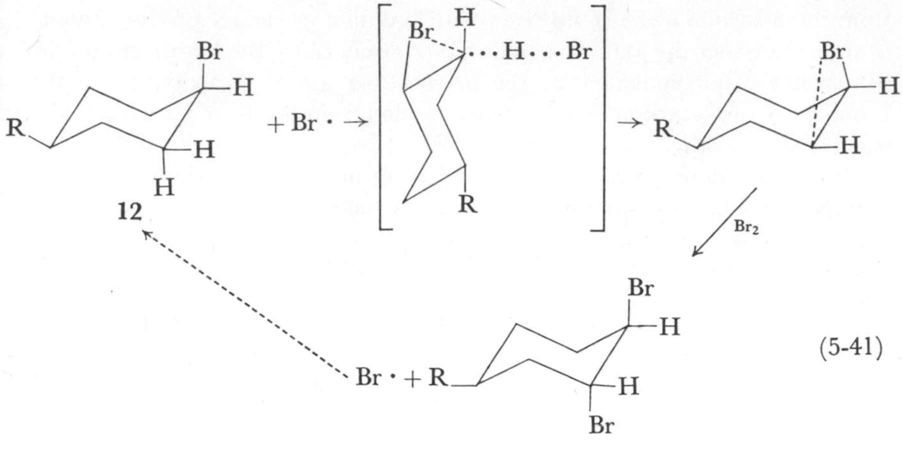

$$\text{12} + \text{Br} \cdot \; \rightarrow \; \left[\quad \right] \; \rightarrow \; R \qquad \qquad (5\text{-}41)$$

$$\text{Br} \cdot + R \qquad \qquad$$

$$R = C(CH_3)_3 \qquad \textbf{13}$$

of the *trans*-isomer. In the latter case, the neighboring group participation could occur only with the conformer having the *t*-butyl group in the very unfavorable axial position or if the ring assumed a twist-boat conformation.

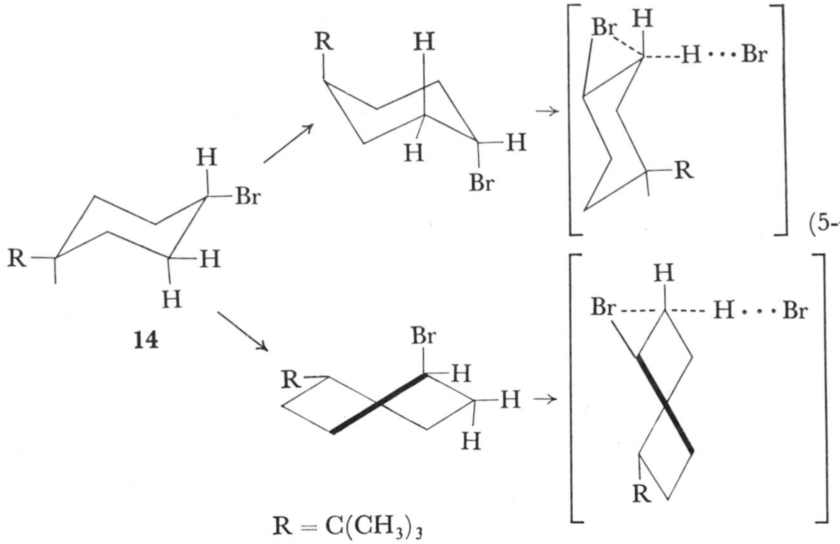

$$\textbf{14} \qquad \rightarrow \left[\quad \right] \qquad (5\text{-}42)$$

$$R = C(CH_3)_3$$

The high degree of stereospecificity observed in the bromination of **12** can be explained if the chain-carrying radical has the bridged structure shown in 5-41 and this radical reacts with the bromine from its least hindered side. This would result in bonding of the bromine to the 3-carbon in an axial position as shown in 5-41. The nonbridged radical formed by abstraction of a hydrogen

from the 3-carbon of **14** could react with bromine to yield a product havin
bromine in either the axial or equatorial position. Since the hydrogens on the
3-carbon are not activated by the neighboring group participation of the
bromine on the 4-carbon, hydrogen abstraction from the 1- and 2-carbon also
occurs yielding other isomeric dibromides.

Compelling evidence favoring the bridged radical comes from the bromination of (+)-1-bromo-2-methylbutane (**15**) which yields (−)-1,2-dibromo-2-

$$\text{(5-43)}$$

15   16   17

methylbutane (**18**) having a high degree of optical purity.[16] The nonbridged
intermediate radical **16** would be symmetrical and would yield a racemic
product. The bridged radical **17**, on the other hand, is asymmetric and reaction
with bromine as shown in 5-44 would yield the asymmetric product **18**. The

$$\text{(5-44)}$$

17   18

chlorination of **15** yields a racemic product mixture, a result analogous to the
chlorination of (+)-1-chloro-2-methylbutane (**19**).[17] However, bromination
of **19** does yield an active product, namely (−)-2-bromo-1-chloro-2-methyl-
butane (**20**), the optical purity of which depends on the reaction conditions.
Only at high bromine concentrations is the maximum optical activity observed
suggesting a competition reaction between the initially formed bridged radical
with bromine and its conversion to the open-chain radical leading to racemic
products. At low bromine concentrations the optical purity of **18** obtained from
15 also decreases.

$$(5\text{-}45)$$

$$(5\text{-}46)$$

$$(5\text{-}47)$$

20

III. Molecular Chlorine

While in many cases a single product may be produced in the bromination of an alkane having several different replaceable hydrogens, a mixture of chlorinated products may result from the reaction of the same alkane with molecular chlorine. Therefore, while advantage might be gained in the longer kinetic chain lengths of the chlorination reactions compared to bromination, much of this advantage is lost in the lack of specificity. If the species to be halogenated has only a single kind of hydrogen (e.g., the cycloalkanes), chlorination holds a marked advantage over bromination as a means of introducing a functionality into the molecule.

A. MECHANISM AND KINETICS OF CHLORINATIONS

The overall mechanism for the chlorination of an alkane with molecular chlorine induced by light is shown in 5-48 through 5-53. The kinetic chain

$$Cl_2 \xrightarrow{h\nu} 2Cl\cdot \tag{5-48}$$

$$Cl\cdot + RH \longrightarrow HCl + R\cdot \tag{5-49}$$

$$R\cdot + Cl_2 \longrightarrow RCl + Cl\cdot \tag{5-50}$$

$$2Cl\cdot \longrightarrow Cl_2 \qquad\qquad (5\text{-}51)$$

$$2R\cdot \longrightarrow R_2 \qquad\qquad (5\text{-}52)$$

$$R\cdot + Cl\cdot \longrightarrow RCl \qquad\qquad (5\text{-}53)$$

length of the sequence 5-49 and 5-50 is often very long, in some cases well over a million. Both reactions in the chain sequence therefore are quite facile, in contrast to the bromination reaction in which hydrogen abstraction by the halogen atom is the rate-limiting step. Also in contrast to bromination reactions, reversal of the hydrogen abstraction reaction 5-49 is unimportant because of the comparatively low reactivity of hydrogen chloride toward attack by alkyl radicals.

The observed kinetic rate laws for the overall chlorination of an alkane are generally complex since more than a single termination process is likely operative. In contrast to brominations, where the steady-state concentration ratio favored the chain-carrying bromine atoms because of their low reactivity as hydrogen abstractors, such a factor does not pertain in chlorination reactions. As is often the case in reactions having long kinetic chain lengths, the steady-state concentrations of the chain-carrying radicals are comparable, and all three termination reactions may take place to varying degrees depending on the concentration of the reactants.

B. SPECIFICITY IN CHLORINATIONS

1 General Comparison of Chlorination with Bromination. As in the case of bromination, the site of hydrogen abstraction from a molecule by the halogen atom will be the site of the substitution in that molecule. The following general observations can be made concerning the nature of the chlorinated products obtained from molecular chlorine with simple alkanes.[18] Every possible monochloride will be formed indicating appreciable reactivity of primary hydrogens toward abstraction as well as secondary and tertiary hydrogens. The relative reactivities of various hydrogens toward abstraction by chlorine atoms are tertiary > secondary > primary. Increasing the temperature of the reaction causes the reactivity differences of primary, secondary, and tertiary hydrogens to diminish.

The enthalpies of reaction of several different types of hydrogen atoms toward abstraction both by chlorine atoms and by bromine atoms, along with the activation energies for the reactions, are listed in Table 5-4. All of the hydrogen abstractions by bromine atoms are endothermic, and the activation energies are at least that of the endothermicity of the reaction. The consequence is rather large differences in the activation energies for the abstractions of the various kinds of hydrogens by bromine atoms and, hence, a significant degree of selectivity (see Table 5-5). The abstractions by chlorine atoms are all exo-

TABLE 5-4   Energetics of Hydrogen Abstractions by Halogen Atoms[19]

		$(X\cdot + RH \rightarrow HX + R\cdot)$	
$R—H$	$X\cdot$	ΔH	$\Delta E_{\ddagger}^{\ddagger}$ $(kcal/mole)$
$CH_3—H$	$Cl\cdot$	-1	3.83
$CH_3—H$	$Br\cdot$	$+16.5$	18.25
$C_2H_5—H$	$Cl\cdot$	-5	1.02
$C_2H_5—H$	$Br\cdot$	$+10.5$	13.39
$(CH_3)_2CH—H$	$Cl\cdot$	-9	0.66
$(CH_3)_2CH—H$	Br	$+7.0$	10.1
$(CH_3)_3C—H$	$Cl\cdot$	-13	0.1
$(CH_3)_3C—H$	$Br\cdot$	$+3.5$	7.51

thermic and the differences in the activation energies need not meet a minimum requirement (namely the differences in the enthalpies of the reactions). Hence, the differences in activation energy can be comparatively small and, consequently, the relative reactivities of various types of hydrogens are correspondingly small. The observation that the reactivities of primary, secondary, and tertiary hydrogens become comparable at elevated temperatures indicates that real differences in the activation energies for abstraction of the hydrogens by chlorine atoms do, however, exist.

While the transition state for an endothermic abstraction of a hydrogen atom by a bromine atom has much character of products contributing to it, the exothermic hydrogen abstractions by chlorine atoms have transition states that resemble the reactants more than the products (Hammond postulate). Therefore, there is little bond-making and bond-breaking in the transition

TABLE 5-5   Relative Reactivities of Hydrogens Toward Abstraction by Halogen Atoms[19]

Type	RH	$Br\cdot$ $(100°C)$	$Cl\cdot$ $(25°C)$
Primary	$C_2H_5—H$	1.00	1.00
Secondary	$(CH_3)_2CH—H$	160	4.4
Tertiary	$(CH_3)_3C—H$	4713	6.7

state of the hydrogen-atom abstraction by a chlorine atom. This is indicated by the primary isotope effects reported for abstraction of benzylic hydrogens from α-duetero toluene (21). In carbon tetrachloride at 77°C, k_H/k_D for X $\cdot$ = Br

$$X \cdot + C_6H_5CH_2D \xrightarrow{k_H} HX + C_6H_5\dot{C}HD \qquad (5\text{-}54a)$$
$$\mathbf{21}$$

$$X \cdot + C_6H_5CH_2D \xrightarrow{k_D} DX + C_6H_5\dot{C}H_2 \qquad (5\text{-}54b)$$
$$\mathbf{21}$$

X = Cl or Br

is 4.59 while in the same solvent at the same temperature, $k_H/k_D = 1.30$ when X $\cdot$ = Cl $\cdot$.[20] The higher value observed for the bromine atom abstraction is indicative of considerably more bond-breaking of the carbon-hydrogen bond in the transition state of the reaction than is observed in the chlorine atom reactions. The k_H/k_D for C_2H_6 compared to C_2D_6 is 2.68 at 27°[21] and 1.3 to 1.5 at -15°C for the tertiary hydrogen in isobutane.[22] Figure 5.2 shows the energy profiles for the abstractions of hydrogen atoms from isobutane and ethane by both chlorine atoms and bromine atoms. The transition states for the hydrogen atom abstractions by chlorine atoms have more reactant character than product character whereas the opposite is true for the bromine atom reactions. In the bromine atom reactions, factors such as resonance stabilization and the donor character of the radical formed are important in determining the activation energy requirement. These factors play a lesser role in determining the activation energy requirement for the hydrogen abstraction by a chlorine atom.

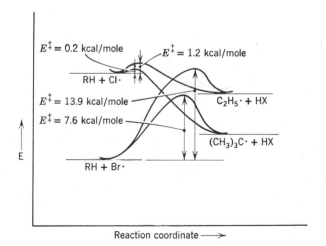

Fig. 5.2.   Hydrogen atom abstraction by halogen atoms.

2. Polar Effects. Most of the selectivity displayed by the chlorine atom as a hydrogen abstractor can be attributed to its electronegativity causing it to react more readily with hydrogens that are bonded to carbons having a high electron density than with those having a low electron density. The observed order of relative reactivities of alkyl hydrogens being in the order primary < secondary < tertiary is the consequence of an increase in electron density at the reaction site resulting from the electron-releasing properties of alkyl groups.

$$CH_3\overset{\longrightarrow}{CH_2}-H \qquad \overset{CH_3}{\underset{CH_3}{>}}CH-H \qquad \overset{CH_3}{\underset{CH_3}{>}}\overset{}{C-H}_{CH_3}$$

The reactivities of the hydrogens in compounds containing strong electron-attracting functionalities also display the electrophylicity of the chlorine atom. For example, chlorination of *n*-butyronitrile (**22**) yields mainly the β-chloro-

$$
\begin{array}{c}
CH_3CHClCH_2CN \quad (69\%) \\
+ \\
\mathbf{23} \\
CH_3CH_2CH_2CN + Cl_2 \xrightarrow{h\nu} ClCH_2CH_2CH_2CN \quad (31\%) \quad (5\text{-}55) \\
\mathbf{22} \\
+ \\
\mathbf{24} \\
CH_3CH_2CHClCN \quad (0\%) \\
\mathbf{25}
\end{array}
$$

isomer (**23**), some of the γ-chloroisomer (**24**) but none of the α-chloroisomer (**25**).[23] If resonance stabilization of the radical produced by the hydrogen abstraction were an important factor in the transition state of the reaction, the α-chloroisomer **25** would have been formed in significant amounts owing to the ability of the cyano group to stabilize the resulting free radical. Finding no **25** is indicative not only of the lack of resonance stabilization playing some significant role in the reaction but also that the reactivities of the hydrogens in the vicinity of the electronegative cyano group are decreased markedly relative to the other hydrogens in the molecule. Similar effects of decreasing reactivity of hydrogens bonded to carbons that have strong electron-withdrawing groups have been noted for esters[24], acids[24, 25], acid chlorides[24, 25], and alkyl halides[26] (Table 5-6).

An interesting observation that illustrates the peculiar nature of the chlorine atom compared to most other radicals is that toluene is less reactive than cyclohexane toward hydrogen abstraction by chlorine atoms. Although the benzyl radical produced in the hydrogen abstraction by a chlorine atom from

TABLE 5-6   Reactivities of Hydrogen toward Abstraction by Chlorine Atoms [24-26]

Compound (% Chlorination)	Chlorinating Agent	Temperature (°C)
(80) CH_3CH_2Cl (20)	Cl_2	208
(47) (7) $CH_3CH_2CH_2CH_2Cl$ (24) (22)	SO_2Cl_2	70
(49) (2) $CH_3CH_2CH_2CHCl_2$ (37) (12)	SO_2Cl_2	70
(45) $CH_3CH_2CH_2CF_3$ (55) (0)	Cl_2	Room temperature
(35) $CH_3CH_2CO_2H$ (65)	SO_2Cl_2	—
(31) (5) $CH_3CH_2CH_2CO_2H$ (14)	Cl_2	—
(30) (5) $CH_3CH_2CH_2COCl$ (65)	Cl_2	—
(25) $CH_3CH_2CH_2CH_2CN$ (75)	Cl_2	—
(29) $CH_3CH_2CH_2CH_2CO_2CH_3$ (71)	Cl_2	—

toluene has considerably more resonance stabilization than the cyclohexyl radical, the hydrogens of cyclohexane are 1.85 times more reactive than the benzylic hydrogens of toluene in the liquid phase (40°C) [27] and 2.1 times more reactive in the gas phase at 70°C. [28] The comparatively low reactivity of the benzylic hydrogens of toluene toward abstraction by chlorine is indicative of the lack of any significant contribution of the resonance stabilization of the benzyl radical to the activation energy requirement of the reaction as well as the electron-withdrawing ability of the aromatic ring.

The reactivities of benzylic hydrogens of substituted toluenes toward abstraction by chlorine atoms are affected by *meta* and *para* substituents that alter the electron density at the reaction site. A linear correlation between $\log k/k_0$ and the σ-values for the substituents was reported by Walling and Miller who found that $\rho = -0.76$ at 70°C.[28] Subsequent studies by Russell and Williamson[29] showed that data obtained by competitive chlorinations at 40° gave a somewhat better correlation when plotted against σ^+ ($\rho = -0.66$, std. dev. 0.02) than when plotted against σ ($\rho = -0.76$, std. dev. 0.06). Unfortunately, in neither study were competition reactions performed using substituents that have σ^+ values markedly different from the σ-values since most such substituents (hydroxy, methoxy, *t*-butyl, amino) activate the aromatic ring toward electrophylic substitution making determination of the reactivity of the benzylic hydrogens difficult by simple competitive experiments.

3. Solvent Effects. The reactivity of chlorine atoms as hydrogen abstractors is markedly affected in many instances by the solvent in which the reaction is performed. Russell showed that the reactivity ratio of the abstraction of a tertiary hydrogen relative to a primary hydrogen (k_{ter}/k_{pri}) from 2,3-dimethylbutane (**3**) was influenced by the solvent in which the reaction was performed[30] (see Table 5-7). Certain solvents have no effect on the ratio of reactivities of the hydrogens, k_{ter}/k_{pri} being about the same in these solvents as it is in the pure hydrocarbon. Other solvents apparently increase the selectivity of the chlorine atom as a hydrogen abstractor. This increase in selectivity can be explained in terms of complexing of the chlorine atom by the solvent, a process that lowers the energy of the chlorine atom. Reactions of a complexed chlorine atom are, therefore, somewhat less exothermic and resemble to some degree those of the less energetic and more selective bromine atom. Figure 5.3 illustrates how complexing of the chlorine atom results in an increase in the difference in the activation energies for the abstraction of primary and tertiary hydrogens provided, of course, that the transition states of the reaction involve the solvated chlorine atom. If the latter were not the case, the activation energy difference would be the same as that observed when there is no complexing of the chlorine atom.

The nature of the interaction of the chlorine atom with the complexing solvent depends very much on the nature of the solvent. It is of interest that some solvents (i.e., Cl_4C, $Cl_2C{=}CHCl$) apparently do not complex chlorine atoms to any appreciable extent. The solvent must possess some functionality that will interact with the chlorine atom. If it is assumed that a complexed chlorine atom has a very high degree of selectivity and abstracts only tertiary hydrogens, then finding appreciable amounts of primary hydrogen abstraction is indicative of free chlorine atoms also participating in the hydrogen atom abstraction reaction.

TABLE 5-7 **Effects of Solvent on Chlorination of 2,3-Dimethylbutane[30]**

Solvent	Solvent Concentration (*Molar*)	k_{ter}/k_{pri} 25°C	55°C
2,3-Dimethylbutane	7.6		3.7
Carbon tetrachloride	4.0		3.5
Trichloroethylene	4.0		3.6
t-Butyl alcohol	4.0		4.8
Dioxane	4.0		5.6
n-Butyl ether	4.0		7.2
CS$_2$	2.0	15	
CS$_2$	4.0	33	
CS$_2$	10.0	161	
CS$_2$	12.0	225	
Nitrobenzene	4.0		4.9
Chlorobenzene	2.0	9.0	
	4.0	17.1	
	6.0	27.5	
Fluorobenzene	4.0		10.3
Benzene	2.0	11.0	
	4.0	20.0	
	8.0	49.0	
Anisole	4.0		18.4

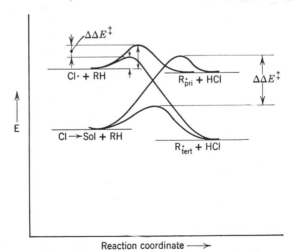

Fig. 5.3. Effect of solvating a chlorine atom on difference in activation energies ($\Delta\Delta E\ddagger$) of primary and tertiary hydrogen atoms.

The degree of selectivity, therefore, serves as a measure of the equilibrium between the free chlorine atom and the complexed chlorine atom. Increasing selectivity with increasing concentration of the complexing solvents is further indication of an equilibrium between the free and complexed chlorine atom.

Alcohols and ethers show that there is some degree of complexing of the chlorine atom with oxygen-containing molecules, the strongly electrophilic chlorine atoms possibly interacting with the nonbonding electrons of the oxygen atoms in the solvent molecules. Carbon disulfide apparently complexes quite readily with chlorine atoms and, at high concentrations of carbon disulfide, the selectivity of tertiary hydrogen abstraction over primary hydrogen abstraction, for all practical purposes, approaches that of bromine atoms. The chlorine in this case may be bonded in a σ-complex with the carbon disulfide and would

$$Cl\cdot + O\!\!-\!\!R \;\rightarrow\; \left(\begin{array}{c} Cl\!\!-\!\!O\!\!-\!\!R \\ | \\ R \end{array}\right) \;\xrightarrow{\text{RH}}\; HCl + R\cdot + O\!\!-\!\!R \qquad (5\text{-}56)$$

with R' below the first O and the last O.

$$Cl\cdot + S\!\!=\!\!C\!\!=\!\!S \;\rightarrow\; (Cl S\dot{C}\!\!=\!\!S) \;\xrightarrow{\text{RH}}\; HCl + R\cdot + CS_2 \qquad (5\text{-}57)$$

$R = $ alkyl, $R' = $ alkyl or H

be less energetic than the free chlorine atom and, therefore, more selective as a hydrogen atom abstractor. The equilibrium favoring the complexed species is apparently greater than that between chlorine atoms and oxygen-containing molecules.

The solvent effects displayed by aromatic compounds indicate a relationship between the electron density of the aromatic ring and its ability to complex the chlorine atom. Russell has shown that a direct correlation exists between $\log k_{ter}/k_{pri}$ and the Hammett σ-(meta) constants for the substituent on the benzene ring indicating the solvated chlorine atom in these solvents is a π-complex consisting of the chlorine atom and the aromatic ring.

$$Cl\cdot + \text{(benzene ring with Y)} \;\rightarrow\; Cl\text{—} \text{(ring with Y)} \;\xrightarrow{\text{RH}}\; HCl + R\cdot + \text{(benzene ring with Y)} \qquad (5\text{-}58)$$

An increased selectivity toward abstraction of secondary hydrogens relative to primary hydrogens is also noted in the chlorinations of n-heptane in benzene and in carbon disulfide relative to reactions performed in the pure hydrocarbon (see Table 5-8).[31] Competitive chlorinations of cyclic alkanes also show an increase in selectivity when the reactions are performed in complexing solvents (Table 5-9).[32] The order of stability of cycloalkyl radicals is cyclooctyl > cyclo-heptyl > cyclopentyl > cyclohexyl[33] and the increased selectivity in complexing solvents is indicative of a greater extent of bond-breaking in the transition state of the reactions involving the less energetic complexed chlorine atoms.

TABLE 5-8   Chlorination of n-Heptane at 20°C[31]

Conditions	Relative Reactivities of Hydrogen
Neat	1.00 3.3 3.3 1.0 $CH_3CH_2CH_2CH_2CH_2CH_2CH_3$ 3.5 3.3 3.5
Benzene solution	1.00 9.6 9.6 1.00 $CH_3CH_2CH_2CH_2CH_2CH_2CH_3$ 8.1 9.1 8.1
Carbon disulfide solution	1.00 23 23 1.00 $CH_3CH_2CH_2CH_2CH_2CH_2CH_3$ 18 10 18

The chlorinations of alkyl aromatics present an interesting situation in that the reactant itself may act as a complexing agent of the chlorine atom. Chlorination of ethylbenzene without a solvent yields a mixture of α-chloroethylbenzene (26) and β-chloroethylbenzene (27). At 40°C the ratio k_α/k_β is 14.5 (see Table 5-10). Similarly, the α-isomer (28) predominates over the β-isomer (29) in the

TABLE 5-9   Relative Reactivities of Cycloalkanes toward Chlorination[32]

Solvent	Cyclohexane	Cyclopentane	Cycloheptane	Cyclooctane
Neat	1.00	1.04	1.11	1.60
t-Butylbenzene(4M)	1.00	1.06	1.45	2.30
Carbon disulfide	1.00	1.15	2.00	3.75

$$C_6H_5\overset{\cdot}{C}HCH_3 \xrightarrow{Cl_2} C_6H_5CHClCH_3 + Cl\cdot \quad (5\text{-}59)$$
$$26$$

$$Cl\cdot + C_6H_5CH_2CH_3$$

$$\overset{k_\alpha}{\underset{-HCl}{\nearrow}} \qquad \overset{k_\beta}{\underset{-HCl}{\searrow}}$$

$$C_6H_5CH_2\overset{\cdot}{C}H_2 \xrightarrow{Cl_2} C_6H_5CH_2CH_2Cl + Cl\cdot \quad (5\text{-}60)$$
$$27$$

chlorination of cumene ($k_\alpha/k_\beta = 42$). However, if these chlorinations are performed in solvents that do not effectively complex chlorine atoms, the selectivity in both cases is decreased as shown by the data in Table 5-10. The most logical explanation is that as the concentration of either ethylbenzene or

TABLE 5-10   Chlorination of Alkyl Aromatics at 40°C[27c]

Alkyl Aromatic	Concentration	Solvent	k_α/k_β[a]
Ethylbenzene	8.10	None	14.5
Ethylbenzene	5.14	Nitrobenzene	6.15
Ethylbenzene	3.07	Nitrobenzene	3.63
Ethylbenzene	1.49	Nitrobenzene	2.70
Ethylbenzene	4.98	Cyclohexane	5.65
Ethylbenzene	2.93	Cyclohexane	3.36
Ethylbenzene	1.49	Cyclohexane	2.42
Ethylbenzene	0.00	Cyclohexane or Nitrobenzene	2.0[b]
Cumene	7.05	None	42.2
Cumene	5.04	Nitrobenzene	20.8
Cumene	3.00	Nitrobenzene	10.6
Cumene	1.49	Nitrobenzene	5.24
Cumene	0.00	Nitrobenzene	3.5[b]

[a] Statistically corrected.
[b] Extrapolated value.

$$(5\text{-}61)$$

cumene, both solvents that are capable of complexing with chlorine atoms, is decreased the amount of hydrogen abstraction by noncomplexed chlorine atoms increases. Extrapolation to infinite dilution of the substrate in the inert solvents gives a value for the reactivities of the hydrogens on the alkyl side chains toward abstraction by noncomplexed chlorine atoms. Interestingly, there is no dilution effect noted on the relative reactivities of cyclohexane and toluene toward chlorination suggesting that toluene apparently does not complex chlorine atoms effectively enough to play a similar role in the reactions of the benzylic hydrogens of toluene relative to those of cyclohexane.[29]

IV. Molecular Fluorine and Iodine

A. REACTIONS OF MOLECULAR FLUORINE

Molecular fluorine is a strong oxidizing agent and reacts with many alkanes yielding carbon tetrafluoride as the ultimate reaction product. Examination of the energetics of the chain-propagating reactions involved in the fluorination of methane shows that both reactions in the chain sequence are quite exothermic. Part of the problem encountered in reactions of molecular

$$F \cdot + CH_4 \rightarrow HF + CH_3 \cdot \quad (\Delta H = -31.2 \text{ kcal/mole}) \quad (5\text{-}62)$$

$$CH_3 \cdot + F_2 \rightarrow CH_3F + F \cdot \quad (\Delta H = -70.0 \text{ kcal/mole}) \quad (5\text{-}63)$$

fluorine with alkanes is dissipation of the heat evolved in the reaction. This can be accomplished in gas phase reactions by diluting the mixture with an inert gas such as nitrogen or in the liquid phase with inert solvents such as fluorinated halocarbons.

A spectrum of products is obtained in the reaction of fluorine with methane, among them the various expected fluorinated methanes as well as C_2F_6 and C_3F_8 suggesting that coupling reactions of trifluoromethyl radicals and other complex reactions occur to a significant extent.[34] The reaction proceeds readily in the dark and at temperatures as low as $-80°C$. The occurrence of appreciable amounts of termination products and the lack of necessity of special means of initiation suggest a comparatively facile means of introducing chain-carrying radicals into the reaction mixture. Although the decomposition energy of F_2 (5-64) is in the range that would allow for formation of fluorine atoms, the

$$F_2 \rightarrow 2F \cdot \quad (\Delta H = +38 \text{ kcal/mole}) \quad (5\text{-}64)$$

kinetic chain lengths of the reactions would necessarily have to be very long at lower temperatures to account for the rapid rate of reaction. However, formation of appreciable amounts of dimeric and trimeric products, which very likely result from termination reactions, indicates that the kinetic chain lengths are comparatively short. A bimolecular reaction between methane and fluorine (5-65), a reaction which results in formation of a pair of chain-carrying radicals, is only slightly endothermic. This bimolecular initiation reaction has been suggested[35] to be a facile means of introducing a sufficient number of chain-carrying radicals at low temperatures to account for both the rate of reaction

$$F_2 + CH_4 \rightarrow HF + CH_3 \cdot + F \cdot \quad (\Delta H = +6.2 \text{ kcal/mole}) \quad (5\text{-}65)$$

and the formation of dimeric products in comparatively large amounts.

The fragmentation of longer alkane chains with the ultimate formation of CF_4 is likely the result of direct displacement by fluorine atoms on carbon. This reaction may be in competition with displacement on hydrogen. Comparison of the energetics for the displacement reactions on both hydrogen and carbon of ethane shows that, in contrast to similar reactions with chlorine atoms, the displacement on carbon by fluorine is appreciably exothermic. The displacement reaction on carbon would still be subjected to steric problems (see Chapter 4)

$$F \cdot + CH_3CH_3 \ \rightarrow\ HF + \cdot CH_2CH_3 \qquad (\Delta H = -37.8 \text{ kcal/mole}) \qquad (5\text{-}66)$$

$$F \cdot + CH_3CH_3 \ \rightarrow\ CH_3F + CH_3 \cdot \qquad (\Delta H = -20 \quad \text{kcal/mole}) \qquad (5\text{-}67)$$

$$Cl \cdot + CH_3CH_3 \ \rightarrow\ HCl + \cdot CH_2CH_3 \qquad (\Delta H = -\ 4 \quad \text{kcal/mole}) \qquad (5\text{-}68)$$

$$Cl \cdot + CH_3CH_3 \ \rightarrow\ CH_3Cl + CH_3 \cdot \qquad (\Delta H = +\ 4 \quad \text{kcal/mole}) \qquad (5\text{-}69)$$

and may not compete with the hydrogen abstraction if hydrogens are still available for reaction. More likely the displacement on carbon occurs after the alkane is completely fluorinated and hydrogens are no longer available to compete with the energetically less favorable displacement on carbon.

B. MOLECULAR IODINE

The energetics of the chain-propagating reactions for the free-radical iodination of methane are shown in 5-70 and 5-71. The activation energy

$$I \cdot + CH_4 \ \rightarrow\ HI + CH_3 \cdot \qquad (\Delta H = +33.7 \text{ kcal/mole}) \qquad (5\text{-}70)$$

$$CH_3 \cdot + I_2 \ \rightarrow\ CH_3I + I \cdot \qquad (\Delta H = -20.2 \text{ kcal/mole}) \qquad (5\text{-}71)$$

requirement for the hydrogen abstraction from methane by iodine atoms would be at least 33.7 kcal/mole and, consequently, the rate would be too slow for the reaction to be part of a chain sequence having a long kinetic chain length. The abstraction of a benzylic hydrogen by iodine atoms is also endothermic to

$$I \cdot + C_6H_5CH_3 \ \rightarrow\ HI + C_6H_5\dot{C}H_2 \qquad (\Delta H = +13.7 \text{ kcal/mole}) \qquad (5\text{-}72)$$

a significant extent but, in this case, a chain reaction may be possible at elevated temperatures. Any chain sequence leading to alkyl or benzylic iodides involving iodine atoms will, however, be complicated by the interaction of the iodine atoms with the products. Displacement of an alkyl or benzyl radical by iodine atom attack on the iodine of the alkyl or benzyl iodide appears to be a facile reaction. Exchange reactions on benzyl iodide[36] and on alkyl iodides[37] with radioactive iodine are known to occur readily presumably by the sequence shown in 5-73 and 5-74. The light-induced deiodination of ethylene diiodide

$$I \cdot + RI \ \rightarrow\ I_2 + R \cdot \qquad (5\text{-}73)$$

$$R \cdot + I_2^* \ \rightarrow\ RI^* + I \cdot \qquad (5\text{-}74)$$

(30) sensitized by molecular iodine likely involves a similar displacement on the iodine atom of an alkyl iodide.[38]

$$I_2 \ \xrightarrow{\ h\nu\ }\ 2I \cdot \qquad (5\text{-}75)$$

$$I \cdot + ICH_2CH_2I \longrightarrow I_2 + \cdot CH_2CH_2I \qquad (5\text{-}76)$$
$$\underset{30}{}$$

$$\cdot CH_2CH_2I \longrightarrow CH_2{=}CH_2 + I \cdot \qquad (5\text{-}77)$$

$$2I \cdot \longrightarrow I_2 \qquad (5\text{-}78)$$

V. Other Halogenating Agents

A. SULFURYL CHLORIDE

Benzoyl peroxide-induced chlorinations of alkanes and alkyl aromatics with sulfuryl chloride[26] proceed by a free-radical chain reaction involving the sequence of chain-propagating steps shown in 5-79 through 5-82. The selectivity

$$R \cdot + SO_2Cl_2 \rightarrow RCl + \cdot SO_2Cl \qquad (5\text{-}79)$$

$$\cdot SO_2Cl + RH \rightarrow HCl + SO_2 + R \cdot \qquad (5\text{-}80)$$

or

$$\cdot SO_2Cl \rightarrow SO_2 + Cl \cdot \qquad (5\text{-}81)$$

$$Cl \cdot + RH \rightarrow HCl + R \cdot \qquad (5\text{-}82)$$

of sulfuryl chloride as a chlorinating agent depends on whether reaction 5-80 or 5-82 is the hydrogen abstracting process—more selectivity being expected if the $SO_2Cl \cdot$ radical rather than the chlorine atom is the hydrogen abstractor. The similarity in reactivity of substituted toluenes toward reaction by molecular chlorine and sulfuryl chloride[39] suggests a similar hydrogen abstractor is involved in the chlorination by both reagents. The chlorination of alkanes, on the other hand, indicates a degree of selectivity of sulfuryl chloride greater than that of molecular chlorine. Table 5-11 lists the relative reactivities of the various hydrogens of several alkanes toward reaction both with molecular chlorine and sulfuryl chloride. The increase in the reactivity ratios of the tertiary hydrogens with respect to primary hydrogens (k_{ter}/k_{pri}) is indicative of a less energetic species acting as the hydrogen abstractor in reactions of sulfuryl chloride.[40, 41]

A plausible explanation for the discrepancy in the reported behavior of sulfuryl chloride as a selective chlorinating agent was presented by Russell who

$$
Cl \cdot (or \cdot SO_2Cl) + H{-}\underset{\underset{R}{|}}{\overset{\overset{R}{|}}{C}}CH_3
\begin{array}{l}
\xrightarrow{k_{ter}} \quad HCl \qquad + \cdot \underset{\underset{R}{|}}{\overset{\overset{R}{|}}{C}}CH_3 \qquad (5\text{-}83) \\
\qquad\qquad (or\ HCl + SO_2) \\[2em]
\xrightarrow{k_{pri}} \quad HCl \qquad + H\underset{\underset{R}{|}}{\overset{\overset{R}{|}}{C}}CH_2 \cdot \quad (5\text{-}84) \\
\qquad\qquad (or\ HCl + SO_2)
\end{array}
$$

TABLE 5-11   Comparison of Selectivity of Cl_2 and SO_2Cl_2

Alkane	Agent	Conditions	Temperature ($°C$)	k_{ter}/k_{pri}
$(CH_3)_2CHCH(CH_3)_2$	Cl_2	Neat	55	3.7
$(CH_3)_2CHCH(CH_3)_2$	SO_2Cl_2	Neat	55	10.0
$(CH_3)_2CHCH(CH_3)_2$	Cl_2	Neat	25	4.2
$(CH_3)_2CHCH(CH_3)_2$	SO_2Cl_2	Neat	25	12.0
$(CH_3)_2CHC_5H_{11}$-n	Cl_2	Neat	20	2.9
$(CH_3)_2CHC_5H_{11}$-n	SO_2Cl_2	Neat	85	5.6

$(CH_3)_2CHCHCH(CH_3)_2$
   $\mid$
   CH_3
	Cl_2	Neat	20	2.4

H
$(CH_3)_2CHCHCH(CH_3)_2$
   $\mid$
   CH_3
	SO_2Cl_2	Neat	85	3.6

found that the selectivity of this reagent was influenced by the solvent in which the reactions were performed. The relative reactivities, ratio k_{ter}/k_{pri}, for 2,3-dimethylbutane become more alike in reactions of both chlorine and sulfuryl chloride when the chlorinations were performed in solvents that complex chlorine atoms effectively (see Table 5-12). The stability of the σ-complexed

TABLE 5-12   Effects of Solvent on Chlorinations of 2,3-Dimethylbutane with SO_2Cl_2 and Cl_2[41]

Solvent	Concentration	Temperature ($°C$)	k_{ter}/k_{pri} Cl_2	SO_2Cl_2
Neat	—	25	4.2	12
Benzene	8M	25	49	53
Neat	—	55	3.7	10
Benzene	4M	55	14.5	27.8
Benzene	8M	55	32	36
t-Butylbenzene	4M	55	24	32

$\cdot SO_2Cl$ radical is apparently great enough for this species to exist and react as such with alkanes. However, the π-complex of the aromatic species with the chlorine atom is quite possibly more stable than the $SO_2Cl\cdot$ radical. Consequently, reactions of sulfuryl chloride in aromatic compounds involve the aromatic π-complexed chlorine atom as the predominant hydrogen abstractor even in sulfuryl chloride reactions. This being so, the degree of selectivity for both chlorinating agents would be similar in aromatic solvents since the same hydrogen abstractor is involved to a significant degree in both reactions.

$$\text{SO}_2\text{Cl} \rightleftharpoons \text{SO}_2 + \text{Cl} \cdot \quad \rightleftharpoons \quad \text{Cl} \qquad (5\text{-}85)$$

B. *t*-BUTYL HYPOCHLORITE

The halogenation of hydrocarbons with *t*-butyl hypochlorite (**31**) was shown by Walling and Jacknow to be a free-radical chain reaction induced both by chemical initiators and by light.[42] The chain sequence of propagating reactions for the chlorination of ethane is shown in 5-86 and 5-87 with the energetics of the radical-propagating reactions.

$$(\text{CH}_3)_3\text{CO} \cdot + \text{CH}_3\text{CH}_3 \rightarrow (\text{CH}_3)_3\text{COH} + \cdot\text{CH}_2\text{CH}_3 \qquad (5\text{-}86)$$

$$(\Delta H = {\sim}{-}5 \text{ kcal/mole})$$

$$\cdot\text{CH}_2\text{CH}_3 + (\text{CH}_3)_3\text{COCl} \rightarrow \text{ClCH}_2\text{CH}_3 + (\text{CH}_3)_3\text{CO} \cdot \qquad (5\text{-}87)$$
$$\mathbf{31}$$

$$(\Delta H = {\sim}{-}20 \text{ kcal/mole})$$

Kinetic analyses of the reactions of *t*-butyl hypochlorite with toluene and with cyclohexane indicate that the reactions can have comparatively long kinetic chain lengths ($>10^4$)[42] and that various modes of termination may be

$$(\text{CH}_3)_3\text{COCl} \xrightarrow{h\nu} (\text{CH}_3)_3\text{CO} \cdot + \text{Cl} \cdot \qquad (5\text{-}88)$$

$$\text{Cl} \cdot + \text{RH} \longrightarrow \text{HCl} + \text{R} \cdot \qquad (5\text{-}89)$$

$$(\text{CH}_3)_3\text{CO} \cdot + \text{RH} \longrightarrow (\text{CH}_3)_3\text{COH} + \text{R} \cdot \qquad (5\text{-}90)$$

$$\text{R} \cdot + (\text{CH}_3)_3\text{COCl} \longrightarrow \text{RCl} + (\text{CH}_3)_3\text{CO} \cdot \qquad (5\text{-}91)$$

$$2(\text{CH}_3)_3\text{CO} \cdot \longrightarrow (\text{CH}_3)_3\text{COOC}(\text{CH}_3)_3 \qquad (5\text{-}92)$$

$$(\text{CH}_3)_3\text{CO} \cdot + \text{R} \cdot \longrightarrow (\text{CH}_3)_3\text{COR} \qquad (5\text{-}93)$$

$$2\text{R} \cdot \longrightarrow \text{R}_2 \qquad (5\text{-}94)$$

operative.[43] Three termination reactions are possible for a reaction involving two chain-carrying free radicals. Ingold and his co-workers (see Chapter 3) found that the kinetic rate law for the light-induced reaction of toluene ($\text{R} = \text{C}_6\text{H}_5\text{CH}_2 \cdot$) indicated termination by coupling of two *t*-butoxy radicals (reaction 5-92) in reactions performed using low ratios of toluene to hypochlorite. Walling and Kurkov found at higher ratios of toluene to hypochlorite that the cross-termination 5-93 was the predominant mode of termination. The two results do not contradict each other since the ratio of t-butoxy radicals with respect to benzyl radicals should be higher at low toluene concentrations and, consequently, favor the coupling reaction 5-92. At low ratios of cyclohexane ($R = C_6H_{11}$) to hypochlorite, cross-termination reaction 5-93 occurs, while at higher ratios of cyclohexane to hypochlorite, the ratio of cyclohexyl radicals to t-butoxy radicals becomes larger and termination by coupling of cyclohexyl radicals (reaction 5-94) becomes operative. Assuming that the reactions of both cyclohexyl and benzyl radicals with **31** are facile and that hydrogen atom abstraction is the rate-limiting step in the chain sequence, then the difference in behavior of the two hydrocarbons is understandable in terms of the greater reactivity of cyclohexane than toluene toward hydrogen abstraction by the t-butoxy radical (see below). A smaller ratio of t-butoxy radicals to cyclohexyl radical exists at the steady state at a given concentration ratio of cyclohexane and **31** than t-butoxy radicals to benzyl radicals at the same concentration ratio of toluene to **31**.

Consideration must be given to one possible complicating reaction encountered when using t-butyl hypochlorite to study reactions of the t-butoxy radical as a hydrogen abstractor. This is the fragmentation of the t-butoxy radical yielding acetone and a methyl radical. The fate of the methyl radical, which is also capable of acting as a hydrogen abstractor, is of some consequence to any investigation with the view of studying the reactions of t-butoxy radicals as hydrogen abstractors. The activation energy of the fragmentation reaction

$$(CH_3)_3CO \cdot \; \rightarrow \; CH_3 \cdot + CH_3COCH_3 \qquad (5\text{-}95)$$

$$CH_3 \cdot + (CH_3)_3COCl \; \rightarrow \; CH_3Cl + (CH_3)_3CO \cdot \qquad (5\text{-}96)$$

is higher than that of most hydrogen abstraction reactions by the t-butoxy radical and, consequently, little fragmentation occurs except at elevated reaction temperatures. Even at that, the reactivity of t-butyl hypochlorite toward the eliminated methyl radical (reaction 5-96) is very high and little, if any, hydrogen abstraction from the substrate by the methyl radical occurs. Another complicating aspect is chlorination by reaction of elemental chlorine formed by reaction of hydrogen chloride with **31**. This complication becomes

$$HCl + (CH_3)_3COCl \; \rightarrow \; Cl_2 + (CH_3)_3COH \qquad (5\text{-}97)$$

important if the chain sequence 5-90 and 5-91 is short and appreciable amounts of initiation (5-88) are required leading to formation of hydrogen chloride (5-89).

A characteristic feature of the t-butoxy radical as a hydrogen abstractor is its high reactivity. Its selectivity as a hydrogen abstractor, while greater than that of the chlorine atom, does resemble this radical more than the bromine atom. For example, cyclohexane is more reactive than toluene toward hydrogen abstraction by t-butoxy radicals. The electrophylicity of the t-butoxy radical is also apparent in its reactions with substituted toluenes. A reasonably good linear relationship is observed when $\log k/k_0$ is plotted against the Hammett σ-values ($\rho = -0.83$ at 40°C).[42]

t-Butoxy radicals show a propensity for abstracting allylic hydrogens from alkenes causing the formation of considerable amounts of allylic chlorination

$$(CH_3)_3CO\cdot + H\overset{|}{C}-\overset{|}{C}=C\overset{\diagup}{\diagdown} \rightarrow (CH_3)_3COH + \cdot\overset{\diagdown}{C}-\overset{|}{C}=C\overset{\diagup}{\diagdown} \qquad (5\text{-}98)$$

$$\overset{\diagdown}{C}-\overset{|}{C}=C\overset{\diagup}{\diagdown} + (CH_3)_3COCl \rightarrow Cl-\overset{|}{C}-\overset{|}{C}=C\overset{\diagup}{\diagdown} + (CH_3)_3CO\cdot \qquad (5\text{-}99)$$

products by the chain sequence 5-98 and 5-99.[44] Reactions of simple alkenes (see Table 5-13) show that a small amount of addition of the t-butoxy radical to the unsaturated linkage occurs in competition with the allylic hydrogen abstraction. Particularly striking is the observation that the allylic radicals retain their cis-$trans$ stereochemistry until they react with t-butyl hypochlorite yielding the allylic chloride. For example, the allylic radical obtained by

$$(CH_3)_3CO\cdot + \quad \underset{\textbf{32}}{\begin{matrix} CH_3 \; H \\ C \\ \| \\ C \\ H \quad CH_3 \end{matrix}} \quad \rightarrow (CH_3)_3COH + \underset{\textbf{33}}{\begin{matrix} \cdot CH_2 \; H \\ C \\ \| \\ C \\ H \quad CH_3 \end{matrix}} \leftrightarrow \begin{matrix} CH_2 \; H \\ C \\ | \\ C \\ H \quad CH_3 \end{matrix} \qquad (5\text{-}100)$$

$$\underset{\textbf{33}}{\begin{matrix} CH_2 \; H \\ C \\ \| \\ C \\ H \quad CH_3 \end{matrix}} + (CH_3)_3COCl \rightarrow \underset{\textbf{34}}{\begin{matrix} ClCH_2 \; H \\ C \\ \| \\ C \\ H \quad CH_3 \end{matrix}} + \underset{\textbf{35}}{\begin{matrix} CH_2 \; H \\ C \\ | \\ CHClCH_3 \end{matrix}} \qquad (5\text{-}101)$$

abstraction of an allylic hydrogen from $trans$-2-butene (**32**) retains its $trans$-stereochemistry as the hybrid radical and reacts to yield only the 1-chloro-2-butene (**34**) with the $trans$-configuration along with the isomeric 3-chloro-1-butene (**35**). The ratios of the 1-chloro-2-alkene to 3-chloro-1-alkenes formed in

TABLE 5-13 Reactions of Alkenes with *t*-Butyl Hypochlorite (40°C)[44]

Alkene	Allylic Products (%)	Addition Products (%)
trans-2-Butene	*trans*-1-Chloro-2-butene (73.2)	2-*t*-Butoxy-3-chloro-butane (3.4)
	3-Chloro-1-butene (26.8)	
cis-2-Butene	*cis*-1-Chloro-2-butene (63.2)	2-*t*-Butoxy-3-chloro-butane (16.9)
	3-Chloro-1-butene (36.8)	
1-Butene	*cis* and *trans* 1-Chloro-2-butene (69.1)	1-*t*-Butoxy-2-chloro-butane (3.3)
	3-Chloro-1-butene (30.9)	
trans-2-Pentene	*trans*-1-Chloro-2-pentene (17.2)	—
	3-Chloro-1-pentene (6.8)	
	cis and *trans*-4-Chloro-2-pentene (76)	
cis-2-Pentene	*cis*-1-Chloro-2-pentene (16)	—
	3-Chloro-1-pentene (9.7)	
	cis and *trans* 4-Chloro-2-pentene	
trans-4,4-Dimethyl-2-pentene	*trans*-1-Chloro-4,4-dimethyl-2-pentene (88.9)	—
	3-Chloro-4,4-dimethyl-1-pentene (11.1)	
cis-4,4-Dimethyl-2-pentene[a]	*cis*-1-Chloro-4,4-dimethyl-2-pentene (76.3)	—
	3-Chloro-4,4-dimethyl-1-pentene (23.7)	

[a] At −78.5°C.

121

these reactions vary for the *cis-* and *trans*-alkenes, an observation that indicates the existence of different hybrid allylic radicals reacting with *t*-butyl hypochlorite in each case.

The reactions of *t*-butyl hypochlorite with alkanes are subject to solvent effects as evidenced by the difference in reactivities of various hydrogens toward abstraction by *t*-butoxy radicals when the reactions are performed in different media. Table 5-14 lists the relative reactivities of primary and tertiary

TABLE 5-14   Reactions of 2,3-Dimethylbutane and *n*-Butane with *t*-Butyl Hypochlorite[45]

			k/k^a				
Alkane	*Solvent*	*100°C*	*70°C*	*40°C*	*25°C*	*0°C*	$E_1^{\ddagger}-E_2^{\ddagger\ b}$
n-Butane	Cl_4C	—	—	10.2	10.9	12.8	0.93
n-Butane	C_6H_5Cl	—	8.85	10.8	11.5	14.6	1.32
n-Butane	CH_3CN	—	4.83	6.52	8.30	10.9	2.19
	CH_3CO_2H	—	3.99	5.88	7.23	10.6	2.59
2,3-Dimethyl-butane	None	—	—	44	54	68	1.85
	C_6H_6	—	—	55	70	89	1.99
	C_6H_5Cl	—	35	54	66	94	2.58
	CH_3COCH_3	20	30	51	76	128	3.77
	CH_3CN	10	17	33	47	—	4.57

[a] $k/k = k_{sec}/k_{pri}$ for *n*-butane and k_{ter}/k_{pri} for 2,3-dimethylbutane.
[b] $E_1^{\ddagger}-E_2^{\ddagger} = E_{pri}^{\ddagger}-E_{sec}^{\ddagger}$ for *n*-butane and $E_{ter}^{\ddagger}-E_{pri}^{\ddagger}$ for 2,3-dimethylbutane.

hydrogens of 2,3-dimethylbutane and primary and secondary hydrogens of *n*-butane observed for the reactions of these alkanes with *t*-butyl hypochlorite in various solvents. Listed with the relative reactivity ratios are the differences in the activation energies for the reactions. While there are no large differences in the relative reactivity ratios at any given temperature, there are appreciable differences in the activation energies for the abstraction reactions. These solvent effects can be explained in terms of complexing the *t*-butoxy radical in the ground state yielding a species that must undergo some degree of desolvation in the transition state of the hydrogen abstraction reaction. Desolvation of the radical is necessary since the solvent molecule is very likely at the reaction site of the radical when it functions as a hydrogen abstractor. This contrasts with the reactions of the chlorine atom which can be solvated in both the ground

$$(CH_3)_3CO{\rightarrow}Sol + RH \; \rightarrow \; \left[(CH_3)_3CO \overset{Sol}{\cdots} H \cdots R\right] \; \rightarrow$$

$$(CH_3)_3COH + R\cdot + Sol \quad (5\text{-}102)$$

state and transition state. The extent of desolvation depends on the nature of the transition state of the hydrogen abstraction. More desolvation will be required for abstraction of the less reactive primary hydrogens where more bond-breaking is involved in the transition state than for abstraction of the more reactive secondary or tertiary hydrogens. Solvents that complex the t-butoxy radical effectively (i.e., acetonitrile and acetic acid) display a larger difference in the activation energies of these hydrogen atom abstraction reactions.

Solvent effects are observed in the relative rates of hydrogen abstraction from cyclohexane (5-103) with respect to fragmentation of the t-butoxy radical (5-104). Table 5-15 lists the differences in the activation energies for the

TABLE 5-15 Activation Energy Differences for Hydrogen Abstraction from Cyclohexane and β-Elimination of t-Butoxy Radicals[46]

Solvent	$E_{104}-E_{103}$ (kcal/mole)
None	10.80
CH_3CN	9.65
C_6H_6	8.66
C_6H_5Cl	7.21
$C_2H_2Cl_2$ (trans)	7.69
$C_2H_2Cl_2$ (cis)	7.04
CH_3CO_2H	5.80

hydrogen abstraction (E_{103}) and β-elimination (E_{104}). If it is assumed that there is complete desolvation in the hydrogen atom abstraction and the extent of solvation of the ground and transition states of the β-elimination are very

$$(CH_3)_3CO{\rightarrow}Sol + C_6H_{12} \rightarrow \left[(CH_3)_3\overset{Sol}{CO}\cdot\cdot H\cdot\cdot C_6H_{11}\right] \rightarrow$$

$$(CH_3)_3COH + C_6H_{11}\cdot + Sol \quad (5\text{-}103)$$

$$(CH_3)_3CO{\rightarrow}Sol \rightarrow \left[\begin{matrix} CH_3 \\ | \\ CH_3\cdot\cdot \overset{}{C}\cdots O{\rightarrow}Sol \\ | \\ CH_3 \end{matrix} \right] \rightarrow$$

$$CH_3\cdot + CH_3COCH_3 + Sol \quad (5\text{-}104)$$

nearly the same, it can be concluded that the solvation of the t-butoxy radical may be as high as 5 kcal/mole.

C. N-BROMOSUCCINIMIDE

Allylic bromination of alkenes with N-bromoamides was first reported in 1919 by Wohl[47] but the process was not fully appreciated as a synthetic method until attention was called to the reaction by Ziegler and his co-workers[48] nearly 25 years later. These studies showed that N-bromosuccinimide (**36**) was a particularly good reagent for this purpose. The synthetic aspects of the

$$(5\text{-}105)$$

36

reaction have been reviewed several times[49] and the mechanism of the reaction has been subjected to an extensive and critical investigation.

A radical chain mechanism proposed by Hey[50] and by Bloomfield[51] consisted of the sequence of reactions shown in 5-106 and 5-107. Support for the radical

$$(5\text{-}106)$$

$$(5\text{-}107)$$

chain character of the reaction came from observation that its rate is increased by radical initiators (peroxides, oxygen, and light).[52] An alternate radical chain mechanism that was subsequently demonstrated to be more probable was suggested by Goldfinger and his co-workers.[53] The Goldfinger mechanism differs markedly from the Bloomfield and Hey mechanism in that the allylic bromination results from a free-radical chain reaction of the alkene with molecular bromine. Bromine is introduced in very small concentrations into the reaction medium by the interaction of **36** with the hydrogen bromide

$$\text{Br} \cdot + \overset{|}{\text{H}}\overset{|}{\text{C}}\!-\!\overset{|}{\text{C}}\!=\!\overset{/}{\underset{\backslash}{\text{C}}} \;\rightleftarrows\; \text{HBr} + \cdot\overset{\backslash}{\underset{/}{\text{C}}}\!-\!\overset{|}{\text{C}}\!=\!\overset{/}{\underset{\backslash}{\text{C}}} \tag{5-108}$$

$$\cdot\overset{\backslash}{\underset{/}{\text{C}}}\!-\!\overset{|}{\text{C}}\!=\!\overset{/}{\underset{\backslash}{\text{C}}} + \text{Br}_2 \;\rightarrow\; \text{Br}\overset{|}{\text{C}}\!-\!\overset{|}{\text{C}}\!=\!\overset{/}{\underset{\backslash}{\text{C}}} + \text{Br} \cdot \tag{5-109}$$

$$\text{HBr} + \underset{O}{\overset{O}{\text{NBr}}} \;\rightarrow\; \underset{O}{\overset{O}{\text{NH}}} + \text{Br}_2 \tag{5-110}$$

formed in the free-radical chain reaction. The key steps of the Goldfinger mechanism are shown in 5-108 through 5-110. The reaction in which the bromine atom abstracts an allylic hydrogen competes with the reaction in

$$\text{Br} \cdot + \text{H}\!-\!\overset{|}{\text{C}}\!-\!\overset{|}{\text{C}}\!=\!\overset{/}{\underset{\backslash}{\text{C}}} \;\rightleftarrows\; \text{H}\overset{|}{\text{C}}\!-\!\overset{\cdot}{\underset{|}{\text{C}}}\!-\!\overset{|}{\text{C}}\!-\!\text{Br} \tag{5-111}$$

37

which the bromine atom adds to the unsaturated linkage. The adduct radical **37** formed in this reaction could react with molecular bromine yielding the alkene dibromide. However, the latter reaction requires a sufficiently high

$$\text{H}\overset{|}{\text{C}}\!-\!\overset{\cdot}{\underset{|}{\text{C}}}\!-\!\overset{|}{\text{C}}\!-\!\text{Br} + \text{Br}_2 \;\rightarrow\; \text{H}\overset{|}{\text{C}}\!-\!\overset{|}{\text{C}}\!-\!\overset{\overset{\text{Br}}{|}}{\text{C}}\!-\!\text{Br} + \text{Br} \cdot \tag{5-112}$$

37

concentration of molecular bromine, apparently much higher than that present under the circumstances, to compete with the fragmentation of the adduct radical yielding the alkene and the free bromine atom. Although reaction 5-108 is also reversible, the rate of the reaction of the allylic radical with the molecular bromine, even if the latter is present in only small concentrations, is faster than the reaction of the allylic radical with hydrogen bromide.

The crucial aspect of the Goldfinger mechanism is the establishment of a concentration of molecular bromine low enough to prevent its reaction with the adduct radical **37** yet high enough for it to react with the allylic radical. The maintenance of the proper bromine concentration results from formation of bromine only in the rapid reaction of the hydrogen bromide, a product of the chain sequence 5-108 and 5-109, with **36**.

Support for the Goldfinger mechanism comes from the observation that allylic bromination of cyclohexene predominates over the addition reaction if molecular bromine is present in very small quantities.[54] Other supporting evidence is that the nonbrominated portion of the alkenes undergo *cis-trans* isomerization in reactions with **36** owing to the loss of stereoconfiguration in the adduct radical **37**. Thus, reaction of *cis*-3-hexene (**38**) with **36** results in isomerization of the "unreacted" alkene to the *trans* isomer **39**.[53] Similar isomerizations occur with oleic and elaidic acids.[55]

Furthermore, the relative reactivities of various alkanes, alkylaromatics, and substituted toluenes toward bromination by NBS are essentially the same as those observed for bromination with molecular bromine.[56] If the Bloomfield and Hey mechanism were operative, the hydrogen abstracting species would be the *N*-succinimidyl radical, a species expected to have hydrogen abstracting properties different to some degree from those of bromine atoms.

D. POLYHALOALKANES

1. Polyhalomethanes. Peroxide-induced reactions of carbon tetrachloride and carbon tetrabromide with aliphatic hydrocarbons result in an exchange

$$X_4C + RH \rightarrow HCX_3 + RX \qquad (5\text{-}114)$$

of an alkyl hydrogen and a halogen atom.[57] A free-radical chain reaction involving hydrogen abstraction by the trihalomethyl radical is likely operative in these exchange reactions. In many instances both the hydrogen abstraction reaction by the trihaloalkyl radical and the reaction of the alkyl radical with the polyhalomethane are not facile enough to permit the existence of a long kinetic chain length for the reaction. Consider, for example, the energetics of

$$X_3C \cdot + RH \rightarrow HCX_3 + R \cdot \qquad (5\text{-}115)$$

$$R \cdot + X_4C \rightarrow RX + X_3C \cdot \qquad (5\text{-}116)$$

the reactions of carbon tetrachloride with toluene and with cyclohexane.

$$Cl_3C \cdot + C_6H_5CH_3 \rightarrow HCCl_3 + C_6H_5\dot{C}H_2 \qquad (5\text{-}117)$$

$$(\Delta H = -9.3 \text{ kcal/mole})$$

$$C_6H_5\dot{C}H_2 + Cl_4C \rightarrow C_6H_5CH_2Cl + Cl_3C\cdot \qquad (5\text{-}118)$$
$$(\Delta H = +5 \text{ kcal/mole})$$

$$Cl_3C\cdot + C_6H_{12} \rightarrow HCCl_3 + C_6H_{11}\cdot \qquad (5\text{-}119)$$
$$(\Delta H = -1.7 \text{ kcal/mole})$$

$$C_6H_{11}\cdot + Cl_4C \rightarrow C_6H_{11}Cl + Cl_3C\cdot \qquad (5\text{-}120)$$
$$(\Delta H = \sim +2 \text{ kcal/mole})$$

Both reactions involve a chain-propagating reaction that is endothermic. Although the reaction can be made to occur with carbon tetrachloride, elevated temperatures are required.[57, 58]

Bromotrichloromethane has been used for bromination of alkyl aromatics in the benzylic position. Although reaction of the benzyl radical with bromotrichloromethane is also endothermic, the activation energy requirement apparently is less than in the reaction of this radical with carbon tetrachloride

$$C_6H_5\dot{C}H_2 + BrCCl_3 \rightarrow C_6H_5CH_2Br + Cl_3C\cdot \qquad (5\text{-}121)$$
$$(\Delta H = +3 \text{ kcal/mole})$$

as evidenced by the longer chain lengths of the reactions with the former. Reactions of alkyl radicals with bromotrichloromethane are comparatively exothermic and bromination of alkanes with this reagent proceed readily.

$$C_6H_{11}\cdot + BrCCl_3 \rightarrow C_6H_{11}Br + Cl_3C\cdot \qquad (5\text{-}122)$$
$$(\Delta H = -14 \text{ kcal/mole})$$

The fairly long kinetic chain lengths obtained in brominations of both alkyl aromatics and certain alkanes with bromotrichloromethane allow for investigations of the trichloromethyl radical as a hydrogen-abstracting radical. Competition reactions of substituted toluenes toward bromination by bromotrichloromethane showed that *meta* and *para* substituents affected the reactivities of benzylic hydrogens toward abstraction by the trichloromethyl radical. A correlation was observed between the logarithm of the reactivity ratio of the substituted toluene with respect to toluene and the σ^+-values[59] of the substituent. The transition state for the hydrogen abstraction reaction for toluene, therefore, has polar character much like that suggested for hydrogen abstraction by bromine atoms.

$$C_6H_5CH_3 + \cdot CCl_3 \rightarrow \left[C_6H_5\overset{\delta+}{CH_2}\cdots H\cdots \overset{\delta-}{CCl_3} \right] \rightarrow C_6H_5\dot{C}H_2 + HCCl_3$$
$$(5\text{-}123)$$

Various reactions of polymethanes with alkanes display the selectivity of the trichloromethyl radical as a hydrogen abstractor. The halogenation of norbornane (**5**) occurs exclusively at the 2-position with both carbon tetrachloride and bromotrichloromethane[60] indicating the same degree of selectivity for abstraction of the hydrogens on the 2-carbon displayed by bromine atoms.

$$5 \qquad + \text{ XCCl}_3 \;\rightarrow\; \text{HCX}_3 + \qquad \qquad \text{X} \;+\; \qquad \text{H} \qquad (5\text{-}124)$$

$$(95\text{–}98\%) \qquad (<5\%)$$

$$\text{via} \qquad \text{Cl}_3\text{C}\cdot + \qquad \text{H} \;\rightarrow\; \text{HCCl}_3 + \qquad \qquad (5\text{-}125)$$

via

$$+ \text{XCCl}_3 \qquad \qquad \text{X} + \text{Cl}_3\text{C}\cdot \qquad (5\text{-}126)$$

$$\text{H} + \text{Cl}_3\text{C}\cdot \qquad (5\text{-}127)$$

$$\text{X} = \text{Br or Cl}$$

The predominant formation of the *exo*-isomer in each case is indicative of the steric interference encountered when the 2-norbornyl radical approaches the polyhalomethane from the *endo*- side of the radical in reaction 5-127.

Reactions of cycloalkanes with bromotrichloromethane also display the selectivity of the trichloromethyl radical as a hydrogen abstractor. The relative reactivities of the C_6, C_5, C_7, and C_8 cycloalkanes toward bromination with bromotrichloromethane at 80° are 1.00:1.57:3.30:9.20, respectively.[61]

The reaction of secondary alcohols with bromotrichloromethane and carbon tetrachloride[62] results in oxidation of the alcohol. Both steps in the chain sequence involve transition states in which there is a significant lowering of the activation energy requirement because of the donor qualities of the alcohol and alcohol-derived radical coupled with the acceptor qualities of the trichloromethyl radical and polyhalomethane. The same is true in the reactions of aldehydes with carbon tetrachloride, a reaction leading to the formation of an acid chloride by the chain sequence 5-132 and 5-133.[63] The reaction of the acyl

$$XCCl_3 + R_2CHOH \rightarrow R_2C{=}O + HCCl_3 + HX \qquad (5\text{-}128)$$

via

$$Cl_3C\cdot + R_2\underset{\underset{OH}{|}}{CH} \rightarrow \left[R_2\overset{\delta+}{\underset{\underset{OH}{|}}{C}}\cdots H\cdots\overset{\delta-}{CCl_3} \right] \rightarrow R_2\dot{C}OH + HCCl_3 \qquad (5\text{-}129)$$

$$R_2\dot{C}OH + XCCl_3 \rightarrow \left[R_2\overset{\delta+}{C}{\cdots}O\cdot\cdot H\cdot\cdot X\cdot\cdot\overset{\delta-}{CCl_3} \right] \rightarrow R_2C{=}O + HX + \cdot CCl_3$$
$$(5\text{-}130)$$

radical with carbon tetrachloride competes with the decarbonylation reaction of this radical, the latter reaction becoming more important with increasing

$$RCHO + Cl_4C \rightarrow R\underset{\underset{O}{\|}}{C}Cl + HCCl_3 \qquad (5\text{-}131)$$

via

$$Cl_3C\cdot + R\underset{\underset{O}{\|}}{C}{-}H \rightarrow \left[R\overset{\delta+}{\underset{\underset{O}{\|}}{C}}\cdot\cdot H\cdot\cdot\overset{\delta-}{CCl_3} \right] \rightarrow R\underset{\underset{O}{\|}}{\overset{}{C}}\cdot + HCCl_3 \qquad (5\text{-}132)$$

$$R\underset{\underset{O}{\|}}{C}\cdot + Cl_4C \rightarrow \left[R\overset{\delta+}{\underset{\underset{O}{\|}}{C}}\cdot\cdot Cl\cdot\cdot\overset{\delta-}{CCl_3} \right] \rightarrow R\underset{\underset{O}{\|}}{C}Cl + Cl_3C\cdot \qquad (5\text{-}133)$$

stability of the alkyl radical.[64] The use of bromotrichloromethane, a more reactive halogen source than carbon tetrachloride, would seem to be an inviting route around the difficulties of decarbonylation. However, a new complication arises, namely the interaction of the acid bromide with unreacted aldehyde yielding an α-bromoalkyl ester (40) [65].

$$R\underset{\underset{O}{\|}}{C}Br + R\underset{\underset{O}{\|}}{C}H \rightleftarrows R\underset{\underset{Br}{|}}{C}H O\underset{\underset{O}{\|}}{C}R \qquad (5\text{-}134)$$
$$\textbf{40}$$

2. Polyhaloethanes. The peroxide-induced reactions of hexachloroethane (**41**) with alkanes yield the halogenated alkane, hydrogen chloride, tetrachloroethylene, and pentachloroethane.[57] The chain sequence 5-135 through 5-138 shows that hydrogen atom abstraction can be performed either by a free chlorine atom or the pentachloroethyl radical. The relative amounts of pentachloroethane with respect to tetrachloroethylene formed increases with decreasing temperature and increasing reactivity of the alkane (Table 5-16).[66]

TABLE 5-16   Reactions of Cl_3CCCl_3 with Alkanes

Alkane	Temperature (°C)	$k_{136}[RH]/k_{137}$
Cyclohexane	40	0.20
Cyclohexane	61	0.095
Cyclohexane	80	0.083
2,3-Dimethylbutane	58	2.86

$$R \cdot + Cl_3CCCl_3 \longrightarrow RCl + Cl_2\dot{C}CCl_3 \qquad (5\text{-}135)$$
$$\mathbf{41}$$

$$Cl_2\dot{C}CCl_3 + RH \xrightarrow{k_{136}} Cl_2CHCCl_3 + R \cdot \qquad (5\text{-}136)$$

$$Cl_2\dot{C}CCl_3 \xrightarrow{k_{137}} Cl_2C{=\!=}CCl_2 + Cl \cdot \qquad (5\text{-}137)$$

$$Cl \cdot + RH \longrightarrow HCl + R \cdot \qquad (5\text{-}138)$$

Hexachloroethane does not chlorinate toluene by the previously outlined chain sequence presumably because of the low reactivity of the chlorinating agent toward reaction with benzyl radicals. Bromopentachloroethane (**42**), however, does react with toluene as well as cyclohexane yielding the brominated hydrocarbon, hydrogen chloride, pentachloroethane and tetrachloroethane. In these brominations, the selectivity is determined in part by the chlorine atom.

$$R \cdot + BrCCl_2CCl_3 \longrightarrow RBr + Cl_2\dot{C}CCl_3 \qquad (5\text{-}139)$$
$$\mathbf{42}$$

(Followed by 5-136 through 5-138)

Dibromotetrachloroethane (**43**) brominates alkanes and alkyl aromatics by a chain mechanism in which a bromine atom is the hydrogen abstractor. In

$$Br \cdot + RH \rightarrow HBr + R \cdot \qquad (5\text{-}140)$$

$$R \cdot + BrCCl_2CCl_2Br \rightarrow RBr + Cl_2\dot{C}CCl_2Br \qquad (5\text{-}141)$$
$$\mathbf{43}$$

$$Cl_2\dot{C}CCl_2Br \rightarrow Cl_2C{=\!=}CCl_2 + Br \cdot \qquad (5\text{-}142)$$

this case, the β-elimination of a bromine atom from the pentahaloethyl radical is faster than reaction of the pentahaloethyl radical with the alkanes.

Dibromotetrachloroethane can be used as an allylic brominating agent.[67] The action of this material as an allylic brominating agent resembles that of N-bromosuccinimide in that the allylic radical is formed by hydrogen abstraction by a bromine atom. The adduct radical is also likely formed (5-144) but fragments due to the low reactivity of dibromotetrachloroethane toward reaction with the adduct radical (5-146). With high concentrations of dibromo-tetrachloroethane, some dibromoalkenes are formed.

$$Br \cdot + H\overset{|}{\underset{|}{C}}C{=}C \rightleftarrows HBr + \cdot \overset{|}{C}{-}C{=}C \qquad (5\text{-}143)$$

$$Br \cdot + H\overset{|}{C}{-}\overset{|}{C}{=}C \rightleftarrows H\overset{|}{C}{-}\overset{\cdot}{C}{-}\overset{|}{C}{-}Br \qquad (5\text{-}144)$$

$$\cdot\overset{|}{C}{-}C{=}\overset{|}{C} + BrCCl_2CCl_2Br \rightarrow Br\overset{|}{C}{-}C{=}\overset{|}{C} + Cl_2\dot{C}CCl_2Br \qquad (5\text{-}145)$$

$$H\overset{|}{C}{-}\overset{\cdot}{C}{-}\overset{|}{C}Br + BrCCl_2CCl_2Br \rightarrow H\overset{|}{C}{-}\overset{\overset{Br}{|}}{C}{-}\overset{|}{C}{-}Br + Cl_2\dot{C}CCl_2Br \qquad (5\text{-}146)$$

$$Cl_2\dot{C}CCl_2Br \rightarrow Cl_2C{=}CCl_2 + Br \cdot \qquad (5\text{-}147)$$

E. MISCELLANEOUS HALOGENATING AGENTS

Trichloromethanesulfonyl chloride (44) chlorinates alkanes and alkyl aromatics yielding chloroform and sulfur dioxide as other products.[68] The chain

$$RH + \underset{\mathbf{44}}{Cl_3CSO_2Cl} \rightarrow RCl + SO_2 + HCCl_3 \qquad (5\text{-}148)$$

reaction can be effectively induced by benzoyl peroxide or light and involves hydrogen abstraction by the trichloromethanesulfonyl radical (45) yielding trichloromethanesulfinic acid (46) which decomposes to the observed reaction products. The relative reactivity of cyclohexane with respect to toluene

$$R \cdot + Cl_3CSO_2Cl \rightarrow RCl + \underset{\mathbf{45}}{Cl_3CSO_2} \cdot \qquad (5\text{-}149)$$

$$\underset{\mathbf{45}}{Cl_3CSO_2} \cdot + RH \rightarrow \underset{\mathbf{46}}{Cl_3CSO_2H} + R \cdot \qquad (5\text{-}150)$$

$$\underset{\mathbf{46}}{Cl_3CSO_2H} \rightarrow HCCl_3 + SO_2 \qquad (5\text{-}151)$$

($k_{cyclohexane}/k_{toluene}$) toward chlorination by trichloromethanesulfonyl chloride is different from that observed for competitive bromination of these hydrocarbons with bromotrichloromethane. This relative reactivity ratio at 78°C is 0.20 for bromotrichloromethane, which almost undoubtedly involves the trichloromethyl radical as the hydrogen abstractor, whereas the value is 1.86

at 80°C for trichloromethanesulfonyl chloride. The greater reactivity of cyclohexane with respect to toluene toward hydrogen abstraction is reminiscent of competition reactions of these hydrocarbons toward chlorine atoms and t-butoxy radicals and would suggest that the nature of the hydrogen abstractor in chlorinations with trichloromethanesulfonyl chloride is also a comparatively reactive species. The hybrid trichloromethanesulfonyl radical has some alkoxy radical character and reacts by bonding the hydrogen to the oxygen yielding the unstable trichloromethanesulfinic acid.

$$Cl_3CS \cdot \quad \longleftrightarrow \quad Cl_3CS \quad + RH \rightarrow Cl_3CS-OH + R \cdot \qquad (5\text{-}152)$$

45

The degree of specificity of **45** as a hydrogen abstractor, although less than a bromine atom or a trichloromethyl radical, is greater than a chlorine atom. Only the benzylic positions of alkyl aromatics are chlorinated by this reagent. The relative reactivities of cycloalkanes toward halogenations with various reagents (Table 5 17) show the difference in selectivity of the various hydrogen abstractors involved in the reactions of these reagents.

TABLE 5-17   Relative Reactivities of Cycloalkanes toward Halogenation

Halogenating Agent	Temperature (°C)	Hydrogen Abstractor	Cyclo- pentane	Cyclo- hexane	Cyclo- heptane	Cyclo- octane
Cl_2	40	$Cl \cdot$	1.0	1.0	1.0	1.5
$Cl_2 + 12MCS_2$	40	$Cl + CS_2$	1.2	1.0	2.0	3.8
Cl_3CSO_2Cl	80	$Cl_3CSO_2 \cdot$	1.0	1.0	2.7	4.2
$BrCCl_3$	80	$Cl_3C \cdot$	1.6	1.0	3.3	9.2

Trichloromethanesulfonyl bromide (**47**) behaves differently from trichloromethanesulfonyl chloride in that it decomposes to bromotrichloromethane and sulfur dioxide readily and the former acts as the brominating agent.[69] The relative reactivity ratio $k_{cyclohexane}/k_{toluene}$ is the same for **47** as for bromotri-

$$Cl_3CSO_2Br \rightarrow BrCCl_3 + SO_2 \qquad (5\text{-}153)$$
47

$$BrCCl_3 + RH \rightarrow HCCl_3 + RBr \qquad (5\text{-}154)$$

via

$$Cl_3C \cdot + RH \rightarrow HCCl_3 + R \cdot \qquad (5\text{-}155)$$

$$R \cdot + BrCCl_3 \rightarrow RBr + Cl_3C \cdot \qquad (5\text{-}156)$$

chloromethane indicating the same hydrogen abstractor is probably involved in both reactions.

Bromine-chlorine mixtures contain an equilibrium concentration of bromo-chloride (BrCl). This reagent is reported to brominate trimethylchlorosilane (48), methylene fluoride (49), and fluoroform (50), compounds which normally are not brominated by molecular bromine but do undergo chlorination.[70] The chain reaction suggested for the reaction is one in which the reactive chlorine atom is the hydrogen abstractor and the displacement on the BrCl occurs on the bromine.

$$R \cdot + BrCl \rightarrow RBr + Cl \cdot \qquad (5\text{-}157)$$

$$Cl \cdot + RH \rightarrow HCl + R \cdot \qquad (5\text{-}158)$$

$R = CH_2SiCl(CH_3)_2$ for **48**, CHF_2 for **49**, and CF_3 for **50**.

Iodine trichloride (ICl_3) both chlorinates and iodinates cyclohexane in a reaction induced by light.[71] The following chain reaction involving both chlorine atoms and a dichloroiodo radical accounts for the reaction products. The usefulness of the reaction is somewhat limited by extensive dehydrohalogenation of the reaction products leading ultimately to dihalogenated products.

$$ICl_3 \rightarrow ICl_2 \cdot + Cl \cdot \qquad (5\text{-}159)$$

$$Cl \cdot + RH \rightarrow HCl + R \cdot \qquad (5\text{-}160)$$

$$R \cdot + ICl_3 \rightarrow RCl + ICl_2 \cdot \qquad (5\text{-}161)$$

$$ICl_2 \cdot + RH \rightarrow HCl + ICl + R \cdot \qquad (5\text{-}162)$$

$$R \cdot + ICl \rightarrow RI + Cl \cdot \qquad (5\text{-}163)$$

Iodobenzene dichloride (**51**) reacts with hydrocarbons in light-induced reactions yielding a chlorinated substrate, hydrogen chloride, and iodobenzene.

$$\underset{\textbf{51}}{C_6H_5ICl_2} + RH \rightarrow C_6H_5I + HCl + RCl \qquad (5\text{-}164)$$

There is a high degree of specificity noted in the reaction of **51** with 2,3-dimethylbutane, namely that no primary but only tertiary chloride is produced. A chain mechanism in which $C_6H_5\dot{I}Cl$ is the hydrogen-abstracting radical

$$\underset{\textbf{51}}{R \cdot + C_6H_5ICl_2} \rightarrow RCl + C_6H_5ICl \cdot \qquad (5\text{-}165)$$

$$C_6H_5ICl \cdot + RH \rightarrow C_6H_5I + HCl + R \cdot \qquad (5\text{-}166)$$

accommodates the specificity noted and, therefore, appears more plausible than mechanisms which require decomposition of **51** to iodobenzene and molecular chlorine or fragmentation of $C_6H_5\dot{I}Cl$ to a free chlorine atom and iodobenzene.[72]

Phosphorus pentachloride reacts with hydrocarbons in benzoyl peroxide-induced reactions. The selectivity noted is greater than that observed in chlorinations with molecular chlorine suggesting the hydrogen abstracting species to be $PCl_4 \cdot$ as shown.[73]

$$R \cdot + PCl_5 \rightarrow RCl + PCl_4 \cdot \qquad (5\text{-}167)$$

$$PCl_4 \cdot + RH \rightarrow PCl_3 + HCl + R \cdot \qquad (5\text{-}168)$$

References

GENERAL

M. L. Poutsma, Chapter 3, Vol. I, *Methods in Free-Radical Chemistry*, E. S. Huyser, Ed., Marcel Dekker, New York, 1969.

W. Thaler, Chapter 2, Vol. II, *ibid.*, 1969.

SPECIFIC

1. H. G. Anderson, G. B. Kistiakowsky, and E. R. Van Artsdalen, *J. Chem. Phys.*, **16**, 305 (1942); G. B. Kistiakowsky and E. R. Van Artsdalene, *ibid.*, **12**, 469 (1944).
2. S. W. Benson and J. H. Buss, *J. Chem. Phys.*, **28**, 301 (1958).
3. G. C. Fettis, J. H. Knox, and A. F. Trotman-Dickenson, *J. Chem. Soc.*, **4177** (1960).
4. G. A. Russell and H. C. Brown, *J. Amer. Chem. Soc.*, **77**, 4025 (1955).
5. R. H. Siegmann, M. J. Beers, and H. O. Huisman, *Rec. trav. chim*, **84**, 67 (1964).
6. E. C. Kooyman and G. C. Vegter, *Tetrahedron*, **4**, 382 (1958).
7. B. Blouri, C. Cerceau, and F. E. Fauvet, *Bull. Soc. Chim. France*, **477** (1962); B. Blouri, C. Cerceau, and G. Lancher, *ibid.*, **304** (1963).
8. M. S. Kharasch and M. Z. Fineman, *J. Amer. Chem. Soc.*, **63**, 2776 (1941).
9. G. A. Russell and C. DeBoer, *ibid.*, **85**, 3136 (1963).
10. L. P. Hammett, *Chem. Rev.*, **17**, 125 (1935); H. H. Jaffe, *ibid.*, **53**, 191 (1953).
11. H. C. Brown and Y. Okamoto, *J. Amer. Chem. Soc.*, **79**, 1913 (1957); **80**, 4979 (1958).
12. R. E. Pearson and J. C. Martin, *ibid.*, **85**, 3142 (1965). See also E. C. Kooyman, R. Van Helden, and A. F. Bickel, *Koninkl. Ned. Akad. Wetinschap. Proc.*, **56B**, 75 (1953).
13. G. A. Russell, *J. Org. Chem.*, **23**, 1407 (1958).
14. W. A. Thaler, *J. Amer. Chem. Soc.*, **85**, 2607 (1963).
15. P. S. Skell and P. D. Readio, *ibid.*, **86**, 3334 (1964).
16. P. S. Skell, D. L. Tuleen, and P. D. Readio, *ibid.*, **85**, 2849 (1963).
17. H. C. Brown, M. S. Kharasch, and T. H. Chao, *ibid.*, **62**, 3439 (1940).

18. H. B. Hass, E. T. McBee, and P. Weber, *Ind. Eng. Chem.*, **27**, 1190 (1935);
    28, 333 (1936).
19. See General References.
20. K. B. Wiberg and L. H. Slaugh, *J. Amer. Chem. Soc.*, **80**, 3033 (1958).
21. G. Chiltz, R. Eckling, P. Goldfinger, G. Huybrechts, H. S. Johnson,
    L. Meyers, and G. Verbeke, *J. Chem. Phys.*, **38**, 1053 (1963).
22. G. A. Russell, *J. Amer. Chem. Soc.*, **80**, 4987 (1958).
23. A. Bruylants, M. Tits, C. Dieu, and R. Gauthier, *Bull. soc. chim. Belgec.*,
    61, 266 (1952).
24. A. Bruylants, M. Tits, and R. Danby, *ibid.*, **58**, 210 (1949).
25. M. S. Kharasch and H. C. Brown, *J. Amer. Chem. Soc.*, **62**, 925 (1940);
    H. J. den Hertog and P. Smit, *Proc. Chem. Soc.*, **1959**, 132.
26. G. A. Russell, *Tetrahedron*, **8**, 101 (1960); A. B. Ash and H. C. Brown,
    Record of Chemical Progress, **9**, 81 (1948); H. C. Brown and A. B. Ash, *J.
    Amer. Chem. Soc.*, **77**, 4019 (1955); M. S. Kharasch and H. C. Brown,
    ibid., **61**, 2142 (1939); A. L. Henne and F. B. Hinkamp, *ibid.*, **67**, 1194,
    1197 (1945).
27. (a) H. C. Brown and G. A. Russell, *ibid.*, **74**, 3996 (1952);
    (b) G. A. Russell and H. C. Brown, *ibid.*, **77**, 4578 (1955);
    (c) G. A. Russell, A. Ito, and D. G. Hendry, *ibid.*, **85**, 2976 (1963).
28. C. Walling and B. Miller, *ibid.*, **79**, 4181 (1957).
29. G. A. Russell and R. C. Williamson, Jr., *ibid.*, **86**, 2357 (1964).
30. G. A. Russell, *ibid.*, **79**, 2977 (1957); **80**, 4987 (1958).
31. P. Smit and H. J. den Hertog, *Rec. trav. chim.*, **83**, 891 (1964); B. Fell and
    L. H. Kung, *Chem. Ber.*, **98**, 2871 (1965).
32. G. A. Russell, *J. Amer. Chem. Soc.*, **80**, 4997 (1958).
33. H. C. Brown, R. S. Fletcher, and R. B. Johannesen, *ibid.*, **73**, 212 (1951);
    C. C. Overberger, H. Biletch, A. B. Finestone, J. Lilker, and J. Herbert,
    75, 2078 (1953).
34. E. H. Hadley and L. A. Bigelow, *J. Amer. Chem. Soc.*, **62**, 1172 (1940).
35. W. T. Miller, Jr., S. D. Koch, Jr., and F. W. McLafferty, *ibid.*, **78**, 4992
    (1956).
36. M. Gazith and R. M. Noyes, *ibid.*, **77**, 6091 (1955).
37. J. E. Bujake, M. W. T. Pratt, and R. M. Noyes, *ibid.*, **83**, 1547 (1961).
38. H. J. Schumacher and E. O. Wiig, *Z. physik. Chem.*, **B11**, 45 (1931); H. J.
    Schumacher and G. Stieger, *ibid.*, **B12**, 348 (1931).
39. R. Van Helden and E. C. Kooyman, *Rec. trav. chim.*, **73**, 269 (1954).
40. A. E. Fuller and W. J. Hickinbottom, *J. Chem. Soc.*, 3228 (1965).
41. G. A. Russell, *J. Amer. Chem. Soc.*, **80**, 5002 (1958).
42. C. Walling and B. B. Jacknow, *J. Amer. Chem. Soc.*, **82**, 6108, 6113 (1960).
43. C. Walling and V. P. Kurkof, *ibid.*, **88**, 4727 (1966); **89**, 4895 (1967);
    D. J. Carlson, J. A. Howard, and K. U. Ingold, *ibid.*, **88**, 4725, 4726 (1966).

44. C. Walling and W. Thaler, *ibid.*, **83**, 3877 (1961).
45. C. Walling and P. Wagner, *ibid.*, **86**, 3368 (1964).
46. C. Walling and P. Wagner, *ibid.*, **85**, 2333 (1963).
47. A. Wohl, *Ber.*, **52**, 51 (1919).
48. K. Ziegler, A. Spaeta, E. Schaaf, W. Schumann, and E. Winkelmann, *Ann.*, **551**, 80 (1942).
49. C. Djerassi, *Chem. Rev.*, **43**, 271 (1948); N. P. Buu Hoi, *Rec. Chem. Prog.*, **13**, 30 (1952); L. Horner and E. H. Winkelmann, *Angew. Chem.*, **71**, 349 (1959).
50. D. H. Hey, *Ann. Rpts. Chem. Soc.*, **41**, 84 (1944).
51. G. F. Bloomfield, *J. Chem. Soc.*, **14** (1944).
52. H. Schmid and P. Karrer, *Helv. Chim. Acta.*, **29**, 573 (1946); H. J. Dauben and L. L. McCoy, *J. Amer. Chem. Soc.*, **81**, 4863 (1959); *J. Org. Chem.*, **24**, 1577 (1959).
53. J. Adam, P. A. Gosselain, and P. Goldfinger, *Nature*, **171**, 704 (1953); *Bull. soc. chim. Belg.*, **65**, 523 (1956).
54. F. L. J. Sixma and R. H. Rheim, *Konikl. Ned. Akad. Wetenschap. Proc.*, **61B**, 183 (1958); B. P. McGrath and J. M. Tedder, *Proc. Chem. Soc.*, **80** (1961).
55. E. Ucciani, J. Chouteau, and M. Naudet, *Bull. Chim. Soc. France*, **1960**, 1511.
56. G. A. Russell, C. DeBoer, and K. M. Desmond, *J. Amer. Chem. Soc.*, **85**, 365 (1963); G. A. Russell and K. M. Desmond, *ibid.*, **85**, 3139 (1963); C. Walling and A. L. Rieger, *ibid.*, **85**, 3134 (1963); R. E. Pearson and J. C. Martin, *ibid.*, **85**, 354, 3142 (1963); C. Walling, A. L. Rieger, and D. D. Tanner, *ibid.*, **85**, 3129 (1963).
57. J. P. West and L. Schmerling, *J. Amer. Chem. Soc.*, **72**, 3525 (1950).
58. Y. A. Ol'dekap, *Doklady Akad. Nauk S.S.S.R.*, **93**, 75 (1953).
59. E. S. Huyser, *J. Amer. Chem. Soc.*, **82**, 391, 394 (1960).
60. E. C. Kooyman and G. C. Vegter, *Tetrahedron*, **4**, 382 (1958).
61. E. S. Huyser, H. Schimke, and R. L. Burham, *J. Org. Chem.*, **28**, 2141 (1963).
62. G. A. Razavaev, B. N. Moryganov, and A. S. Volkova, *Zhur. Obschei Khim.*, **23**, 1519 (1953).
63. S. Winstein and F. H. Seubold, *J. Amer. Chem. Soc.*, **87**, 2916 (1947).
64. D. E. Applequist and L. Kaplan, *ibid.*, **87**, 2194 (1965).
65. E. S. Huyser, C. Coon, and J. Finnerty, unpublished results.
66. D. N. DeMott, PhD. Thesis, University of Kansas, 1963. Diss. Abstract **XXV**, 4403 (1964).
67. E. S. Huyser and D. N. DeMott, *Chem. and Ind.*, 1954 (1963)
68. E. S. Huyser, *J. Amer. Chem. Soc.*, **82**, 5246 (1960); E. S. Huyser and B. Giddings, *J. Org. Chem.*, **27**, 3391 (1962).
69. R. P. Pinnell, E. S. Huyser, and J. Kleinberg, *ibid.*, **30**, 38 (1965).

70. J. L. Speier, *J. Amer. Chem. Soc.*, **73**, 826 (1951); R. P. Rue and R. A. Davis, U.S. Pat. 2,658,086 (1953).
71. W. W. Hess, E. S. Huyser, and J. Kleinberg, *J. Org. Chem.*, **29**, 1106 (1964).
72. D. F. Banks, E. S. Huyser, and J. Kleinberg, *ibid.*, **29**, 3692 (1964).
73. D. P. Wyman, J. Y. C. Wang, and W. R. Freeman, *ibid.*, **28**, 3173 (1963); B. Fell and L. H. Kung, *Chem. Ber.*, **98**, 2871 (1965).

Chapter 6

Free-Radical Additions: Chemistry of the Adding Reagent

One of the more familiar reactions encountered in organic chemistry is the addition of a reagent across an unsaturated linkage. Although some of the reagents that add to unsaturated linkages by an ionic mechanism will do so by a free-radical mechanism as well, there are many reagents which add only by a free-radical chain reaction. The mechanistic aspects of free-radical additions covered in this and the next chapter hopefully will point out the synthetic potential (and limitations) of these reactions and the value of an appreciation of the kinetic aspects of a reaction in determining its synthetic potential.

I. General Features of Addition Reactions

A. THE CHAIN SEQUENCE

Addition of a reagent AB across the unsaturated linkage of an alkene $RCH{=}CH_2$ in a free-radical reaction consists of two chain-propagating steps, one an addition reaction and the other a displacement reaction. In free-radical

$$AB + CH_2{=}CHR \rightarrow ACH_2CHBR \qquad (6\text{-}1)$$

via

$$A{\cdot} + CH_2{=}CHR \rightarrow ACH_2\dot{C}HR \qquad (6\text{-}2)$$

$$ACH_2\dot{C}HR + AB \rightarrow ACH_2CHBR + A{\cdot} \qquad (6\text{-}3)$$

additions, the displacement (or abstraction) reaction is often referred to as the chain-transfer step, a name resulting from the appearance of the same reaction in vinyl polymerizations (see Chapter 12). The direction of the addition of the unsymmetrical reagent AB to an unsymmetrically substituted ethylene is dictated by the nature of the addition of the adding radical $A{\cdot}$, a reaction that almost always occurs so as to produce the more stable of the two possible adduct radicals. The displacement reaction on the adding reagent by the adduct radical produces the product of the reaction and regenerates the adding radical.

This chapter is devoted to the nature of the adding reagent and the reaction discussed largely in terms of additions to simple alkenes. The next chapter is concerned with the chemistry of the unsaturated compounds that undergo free-radical addition.

B. TELOMER FORMATION

The severest limitation of the free-radical addition reaction occurs in the displacement reaction of the adduct radical on the adding reagent. In order to obtain good yields of a simple 1 : 1 addition product (one molecule of the alkene and one of the addition product) as shown in 6-1, the displacement reaction 6-3 must compete with the addition of the adduct radical to another molecule of the alkene (reaction 6-4). Reaction 6-4 results in the formation of a 2 : 1 adduct radical (**1**), a species consisting of two molecules of the alkene and one adding radical from the original reagent. Reaction of the 2 : 1 adduct radical

$$ACH_2\dot{C}HR + CH_2{=}CHR \rightarrow \begin{array}{l} ACH_2CHR \\ \quad | \\ \ \ CH_2\dot{C}HR \\ \qquad \mathbf{1} \end{array} \qquad (6\text{-}4)$$

$$\begin{array}{l} ACH_2CHR \\ \quad | \\ \ \ CH_2\dot{C}HR \end{array} + AB \rightarrow \begin{array}{l} ACH_2CHR \\ \quad | \\ \ \ CH_2CHBR \\ \qquad \mathbf{2} \end{array} + A{\cdot} \qquad (6\text{-}5)$$

with the adding reagent (reaction 6-5) yields a 2:1 addition product (**2**). The 2:1 adduct radical is faced with the same choice as the 1:1 adduct radical. Addition to another molecule of alkene yields a 3:1 adduct radical from which a 3:1 addition product is formed. Further reaction of the 3:1 adduct radical with the alkene yields a 4:1 adduct radical. If the addition reaction of the adduct radical to the alkene competes favorably with the abstraction reaction from the adding reagent AB, a series of addition products consisting of 2, 3, 4, etc. molecules of alkene per molecule of adding reagent (referred to as telomers) will be formed. Although in some cases such a series of telomeric addition products may be desired, for most synthetic purposes the simple 1:1 addition product is preferred.

C. CHAIN-TRANSFER CONSTANTS

The free-radical chemistry of the adduct radical with both the alkene and the adding reagent will dictate largely what the distribution of products will be in the addition of a given reagent to a given unsaturated linkage. The reactivity ratio, k_{tr}/k_p, referred to as the chain-transfer constant, is a measure of the reactivity of the adding reagent with respect to the unsaturated compound toward reaction with the adduct radical. An adding reagent-unsaturated system

$$ACH_2\dot{C}HR + AB \xrightarrow{k_{tr}} ACH_2CHBR + A\cdot \qquad (6\text{-}6)$$

$$ACH_2\dot{C}HR + CH_2{=}CHR \xrightarrow{k_p} ACH_2CHR \qquad (6\text{-}7)$$
$$\underset{\displaystyle CH_2\dot{C}HR}{\big|}$$

with a high chain-transfer constant ($k_{tr}/k_p > 10$) results in high yields of the simple 1:1 addition product even when approximately equal amounts of the two reagents are available, whereas a system with a low chain-transfer constant ($k_{tr}/k_p < 1$) will give mainly telomeric products under the same conditions. Some of the general characteristics of these reactions that allow for making predictions concerning chain-transfer constants for particular systems in terms of the adduct radicals and the substrates (adding reagent or unsaturated linkage) are outlined below.

1. Resonance. Resonance can be an important factor in determining the chain-transfer constant of an adding reagent-alkene system. An adduct radical that is stabilized by resonance to a high degree is generally less reactive toward both chain transfer with the adding reagent and addition to the unsaturated compound than one which is not stabilized by resonance. Furthermore, substrates (either the adding reagent or the unsaturated compound) that produce stabilized radicals in their respective chain-propagating reaction are generally (but not always) more reactive toward any adduct radical than substrates that yield radicals that are not stabilized to a significant degree. These generalizations can be demonstrated by examination of chain-transfer constants reported for cyclohexane and toluene with styrene (3) and vinyl acetate (4) (see Table 6-1). Both styrene and vinyl acetate are reactive monomers capable of being

TABLE 6-1   Chain-Transfer Constants Illustrating Resonance Effects at 60°[1]

	k_{tr}/k_p	
Adding Reagent	*Styrene*	*Vinyl Acetate*
Cyclohexane	2.4×10^{-6}	6.6×10^{-4}
Toluene	1.25×10^{-5}	2.1×10^{-3}

converted to high molecular weight polymers because of their reactivity toward radical addition. Although little, if any, 1:1 addition product would result from reactions of these adding reagents with such reactive monomers, the chain-transfer constants, which will be small because of the reactivity of 3 and 4 toward adduct radical addition, are useful in appreciating the role of resonance in these reactions.

Cyclohexane and vinyl acetate are substrates that yield radicals in their reactions with an adding radical that are not resonance-stabilized to any significant degree, whereas toluene and styrene both yield resonance-stabilized benzylic radicals. Comparing the chain-transfer constants of cyclohexane and toluene

Styrene-Cyclohexane

$$\sim CH_2\dot{C}HC_6H_5 + C_6H_{12} \xrightarrow{\ k_{tr}\ } \sim CH_2CH_2C_6H_5 + C_6H_{11} \cdot \quad (6\text{-}8)$$

$$\sim CH_2\dot{C}HC_6H_5 + CH_2{=}CHC_6H_5 \xrightarrow{\ k_p\ } \underset{\underset{\displaystyle CH_2\dot{C}HC_6H_5}{|}}{\sim CH_2CHC_6H_5} \quad (6\text{-}9)$$
$$\mathbf{4}$$

Styrene-Toluene

$$\sim CH_2\dot{C}HC_6H_5 + C_6H_5CH_3 \xrightarrow{\ k_{tr}\ } \sim CH_2CH_2C_6H_5 + C_6H_5\dot{C}H_2 \quad (6\text{-}10)$$

$$\sim CH_2\dot{C}HC_6H_5 + CH_2{=}CHC_6H_5 \xrightarrow{\ k_p\ } \underset{\underset{\displaystyle CH_2\dot{C}HC_6H_5}{|}}{\sim CH_2CHC_6H_5} \quad (6\text{-}11)$$
$$\mathbf{3}$$

Vinyl Acetate-Cyclohexane

$$\sim CH_2\dot{C}HOAc + C_6H_{12} \xrightarrow{\ k_{tr}\ } \sim CH_2CH_2OAc + C_6H_{11} \cdot \quad (6\text{-}12)$$

$$\sim CH_2\dot{C}HOAc + CH_2{=}CHOAc \xrightarrow{\ k_p\ } \sim CH_2CHOAc \qquad (6\text{-}13)$$

$$\underset{4}{}$$

$$\begin{array}{c} | \\ CH_2\dot{C}HOAc \end{array}$$

Vinyl Acetate-Toluene

$$\sim CH_2\dot{C}HOAc + C_6H_5CH_3 \xrightarrow{\ k_{tr}\ } \sim CH_2CH_2OAc + C_6H_5\dot{C}H_2 \quad (6\text{-}14)$$

$$\sim CH_2\dot{C}HOAc + CH_2{=}CHOAc \xrightarrow{\ k_p\ } \sim CH_2CHOAc \qquad (6\text{-}15)$$

$$\underset{4}{}$$

$$\begin{array}{c} | \\ CH_2\dot{C}HOAc \end{array}$$

with each of these monomers shows that toluene, the substrate yielding a resonance-stabilized benzyl radical in the hydrogen abstraction reaction, is more reactive than cyclohexane in each case. Further examination shows that resonance-stabilized adduct radicals are less prone to participate in the abstraction reaction relative to addition than adduct radicals possessing little stabilization. If it is assumed that the reaction rate constants for the addition of a styrene-derived adduct radical to styrene is about the same as that of a vinyl acetate-derived radical to vinyl acetate, the transfer constants show that the less stabilized vinyl acetate-derived radical is more reactive than the styrene-derived radical in abstracting hydrogens from both toluene and cyclohexane. The assumption concerning the similarities of the k_p for styrene and vinyl acetate has merit both in theory and in fact. A reactive adduct radical such as the vinyl acetate-derived radical could be expected to react with a comparatively unreactive substrate such as vinyl acetate at approximately the same rate as a resonance-stabilized styrene-derived adduct radical would react with a very reactive substrate such as styrene. The k_p for vinyl acetate is, in fact, about 20 times greater than that of styrene and finding the chain-transfer constants greater for each of the solvents with vinyl acetate simply augments the argument that a nonresonance-stabilized adduct radical participates in the abstraction reaction more readily than a stabilized adduct radical.

It would appear then that chain-transfer constants can be expected to be high for systems in which the adding reagent produces a resonance-stabilized adding radical and the unsaturated compound to which it adds yields an adduct radical that is not resonance stabilized to a high degree. There are, of course, some limitations to this generalization. The adding reagent must not produce an adding radical that is too highly stabilized or it may be too unreactive to add to the unsaturated compound, or the unsaturated compound must not be too unreactive or it may not react with the adding radical.

2. Polar Effects. The good yields of simple 1:1 addition products obtained in many cases are often the result of a favorable polar effect encountered in the chain-transfer reaction. This point will be amplified in several of the examples given later in this chapter.

Some chain-transfer constants are listed in Table 6-2 that illustrate the significance of polar effects in the chain-transfer reaction. The examples chosen include polymerizable monomers and adding reagents that have either donor or acceptor characteristics. Although it is unlikely that 1:1 addition products will result from all of these reactions, the chain-transfer constants are illustrative of polar effects.

TABLE 6-2 Chain-Transfer Constants Illustrating Polar Effects at $60°$[2,3,4]

Adding Reagent	Styrene	Acrylonitril	Methyl Methacrylate
Toluene	1.25×10^{-5}	5.86×10^{-4}	5.2×10^{-5}
Cl_4C	9.0×10^{-3}	8.5×10^{-5}	2.4×10^{-4}
$n\text{-}C_4H_9SH$	22	—	0.67

The adduct radicals obtained from styrene have donor qualities, whereas those from acrylonitrile (5) and methyl methacrylate (6) are acceptor radicals. Toluene shows donor qualities as a substrate in hydrogen abstraction reactions since a donor benzyl radical is formed in the reaction. Carbon tetrachloride and mercaptans show acceptor qualities since they yield acceptor trichloromethyl and thiyl radicals, respectively.

Better chain-transfer constants can be expected from monomer-adding reagent pairs in which one reagent is a donor and the other an acceptor than from pairs in which both reagents are donors or both are acceptors. The chain-transfer constants for toluene, a donor substrate, are higher with the acceptor monomers acrylonitrile and methyl methacrylate than with styrene. Polar contributions to the transition state of the chain-transfer reactions with the former are possible but not with styrene. By itself, the above comparison does

$$\sim CH_2\overset{\cdot}{C}H + HCH_2C_6H_5 \rightarrow$$
$$\underset{CN}{|}$$

$$\mathbf{5}$$

$$\left[\sim CH_2\overset{\delta-}{C}H \cdots H \cdots \overset{\delta+}{C}H_2C_6H_5 \atop \underset{CN}{|} \right] \rightarrow$$

$$\sim CH_2CH_2CN + \cdot CH_2C_6H_5 \quad (6\text{-}16)$$

$$\underset{\substack{| \\ CO_2CH_3 \\ \textbf{6}}}{\overset{\substack{CH_3 \\ |}}{\sim CH_2C\cdot}} \quad + HCH_2C_6H_5 \rightarrow$$

$$\left[\underset{\substack{| \\ CO_2CH_3}}{\overset{\substack{CH_3 \\ |\delta-}}{\sim CH_2C}} \cdots H \cdots \overset{\delta+}{CH_2C_6H_5} \right] \rightarrow$$

$$\underset{\substack{| \\ CO_2CH_3}}{\overset{\substack{CH_3 \\ |}}{\sim CH_2CH}} \quad + \cdot CH_2C_6H_5 \quad (6\text{-}17)$$

not lend strong support to the concept of polar contributions to the transition state of the chain-transfer reaction since the differences could be attributed to resonance contributions. In the case of the acceptor adding reagents, the chain-transfer constants are higher for styrene than for acrylonitrile and methyl methacrylate, both of which involve acceptor radicals in the chain-transfer reaction.

$$\underset{\substack{| \\ C_6H_5}}{\sim CH_2\dot{C}H} \quad + ClCCl_3 \rightarrow$$

$$\left[\underset{\substack{| \\ C_6H_5}}{\overset{\delta+}{\sim CH_2CH}} \cdots Cl \cdots \overset{\delta-}{CCl_3} \right] \rightarrow$$

$$\underset{\substack{| \\ C_6H_5}}{\sim CH_2CHCl} + \cdot CCl_3 \quad (6\text{-}18)$$

$$\underset{\substack{| \\ C_6H_5}}{\sim CH_2\dot{C}H} \quad + HSC_4H_9\text{-}n \rightarrow$$

$$\left[\underset{\substack{| \\ C_6H_5}}{\overset{\delta+}{\sim CH_2CH}} \cdots H \cdots \overset{\delta-}{SC_4H_9\text{-}n} \right] \rightarrow$$

$$\underset{\substack{| \\ C_6H_5}}{\sim CH_2CH_2} + \cdot SC_4H_9\text{-}n \quad (6\text{-}19)$$

Reactions that have a favorable polar effect in the chain-transfer reaction yield an adding free radical that displays a favorable polar effect in its addition reaction to the monomer. The reaction of the acceptor adduct radicals obtained from either acrylonitrile or methyl methacrylate with toluene yields the donor benzyl radical which adds to either acrylonitrile or methyl methacrylate in

$$C_6H_5\dot{C}H_2 + CH_2{=}CHCN \rightarrow$$

$$\left[C_6H_5\overset{\delta+}{C}H_2\cdot\cdot CH_2\overset{\delta-}{\text{---}}\dot{C}HCN\right] \rightarrow$$

$$C_6H_5CH_2CH_2\dot{C}HCN \quad (6\text{-}20)$$

reactions having favorable polar contributions. Similarly, the acceptor trichloromethyl radical formed in the reaction of the styrene-derived radical with carbon tetrachloride adds to styrene in a reaction in which polar contributions again play a significant role.

$$Cl_3C\cdot + CH_2{=}CHC_6H_5 \rightarrow$$

$$\left[\overset{\delta-}{Cl_3C}\cdot\cdot CH_2\overset{\delta+}{\text{---}}\dot{C}HC_6H_5\right] \rightarrow$$

$$Cl_3CCH_2\dot{C}HC_6H_5 \quad (6\text{-}21)$$

3. Steric Effects. Any factor that retards addition of the adduct radical to the unsaturated compound without affecting the transfer reaction with the adding reagent increases the yields of the simple addition products. The low reactivity of nonterminal unsaturated linkages toward radical addition reactions is such a factor, and free-radical addition reactions to nonterminally unsaturated compounds often yield mainly the 1:1 addition product with very little of the higher telomers being produced. The low reactivity of the nonterminal unsaturation does impede the addition reaction of the adding radical yielding the adduct radical as well and may lead to complicating factors such as the shortening of the kinetic chain length of the chain sequence by a coupling reaction of the adding radicals. The adding radical may participate in other chain-propagating reactions with the alkene (e.g., abstraction of an allylic hydrogen), and a chain sequence yielding products from the adding reagent and unsaturated compound other than the addition product results. Many good addition reactions involving nonterminally unsaturated compounds are known to occur when other factors outweigh the difficulties caused by the steric effect (see Chapter 7).

4. Concentration and Temperature Effects. The ideal situation in an addition reaction for formation of the simple 1:1 addition product would be one in which the chain-transfer constant (k_{tr}/k_p) is considerably greater than unity. A great many of the addition reactions of synthetic interest, however, involve adding reagent-unsaturated compound pairs with chain-transfer constants considerably less than unity. Although there is little that can be done about structural features of the reactants that are responsible for a low chain-transfer constant, altering the experimental conditions can affect the amount of the simple addition product that may be obtained.

The product distribution in a free-radical addition reaction depends on the relative rates of reaction of the adduct radical with the adding reagent and the unsaturated compound. Although the chain-transfer constant does play a role in dictating the relative rates of these two reactions, the concentration of the reactants is also important. If the chain-transfer constant is low, that is $k_{tr} \ll k_p$, and the adding reagent and unsaturated compound are present in comparable amounts, the rate of the reaction of the adduct radical with the adding reagent (6-22) will be slow compared to that of the reaction leading to the 2:1 adduct radical (6-23). The rate of 6-22 can be made comparable or even greater than

$$\text{Rate of addition product formation} = k_{tr}[\text{Adduct radical}][\text{Adding reagent}]$$

$$(6\text{-}22)$$

$$\text{Rate of 2:1 adduct radical formation} = k_p[\text{Adduct radical}][\text{Unsaturate}]$$

$$(6\text{-}23)$$

6-23 if the concentration of the adding reagent is increased relative to that of the unsaturated compound. Experimentally this can be accomplished by adding the unsaturated compound slowly to a large excess of the adding reagent.

If the chain-transfer constant of a given adding reagent-unsaturated compound pair is less than unity, it can generally be increased by increasing the reaction temperature. The Arrhenius equations for the two reaction rate constants are given in 6-24 and 6-25. Both reactions are bimolecular and the

$$k_{tr} = P_{tr} Z_{tr} e^{\dfrac{-E_{tr}^{\ddagger}}{RT}} \tag{6-24}$$

$$k_p = P_p Z_p e^{\dfrac{-E_p^{\ddagger}}{RT}} \tag{6-25}$$

pre-exponential terms for both are likely similar. The difference in the values for the reaction rate constants at any given temperature is, therefore, largely the result of differences in the activation energies $E_{tr}^{\ddagger}$ and $E_p^{\ddagger}$. If $k_{tr} \ll k_p$, which means the reaction has a low chain-transfer constant, it must follow that $E_{tr}^{\ddagger} \gg E_p^{\ddagger}$. An increase in temperature has a greater effect on k_{tr}, the reaction with a higher activation energy, than k_p and, consequently, the chain-transfer constant increases with increasing temperature.

II. Chemistry of Adding Reagents

A. POLYHALOMETHANES[5]

Free-radical additions of carbon tetrachloride to simple alkenes yield simple 1:1 addition products in a chain reaction having favorable polar factors in the transition states of both radical-propagating reactions. A similar polar effect

$$Cl_3C \cdot + CH_2{=}CHR \rightarrow \left[Cl_3\overset{\delta-}{C} \cdot \cdot CH_2 \overset{\delta+}{\text{---}} \dot{C}HR \right] \rightarrow$$

$$Cl_3CCH_2\dot{C}HR \quad (6\text{-}26)$$

$$\underset{\underset{R}{|}}{Cl_3CH_2\dot{C}H} + ClCCl_3 \rightarrow \left[\underset{\underset{R}{|}}{Cl_3CCH_2\overset{\delta+}{C}H} \cdot \cdot Cl \cdot \cdot \overset{\delta-}{C}Cl_3 \right] \rightarrow$$

$$\underset{\underset{R}{|}}{Cl_3CCH_2CHCl} + Cl_3C \cdot \quad (6\text{-}27)$$

might be expected with styrene but the chain-transfer constant in this case is comparatively low (see Table 6-2) because of the low reactivity of the polyhalomethane toward reaction with the benzylic adduct radical. Additions to vinyl ethers involve polar effects in the transition states of both reactions that are more pronounced than in the reactions with simple alkenes, and high yields of the 1:1 addition product are produced.[6]

The yields of the simple addition product are very high in the reactions of bromotrichloromethane with alkenes, indicating that abstraction of the bromine atom from this polyhalomethane is more facile than abstraction of a chlorine from carbon tetrachloride. The enthalpy of the reaction of secondary alkyl radical with bromotrichloromethane is more exothermic than reaction with carbon tetrachloride (see Table 4-2).

$$\underset{\underset{CH_3}{|}}{Cl_3CCH_2\dot{C}H} + BrCCl_3 \rightarrow \underset{\underset{CH_3}{|}}{Cl_3CCH_2CHBr} + Cl_3C \cdot \quad (6\text{-}28)$$

$$\Delta H = -14 \text{ kcal/mole}$$

$$\underset{\underset{Cl_3}{|}}{Cl_3CCH_2\dot{C}H} + ClCCl_3 \rightarrow \underset{\underset{CH_3}{|}}{Cl_3CCH_2CHCl} + Cl_3C \cdot \quad (6\text{-}29)$$

$$\Delta H = -8 \text{ kcal/mole}$$

Addition of chloroform involves reaction of the haloform hydrogen rather than chlorine in the transfer reaction.[7] This reaction is somewhat endothermic whereas the abstraction of a chlorine is possibly exothermic. A more pronounced

$$Cl_3CCH_2\overset{\cdot}{C}H + HCCl_3 \rightarrow Cl_3CCH_2CH_2 + Cl_3C\cdot \qquad (6\text{-}30)$$
$$\underset{CH_3}{|} \qquad\qquad\qquad \underset{CH_3}{|}$$

$$\Delta H = +1.2 \text{ kcal/mole}$$

polar effect is operative in the hydrogen abstraction because the acceptor qualities of the trichloromethyl moiety are greater in the transition state **7** than those of a dichloromethyl species in the transition state of the chlorine abstrac-

$$\left[Cl_3CCH_2\overset{\delta+}{CH} \cdot\cdot H \cdot\cdot \overset{\delta-}{CCl_3} \right] \left[Cl_3CCH_2\overset{\delta+}{CH} \cdot\cdot Cl \cdot\cdot \overset{\delta-}{CHCl_2} \right]$$
$$\underset{\substack{CH_3 \\ \mathbf{7}}}{|} \qquad\qquad \underset{\substack{CH_3 \\ \mathbf{8}}}{|}$$

tions. Additions of bromoform[7] and iodoform,[8] however, proceed by abstraction of a halogen atom rather than a hydrogen atom in the transfer reaction.

Iodotrifluoromethane (**9**) adds to many unsaturated compounds resulting in high yields of the simple addition products.[5] Particularly interesting are the additions to perfluorinated alkenes such as tetrafluoroethylene[9] (**10**). Reactions

$$ICF_3 + F_2C{=}CF_2 \rightarrow ICF_2CF_2CF_3 \qquad (6\text{-}31)$$
$$\mathbf{9} \qquad \mathbf{10}$$

of iodotrifluoromethane with alkenes are somewhat better than carbon tetrachloride under similar conditions but its transfer constant is most likely not as large as that of bromotrichloromethane. The stability of the trifluoromethyl radical is considerably less than that of the trichloromethyl owing to the lack of significant amounts of delocalization of the unpaired electron because there are no low energy d-orbitals in fluorine. The bond dissociation energy of the carbon–iodine bond in iodotrifluoromethane is low enough ($H_{diss} = 54$ kcal/mole) to allow for chain transfer to occur readily enough to propagate the chain reaction. Bromotrifluoromethane does not react readily with alkenes to yield simple addition products.

One of the serious side reactions, aside from telomer formation, that occurs in the reactions of carbon tetrachloride and bromotrichloromethane with alkenes is the halogenation of the allylic position of the molecule by the chain sequence shown in 6-32 and 6-33. Abstraction of an allylic hydrogen by the

$$Cl_3C\cdot + H\overset{|}{C}\overset{|}{C}{=}C\diagup \overset{k_{tr}}{\longrightarrow} HCCl_3 + \cdot\overset{|}{C}C{=}C\diagup \qquad (6\text{-}32)$$

$$\cdot\overset{}{C}C{=}C\diagup + XCCl_3 \rightarrow X\overset{|}{C}C{=}C\diagup + Cl_3C\cdot \qquad (6\text{-}33)$$

$$Cl_3C \cdot + \overset{|}{H}\overset{|}{C}C\!\!=\!\!C\overset{\diagup}{\diagdown} \overset{k_a}{\longrightarrow} \overset{|}{H}C\!\!=\!\!\overset{|}{\underset{|}{\overset{\cdot}{C}}}\!\!-\!\!\overset{|}{\underset{|}{C}}CCl_3 \qquad (6\text{-}26)$$

trichloromethyl radical competes with the addition of this radical to the unsaturated linkage. The amount of addition product relative to allylic halogenation product produced depends on both the reactivities of the double bond toward addition and of the allylic hydrogen toward abstraction. The relative reactivity ratio k_a/k_{tr}, a measure of the relative amounts of addition and allylic hydrogen abstraction performed by the trichloromethyl radical, has been determined for several alkenes using bromotrichloromethane. The data in Table 6-3 show that allylic hydrogen abstraction occurs more readily with

TABLE 6-3 Relative Reactivities of Double Bonds with Respect to Allylic Hydrogens toward Reaction with Trichloromethyl Radicals[9]

Alkene	Temperature (°C)	k_a/k_{tr}
1-Octene	77.8	44
1-Decene	77.8	43
2-Pentene	77.8	5.7
3-Heptene	77.8	3.5
3-Heptene	40.0	5.0
4-Methyl-2-pentene	77.8	1.26
4-Methyl-2-pentene	40.0	1.68
cis-2-Butene	99.0	34
trans-2-Butene	99.0	26
Cyclohexene	77.8	1.20
Cyclopentene	77.8	5.4
Cycloheptene	77.8	5.5

nonterminal alkenes where the addition reaction is slow because of steric effects. Furthermore, the reactivities of allylic hydrogens increases in the order primary ≪ secondary < tertiary as evidenced by the reactivity ratios noted for *cis*- and *trans*-2-butene, 2-pentene and 4-methyl-2-pentene, all of which should be about equally reactive toward addition by the trichloromethyl radical but have primary, secondary, and tertiary allylic hydrogens, respectively, available for abstraction. Conformational effects must also play a role in determining either the addition or abstraction reactions (or both) in the reactions of cycloalkenes. The temperature effect indicates that allylic halogenation of nonterminal alkenes could be suppressed by performing the reactions at low temperatures.

The elements of carbon tetrachloride can be added to alkenes in a free-radical reaction of the alkene with trichloromethanesulfonyl chloride (**11**) rather than by using the polyhalomethane itself.[11] Sulfur dioxide is formed as a by-product

$$Cl_3CSO_2Cl + CH_2{=}CHR \rightarrow Cl_3CCH_2CHClR + SO_2 \qquad (6\text{-}34)$$
11

in these reactions. The nature of the addition product indicates that the tri-chloromethyl radical adds to the alkene suggesting that the chain sequence 6-35 through 6-37 may be responsible for the observed reaction products.

$$Cl_3C\cdot + CH_2{=}CHR \rightarrow Cl_3CCH_2\dot{C}HR \qquad\qquad (6\text{-}35)$$

$$Cl_3CCH_2\dot{C}HR + Cl_3CSO_2Cl \rightarrow Cl_3CCH_2CHClR + Cl_3CSO_2\cdot$$
$$(6\text{-}36)$$

$$Cl_3CSO_2\cdot \rightarrow Cl_3C\cdot + SO_2 \qquad\qquad (6\text{-}37)$$
12

This chain sequence requires the decomposition of the trichloromethanesulfonyl radical (**12**) yielding the trichloromethyl radical and sulfur dioxide, whereas reactions of trichloromethanesulfonyl chloride with alkanes (see Chapter 5) indicate that the trichloromethanesulfonyl radical is comparatively stable and abstracts hydrogens without decomposition by reaction 6-37.[12] The anomalous behavior of this radical can be explained in terms of complexing the radical with the alkene and reaction of the alkene-radical complex (**13**) with another molecule of alkene yielding the adduct radical by displacement of the sulfur dioxide-alkene complex (**14**).[13] Compounds that complex sulfur dioxide more

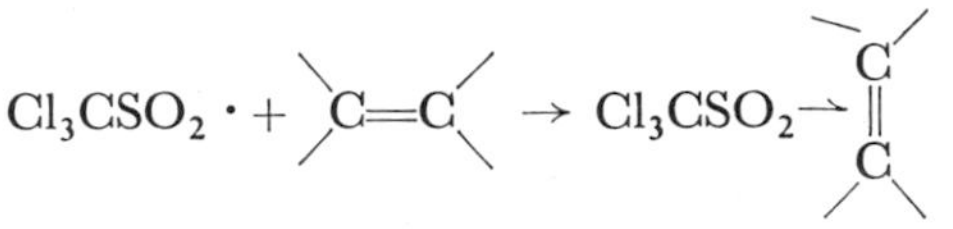

$$(6\text{-}38)$$

13

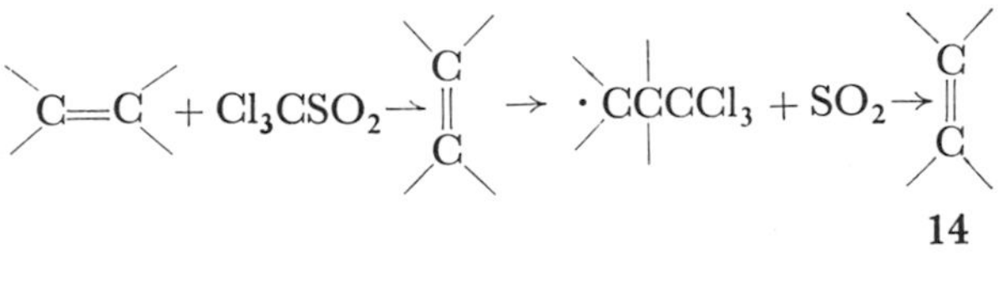

$$(6\text{-}39)$$

14

effectively than simple alkenes (e.g., pyridine) also complex the trichloro-methanesulfonyl radical more effectively than the alkene.

B. ALDEHYDES

Free-radical additions of aldehydes to unsaturated compounds yield ketones. For example, the peroxide-induced reaction of acetaldehyde with 1-octene yields 2-decanone (15) in a chain sequence involving addition of an acetyl radical to the alkene yielding an adduct radical which abstracts the aldehydic hydrogen from the aldehyde.[14]

$$CH_3CHO + CH_2{=}CHC_6H_{13}\text{-}n \xrightarrow{\text{Per.}} CH_3COC_8H_{17}\text{-}n \qquad (6\text{-}40)$$
$$\textbf{15}$$

by

$$CH_3\dot{C}O + CH_2{=}CHC_6H_{13}\text{-}n \longrightarrow CH_3COCH_2\dot{C}HC_6H_{13}\text{-}n \quad (6\text{-}41)$$

$$CH_3COCH_2\dot{C}HC_6H_{13}\text{-}n + CH_3CHO \longrightarrow \textbf{15} + CH_3\dot{C}O \qquad (6\text{-}42)$$

$$(\Delta H = -6.5\ \text{kcal/mole})$$

The reactions of aldehydes with alkenes usually result in some telomer formation. Although the enthalpy of the transfer reaction is favorable for a low activation energy requirement, there is no polar effect operative that might lower the activation energy since both the adduct radical and aldehyde are donor species. The addition of aldehydes to α,β-unsaturated ketones and esters occurs readily, however, since polar effects are operative in the transition states of both the addition reaction and chain-transfer step. Additions to nonterminal α,β-unsaturated carbonyl compounds also occur readily in spite of the steric problems generally encountered in such compounds. Thus, additions of acetaldehyde, n-butyraldehyde and n-heptaldehyde to diethyl maleate (16) result in high yields (75–80%) of the simple addition products.[15] In the addition

$$RC\cdot + CH{=}CHCO_2C_2H_5 \rightarrow$$
$$\Vert \quad \vert$$
$$O \quad CO_2C_2H_5$$
$$\textbf{16}$$

$$\left[\begin{array}{c} \overset{\delta+}{RC}\cdots CH\cdots\overset{\delta-}{C}HCO_2C_2H_5 \\ \Vert \quad \vert \\ O \quad CO_2C_2H_5 \end{array}\right] \rightarrow$$

$$RCCH\dot{C}HCO_2C_2H_5 \xrightarrow{RCHO} RCCHCH_2CO_2C_2H_5 + RC\cdot \quad (6\text{-}43)$$
$$\Vert\ \vert \qquad\qquad\qquad\quad \Vert\ \vert \qquad\qquad\qquad \Vert$$
$$O\ \vert \qquad\qquad\qquad\quad O\ \vert \qquad\qquad\qquad O$$
$$CO_2C_2H_5 \qquad\qquad\qquad CO_2C_2H_5$$

of acetaldehyde and n-butyraldehyde to mesityl oxide (17), the activation energy requirement for the transition state 18 leading to the adduct radical 19 is apparently lower than that of **20** as evidenced by the formation of **21** as the major 1:1 addition product.[16]

$$RC\cdot + (CH_3)_2C{=}CHCOCH_3 \rightarrow$$
$$\underset{\text{O}}{\overset{\parallel}{}}$$
$$\textbf{17}$$

$$\left[\begin{array}{c} \overset{CH_3}{\underset{|}{}} \\ \overset{\delta+}{RC}\cdot\cdot\overset{|}{C}\cdots\overset{\delta-}{C}HCOCH_3 \\ \overset{\parallel}{O}\quad\overset{|}{C}H_3 \end{array}\right] \rightarrow$$
$$\textbf{18}$$

$$\underset{\underset{\underset{CH_3}{|}}{\overset{\parallel}{O}}}{\overset{CH_3}{\underset{|}{}}RCC\dot{C}HCOCH_3} \xrightarrow{\text{RCHO}} \underset{\underset{\underset{CH_3}{|}}{\overset{\parallel}{O}}}{\overset{CH_3}{\underset{|}{}}RCCCH_2COCH_3} + \underset{\overset{\parallel}{O}}{RC\cdot} \quad (6\text{-}44)$$
$$\qquad\qquad \textbf{19} \qquad\qquad\qquad \textbf{21}$$

$$RC\cdot + CH_3COCH{=}C(CH_3)_2 \rightarrow$$
$$\underset{\text{O}}{\overset{\parallel}{}}$$

$$\left[\begin{array}{c} RC\cdot\cdot CH\cdots C(CH_3)_2 \\ \overset{\parallel}{O}\quad\overset{|}{C}OCH_3 \end{array}\right] \rightarrow$$
$$\textbf{20}$$

$$\underset{\underset{\underset{COCH_3}{|}}{\overset{\parallel}{O}}}{RCCH\dot{C}(CH_3)_2} \xrightarrow{\text{RCHO}} \underset{\underset{\underset{COCH_3}{|}}{\overset{\parallel}{O}}}{RCCHCH(CH_3)_2} + \underset{\overset{\parallel}{O}}{RC\cdot} \quad (6\text{-}45)$$
$$\qquad\qquad\qquad\qquad\qquad \textbf{22}$$

Acyl radicals decarbonylate yielding alkyl radicals and carbon monoxide,

$$R\dot{C}O \rightarrow R\cdot + CO \qquad (6\text{-}46)$$

reactions that are accelerated by increasing the temperature of the reaction. Most of the additions of aldehydes have been performed at moderate temperatures ($\sim$80°C) using initiators such as benzoyl peroxide or azobisisobutyronitrile (see Chapter 10). Increasing the reaction temperature may diminish the amount of telomer formation in aldehyde additions, but the decarbonylation reaction would lead to formation of alkanes as side products. The ease of decarbonylation of an acyl radical depends significantly on the stability of the alkyl radical produced in the reaction. At moderately low temperatures the acyl radical obtained from trimethylacetaldehyde (**23**) undergoes complete decarbonylation and only alkanes are obtained in the free-radical reactions of this aldehyde with alkenes.[17]

$$(CH_3)_3C\dot{C}O \rightarrow (CH_3)_3C\cdot + CO \qquad (6\text{-}47)$$

$$(CH_3)_3C\cdot + CH_2{=}CHR \rightarrow (CH_3)_3CCH_2\dot{C}HR \qquad (6\text{-}48)$$

$$(CH_3)_3CCH_2\dot{C}HR + (CH_3)_3CCHO \rightarrow (CH_3)_3CCH_2CH_2R + (CH_3)_3C\dot{C}O$$
$$\underset{\textbf{23}}{}$$

$$(6\text{-}49)$$

C. ALCOHOLS

The free-radical additions of primary and secondary alcohols to alkenes yield secondary and tertiary alcohols, respectively.[18] The nature of the addition product indicates that the transfer step is the abstraction of a hydrogen atom bonded to the carbon bearing the hydroxy group yielding an α-hydroxyalkyl radical as the adding species. Tertiary alcohols having no α-hydrogens do not

$$R_2\dot{C}OH + CH_2{=}CHR \longrightarrow R_2CCH_2\dot{C}HR \qquad (6\text{-}50)$$
$$\qquad\qquad\qquad\qquad\qquad | $$
$$\qquad\qquad\qquad\qquad\qquad OH$$

$$R_2CCH_2\dot{C}HR + R_2CHOH \xrightarrow{k_{tr}} R_2CCH_2CH_2R + R_2\dot{C}OH \qquad (6\text{-}51)$$
$$\quad | \qquad\qquad\qquad\qquad\qquad\qquad | $$
$$\quad OH \qquad\qquad\qquad\qquad\qquad\quad OH$$

$$R_2CCH_2\dot{C}HR + CH_2{=}CHR \xrightarrow{k_p} R_2CCH_2CHR \qquad (6\text{-}52)$$
$$\quad | \qquad\qquad\qquad\qquad\qquad\qquad | \quad | $$
$$\quad OH \qquad\qquad\qquad\qquad\qquad OH \quad CH_2\dot{C}HR$$

participate in this chain reaction. Table 6-4 lists the chain-transfer constants (k_{tr}/k_p) determined for various alcohols with some alkenes and polymerizable monomers.[18,19] The chain-transfer constants are low for styrene and methyl methacrylate compared to the terminal alkenes not only because these compounds polymerize readily but also because they were measured at lower temperatures. The chain-transfer constants for the secondary alcohols with 1-alkenes are marginal for the formation of the 1:1 addition product in significant conversions in t-butyl peroxide-induced reactions at 130°C. It is interesting that the benzoyl peroxide and azobisisobutyronitrile (AIBN) induced reactions of secondary alcohols and 1-alkenes at 80°C yield no simple addition product.[18] The lack of reaction using these initiators may be attributed to induced decomposition of the benzoyl peroxide as described in Chapter 10 or lack of reactivity of the radical fragments from AIBN. The lower reaction temperature may also be a significant factor. Reactions of alcohols with reactive monomers such as acrylic acid and methyl acrylate do yield simple addition products in appreciable quantities in t-butyl peroxide induced reactions at

TABLE 6-4   Chain-Transfer Constants for Alcohols

Alcohol	Unsaturated Compounds	Temperature ($°C$)	k_{tr}/k_p
Methanol	1-Octene	130	1.1×10^{-2}
Ethanol	1-Octene	130	2.3×10^{-2}
2-Propanol	1-Octene	130	5.2×10^{-2}
2-Propanol	Styrene	100	1.7×10^{-4}
2-Propanol	Methyl methacrylate	80	1.4×10^{-4}
1-Butanol	1-Octene	130	2.7×10^{-2}
1-Butanol	Methyl methacrylate	80	2.5×10^{-2}
2-Butanol	1-Octene	130	5.2×10^{-2}
2-Butanol	Methyl methacrylate	80	8.5×10^{-5}
Cyclohexanol	1-Hexene	130	3.9×10^{-2}

higher temperatures (155–170°) yielding γ-hydroxyacids or esters.[19] The addition products react to form the substituted γ-lactones as illustrated for the reaction of 2-octanol (24) with acrylic acid (25) which yields 2-methyl-2-n-hexyl-γ-butyrolactone (27) by reaction of the addition product (26).

$$n\text{-}C_6H_{13}\overset{\underset{\displaystyle CH_3}{|}}{\underset{\underset{\displaystyle OH}{|}}{C}} \cdot + CH_2{=}CHCO_2H \rightarrow n\text{-}C_6H_{13}\overset{\underset{\displaystyle CH_3}{|}}{\underset{\underset{\displaystyle OH}{|}}{C}}CH_2\overset{\displaystyle \cdot}{C}HCO_2H \qquad (6\text{-}53)$$

$$\underset{25}{}$$

$$n\text{-}C_6H_{13}\overset{\underset{\displaystyle CH_3}{|}}{\underset{\underset{\displaystyle OH}{|}}{C}}CH_2\overset{\displaystyle \cdot}{C}HCO_2H + n\text{-}C_6H_{13}CHOHCH_3 \rightarrow$$

$$\underset{24}{}$$

$$(6\text{-}54)$$

$$n\text{-}C_6H_{13}\overset{\displaystyle \cdot}{C}OHCH_3 + n\text{-}C_6H_{13}\overset{\underset{\displaystyle CH_3}{|}}{\underset{\underset{\displaystyle OH}{|}}{C}}CH_2CH_2CO_2H \xrightarrow{-H_2O}$$

26   27

The order of increasing ease of chain transfer is methyl alcohol < primary alcohols < secondary alcohols which is also the order of increasing resonance stabilization of the respective α-hydroxyalkyl radicals resulting from hydrogen abstraction. As donor species, alcohols are more reactive toward hydrogen abstraction by the acceptor methyl methacrylate derived radicals than the donor styrene derived radicals.

D. AMINES

Amines that have at least one hydrogen bonded to an α-carbon add to unsaturated linkages in a free-radical chain reaction as illustrated for the reaction of piperidine (28) with 1-octene yielding 2-n-octyl piperidine (29).[20]

$$(6\text{-}55)$$

$$(6\text{-}56)$$

The chain-transfer constants of amines with 1-octene are somewhat larger than are those of alcohols having comparable structures. The chain-transfer constants for n-butyl amine, isopropyl amine, and piperidine with 1-octene at 125–130°C are reported to be 0.062, 0.095, and 0.26 respectively.

Amines and α-aminoalkyl radicals display donor qualities in their chain-propagating reactions and react readily with unsaturated compounds having acceptor qualities. The chain-transfer constant of dimethyl aniline, for example, is higher with methyl methacrylate than with styrene at 50°C, the reported values being 4.3×10^{-2} and 5.3×10^{-3}, respectively.[21]

The experimental conditions for the additions of amines to unsaturated compounds are similar to those for alcohol additions. Higher temperatures favor the reaction indicating a high temperature coefficient for the chain-transfer constant. The reactivity of amines with acyl peroxides (see Chapter 10) precludes their use as initiators for these reactions and most of the reported additions of amines to alkenes were initiated with t-butyl peroxide.

E. ESTERS AND ACIDS

Esters and acids can be alkylated in the α-position by free-radical addition to an appropriate alkene. The peroxide-induced addition of malonic ester (30) to 1-octene yielding n-octyl malonic ester (31) is illustrative of this reaction.

Ester-alkene combinations have fairly high chain-transfer constants in part

$$CH_2(CO_2C_2H_5)_2 + CH_2{=}CHC_6H_{13}{\text -}n \xrightarrow{\text{Per.}}$$
30

$$n\text{-}C_8H_{17}CH(CO_2C_2H_5)_2 \quad (6\text{-}57)$$
31

via

$$\cdot CH(CO_2C_2H_5)_2 + CH_2{=}CHC_6H_{13}{\text -}n \;\rightarrow$$

$$n\text{-}C_6H_{13}\dot{C}HCH_2CH(CO_2C_2H_5)_2 \quad (6\text{-}58)$$

$$n\text{-}C_6H_{13}\dot{C}HCH_2CH(CO_2C_2H_5) + CH_2(CO_2C_2H_5)_2 \;\rightarrow$$

$$n\text{-}C_6H_{13}CH_2CH_2CH(CO_2C_2H_5)_2 + \cdot CH(CO_2C_2H_5)_2 \quad (6\text{-}59)$$

because of a favorable resonance contribution in the displacement reaction due to the stabilizing effect of a carbonyl function. A strong polar factor is also operative, the adduct radical displaying donor qualities and the ester showing strong acceptor characteristics in the transition state of the transfer reaction. By maintaining a high ratio of ester to alkene and performing the reaction at temperatures in the region of about 150°C, high yields (>90%) of the simple addition product can be obtained. Similarly, additions of acetoacetic ester and ethyl cyanoacetate result in high yields of the 1:1 addition product.[22] Additions of monocarboxylate esters and lactones to alkenes yield α-alkylated esters. The alkylation of the free acids in the α-position also proceeds by a similar free-radical chain reaction.[22]

Methyl formate adds to alkenes in a t-butyl peroxide-induced chain reaction involving displacement on the aldehydic hydrogen of the ester in the chain-transfer step. Reaction with ethylene, for example, yields a series of telomeric methyl esters.[23] The chain-transfer constants for methyl formate with alkenes (0.014 with 1-octene at 125°C) are considerably lower than those of aldehydes as evidenced by the large amounts of telomeric products formed in the reactions.

$$\cdot CO_2CH_3 + CH_2{=}CH_2 \;\rightarrow\; \cdot CH_2CH_2CO_2CH_3 \quad (6\text{-}60)$$

$$\cdot CH_2CH_2CO_2CH_3 + HCO_2CH_3 \;\rightarrow\; CH_3CH_2CO_2CH_3 + \cdot CO_2CH_3$$
$$(6\text{-}61)$$

$$\cdot CH_2CH_2CO_2CH_3 + CH_2{=}CH_2 \;\rightarrow\; \cdot CH_2CH_2CH_2CH_2CO_2CH_3$$
$$(6\text{-}62)$$

$$\cdot CH_2CH_2CH_2CH_2CO_2CH_3 + HCO_2CH_3 \;\rightarrow\; n\text{-}C_4H_9CO_2CH_3 + \cdot CO_2CH_3$$
$$(6\text{-}63)$$

In contrast to the free-radical reactions of aldehydes, no decarbonylation is observed in the reactions of methyl formate even at elevated reaction temperatures. The t-butyl peroxide-induced reactions of ethyl formate with alkenes result in formation of telomeric products that arise from chain-transfer reactions of the α-hydrogens of the alkyl moiety of the ester as well as the aldehydic hydrogen.

F. ETHERS AND ACETALS

The t-butyl peroxide-induced addition n-butyl ether to 1-octene[24] by the sequence shown in 6-64 and 6-65 illustrates the key steps of the addition reaction that yields an α-alkylated ether as the simple addition product. Along with the expected addition product **32**, considerable amounts of 4-dodecanone (**33**) and dodecane (**34**) are also formed, products resulting from the fragmentation of the α-alkoxyalkyl radical yielding the n-butyl radical and n-butyraldehyde (see Chapter 8). Reaction of the n-butyl radical with 1-octene yields dodecane as shown in 6-67 and 6-68 and the aldehyde reacts with 1-octene in the manner described previously producing 4-dodecanone (6-69).

$$n\text{-}C_4H_9O\dot{C}HC_3H_7\text{-}n + CH_2{=}CHC_6H_{13}\text{-}n \;\rightarrow$$

$$
\begin{array}{cc}
n\text{-}C_4H_9OCHC_3H_7\text{-}n & (6\text{-}64) \\
| & \\
CH_2\dot{C}HC_6H_{13}\text{-}n &
\end{array}
$$

$$
\begin{array}{l}
n\text{-}C_4H_9OCHC_3H_7\text{-}n + n\text{-}C_4H_9OCH_2C_3H_7\text{-}n \;\rightarrow \\
| \\
CH_2\dot{C}HC_6H_{13}\text{-}n
\end{array}
$$

$$
\begin{array}{cc}
n\text{-}C_4H_9O\dot{C}HC_3H_7\text{-}n + n\text{-}C_4H_9OCHC_3H_7\text{-}n & (6\text{-}65) \\
& | \\
& C_8H_{17}\text{-}n \\
& \mathbf{32}
\end{array}
$$

$$n\text{-}C_4H_9O\dot{C}HC_3H_7\text{-}n \;\rightarrow\; n\text{-}C_4H_9\cdot + O{=}CHC_3H_7\text{-}n \qquad (6\text{-}66)$$

$$n\text{-}C_4H_9\cdot + CH_2{=}CHC_6H_{13}\text{-}n \;\rightarrow\; n\text{-}C_4H_9CH_2\dot{C}HC_6H_{13}\text{-}n \qquad (6\text{-}67)$$

$$n\text{-}C_4H_9CH_2\dot{C}HC_6H_{13}\text{-}n + n\text{-}C_4H_9OCH_2C_3H_7\text{-}n \;\rightarrow$$

$$
\begin{array}{cc}
n\text{-}C_4H_9CH_2CH_2C_6H_{13}\text{-}n + n\text{-}C_4H_9O\dot{C}HC_3H_7\text{-}n & (6\text{-}68) \\
\mathbf{34} &
\end{array}
$$

$$
\begin{array}{cc}
n\text{-}C_3H_7CHO + CH_2{=}CHC_6H_{13}\text{-}n \;\rightarrow\; n\text{-}C_3H_7COCH_2CH_2C_6H_{13}\text{-}n & (6\text{-}69) \\
\mathbf{33} &
\end{array}
$$

Additions of acetals to α,β-unsaturated esters occur in fairly high yields because of a favorable polar factor in the addition of the donor acetal-derived radical to these acceptor substrates. The chain-transfer reaction is also enhanced by the polar effect. Addition of diethylacetal (**35**) to methyl maleate (**36**) proceeds without complications resulting from fragmentation of the acetal-

$$CH_3\overset{\cdot}{C}(OC_2H_5)_2 + \underset{\substack{||\\ HCCO_2CH_3\\ \textbf{36}}}{HCCO_2CH_3} \rightarrow \underset{\substack{|\ \ \ |\\ C_2H_5O\ \ \underset{}{C}HCO_2CH_3}}{\overset{\overset{C_2H_5O}{|}}{CH_3C}-CHCO_2CH_3} \quad (6\text{-}70)$$

$$\underset{\substack{|\ \ \ |\\ C_2H_5O\ \ \underset{}{C}HCO_2CH_3}}{\overset{\overset{C_2H_5O}{|}}{CH_3C}-CHCO_2CH_3} + CH_3CH(OC_2H_5)_2 \rightarrow$$
$$\textbf{35}$$

$$\underset{\substack{|\ |\\ C_2H_5O\ |\\ CH_2CO_2CH_3}}{\overset{\overset{C_2H_5O}{|}}{CH_3CCHCO_2CH_3}} + CH_3\overset{\cdot}{C}(OC_2H_5)_2 \quad (6\text{-}71)$$

derived radical **37** yielding ethyl acetate and an ethyl radical.[25] Since the

$$\underset{\substack{|\\ \overset{}{O}C_2H_5}}{\overset{\overset{OC_2H_5}{|}}{CH_3\overset{\cdot}{C}}} \rightarrow CH_3\overset{\nearrow O}{\underset{\searrow OC_2H_5}{C}} + C_2H_5\cdot \quad (6\text{-}72)$$

$$\textbf{37}$$

reaction can be induced with benzoyl peroxide, the reactivity of the acetal derived radical with the maleate ester is greater than the reaction with peroxide, a species quite reactive toward ether derived radicals (see Chapter 10). Ethers and acetals also add readily to perfluoroethylene in benzoyl peroxide-induced reactions illustrating the acceptor qualities of this unsaturated species.[26]

G. MERCAPTANS AND OTHER SULFUR COMPOUNDS

Reactions of mercaptans with alkenes induced by light, peroxides, or oxygen, yield sulfides as the free-radical addition products.[27] The yields of sulfides for

$$RSH + CH_2{=}CHR' \rightarrow RSCH_2CH_2R' \quad (6\text{-}73)$$

most mercaptan-alkene systems are high and indicative of high chain-transfer constants for the reactions. Estimations of the energetics of the transfer reactions for the additions of methyl mercaptan and thiophenol to a terminal alkene such as 1-octene show that both reactions are exothermic. Another reason for

$$RS\cdot + CH_2{=}CHC_6H_{13}\text{-}n \rightarrow RSCH_2\overset{\cdot}{C}HC_6H_{13}\text{-}n \quad (6\text{-}74)$$

$$RSCH_2\dot{C}HC_6H_{13}\text{-}n + RSH \rightarrow RSCH_2CH_2C_6H_{13}\text{-}n + RS\cdot \quad (6\text{-}75)$$

$R = CH_3, \quad \Delta H = \sim- 6.5 \text{ kcal/mole}$

$R = C_6H_5, \Delta H = \sim-19.5 \text{ kcal/mole}$

the rapid chain-transfer step is that the donor qualities of the adduct radical and the acceptor qualities of the mercaptan render a favorable polar effect in the transition state of the reaction. The chain-transfer constants of n-butyl mercaptan with some polymerizable monomers are listed in Table 6-5.

TABLE 6-5   Chain-Transfer Constants of n-Butyl Mercaptan at 60°C[28]

Monomer	Radical	k_{tr}/k_p
Vinyl acetate	$CH_2CH\cdot$ | OAc	>48
Styrene	$CH_2CH\cdot$ | C_6H_5	22
Methyl acrylate	$CH_2CH\cdot$ | CO_2CH_3	1.69
Methyl methacrylate	$CH_2\dot{C}CH_3$ | CO_2CH_3	0.67

Both the vinyl acetate-derived radicals and styrene-derived radicals react more rapidly with the mercaptan than with their respective monomers. In the

$$\sim CH_2\dot{C}H + CH_2{=}CHR \xrightarrow{\ k_p\ } \sim CH_2CHCH_2\dot{C}H \quad (6\text{-}76)$$
$$\quad\ \ | \qquad\qquad\qquad\qquad\qquad\quad | \qquad\ \ |$$
$$\quad\ \ R \qquad\qquad\qquad\qquad\qquad\quad R \quad\ \ R$$

case of the vinyl acetate, the adduct radical owes part of its reactivity to its lack of resonance stabilization. The styrene-derived radical, being a benzylic radical, is stabilized to a considerable extent and estimation of the enthalpy of the transfer reaction indicates that this reaction may be slightly endothermic in

$$\sim CH_2\dot{C}H + HSC_4H_9\text{-}n \rightarrow \sim CH_2CH_2R + \cdot SC_4H_9\text{-}n \quad (6\text{-}77)$$
$$\quad\ \ |$$
$$\quad\ \ C_6H_5$$
$$\Delta H = \sim{+}4 \text{ kcal/mole}.$$

contrast to the addition reaction of the adduct radical to styrene which is exothermic (see Chapter 12 for discussion of heats of polymerization). The

$$\sim CH_2\dot{C}H + CH_2{=}CHC_6H_5 \ \rightarrow \ \sim CH_2CHCH_2\dot{C}H \qquad (6\text{-}78)$$

$$\underset{C_6H_5}{|} \qquad\qquad\qquad \underset{C_6H_5}{|} \ \underset{C_6H_5}{|}$$

$$\Delta H = \sim{-}18 \text{ kcal/mole}.$$

ability of the endothermic transfer reaction to compete favorably with the exothermic addition is that a polar factor is operative in the transition state of the hydrogen abstraction in the addition process.

$$\left[\ \sim CH_2\overset{\delta+}{CH}{\cdot}{\cdot}H{\cdot}{\cdot}\overset{\delta-}{S}C_4H_9\text{-}n \ \right]$$
$$\underset{C_6H_5}{|}$$

The energetics of the hydrogen abstraction and addition reaction for the acrylates are about the same as those for styrene. However, the transfer reaction does not have the advantage of a polar effect since the acrylate-derived radicals are acceptor species and the transfer constants are comparatively smaller than in the styrene case.

Note must be taken of the fact that thiyl radical additions to unsaturated linkages are generally reversible reactions (see Chapter 7). Increasing the temperature of mercaptan additions has an interesting effect on the reaction caused by the reversibility of the addition step. In view of the earlier discussions concerning the increase in the chain-transfer constant with increasing temperature for systems with chain-transfer constants less than unity, the opposite might be expected for mercaptan additions, namely increasing the temperature would lower the chain-transfer constant. Although this may be true, the situation can be complicated by the reversibility of the addition step. The reaction of the adduct radical competing with the chain-transfer process is not the addition to the alkene leading to telomers but rather the decomposition of the adduct radical. If equilibrium between the adduct radical and its fragmentation products is attained, the rate of formation of the addition product

$$RS{\cdot} + \underset{/}{\overset{\backslash}{C}}{=}\underset{\backslash}{\overset{/}{C}} \rightleftarrows RS\overset{|}{\underset{|}{C}}{-}\overset{/}{\underset{\backslash}{C}}{\cdot} \qquad (6\text{-}79)$$

$$RS\overset{|}{\underset{|}{C}}{-}\overset{/}{\underset{\backslash}{C}}{\cdot} + RSH \rightarrow RS\overset{|}{\underset{|}{C}}{-}\overset{|}{\underset{|}{C}}{-}H + RS{\cdot} \qquad (6\text{-}80)$$

will depend on the equilibrium concentration of the adduct radical. This concentration, and, consequently, the overall rate of the addition reaction, should be expected to decrease with increasing temperature owing to the increase in entropy on the thiyl radical-alkene side of the equilibrium.[29]

Thio acids add to unsaturated compounds yielding thioesters in high yields.[30] Although the adding radicals in these reactions are ambident hybrids that

$$RCSH + CH_2\!=\!CHR' \rightarrow RCSCH_2CHR' \qquad (6\text{-}81)$$
$$\underset{O}{\overset{\|}{}} \qquad\qquad\qquad \underset{O}{\overset{\|}{}}$$

could react to yield O-alkylated esters, the hydrolysis of the addition products

$$RC\overset{S\cdot}{\underset{O}{\diagdown}} \leftrightarrow RC\overset{S}{\underset{O\cdot}{\diagdown}} + CH_2\!=\!CHR' \rightarrow RCSCH_2\overset{\cdot}{C}HR'$$
$$\underset{O}{\overset{\|}{}} \qquad (6\text{-}82)$$

$$RCSCH_2\overset{\cdot}{C}HR' + RCSH \rightarrow RCSCH_2CH_2R' + RC\overset{S\cdot}{\underset{O}{\diagup}} \qquad (6\text{-}83)$$
$$\underset{O}{\overset{\|}{}} \qquad\quad \underset{O}{\overset{\|}{}} \qquad \underset{O}{\overset{\|}{}}$$

are reported to yield only mercaptans. Hydrogen sulfide adds to alkenes to form mercaptans that undergo further reaction yielding a sulfide as the ultimate

$$H_2S + \overset{\diagdown}{\underset{\diagup}{C}}\!-\!\overset{\diagup}{\underset{\diagdown}{C}} \rightarrow HS\overset{|}{\underset{|}{C}}\!-\!\overset{\diagup}{\underset{\diagdown}{C}}\cdot \qquad (6\text{-}84)$$

$$HS\overset{|}{\underset{|}{C}}\!-\!\overset{\diagup}{\underset{\diagdown}{C}}\cdot + H_2S \rightarrow HS\overset{|}{\underset{|}{C}}\!-\!\overset{|}{\underset{|}{C}}\!-\!H + HS\cdot \qquad (6\text{-}85)$$

$$HS\overset{|}{\underset{|}{C}}\!-\!\overset{|}{\underset{|}{C}}\!-\!H + Rad\cdot \rightarrow \cdot S\overset{|}{\underset{|}{C}}\!-\!\overset{|}{\underset{|}{C}}\!-\!H \qquad (6\text{-}86)$$

$$\cdot S\overset{|}{\underset{|}{C}}\!-\!\overset{|}{\underset{|}{C}}\!-\!H + \overset{\diagdown}{\underset{\diagup}{C}}\!=\!\overset{\diagup}{\underset{\diagdown}{C}} \rightarrow \cdot\overset{\diagdown}{\underset{\diagup}{C}}\!-\!\overset{|}{\underset{|}{C}}S\overset{|}{\underset{|}{C}}\!-\!\overset{|}{\underset{|}{C}}\!-\!H \qquad (6\text{-}87)$$

$$\overset{\diagdown}{\underset{\diagup}{C}}\!-\!\overset{|}{\underset{|}{C}}\!-\!S\!-\!\overset{|}{\underset{|}{C}}\!-\!\overset{|}{\underset{|}{C}}H + H_2S \rightarrow H\!-\!\overset{|}{\underset{|}{C}}\!-\!\overset{|}{\underset{|}{C}}\!-\!S\!-\!\overset{|}{\underset{|}{C}}\!-\!\overset{|}{\underset{|}{C}}H + HS\cdot \quad (6\text{-}88)$$

reaction product. Reactions of hydrogen sulfide with dienes, as might be expected, result in formation of polymeric sulfides.[31]

Aromatic sulfonyl halides add to alkenes in a free-radical chain reaction. Addition of *p*-chlorobenzenesulfonyl chloride (**38**) to ethylene, for example yields, along with telomeric products, the β-haloalkylaryl sulfone (**39**) as the 1:1 addition product.[32] The sulfonyl radical produced in the transfer reaction

$$Cl-\text{(ring)}-S \overset{O}{\underset{O\cdot}{\diagup}} \leftrightarrow Cl-\text{(ring)}-S\cdot \overset{O}{\underset{O}{\diagup}} + CH_2=CH_2 \rightarrow$$

$$Cl-\text{(ring)}-\overset{O}{\underset{O}{S}}CH_2\dot{C}H_2 \qquad (6\text{-}89)$$

$$Cl-\text{(ring)}-\overset{O}{\underset{O}{S}}CH_2\dot{C}H_2 + Cl-\text{(ring)}-SO_2Cl \rightarrow$$
$$\qquad\qquad\qquad\qquad\qquad \mathbf{38}$$

$$Cl-\text{(ring)}-SO_2CH_2CH_3 + Cl-\text{(ring)}-S\cdot\overset{O}{\underset{O}{\diagup}} \qquad (6\text{-}90)$$
$$\mathbf{39}$$

6-90, in contrast to the trichloromethanesulfonyl radical encountered in additions of trichloromethanesulfonyl chloride to alkenes, does not lose sulfur dioxide when adding to the unsaturated linkage. Although this may be due to a greater stability of the arylsulfonyl radical compared to the trichloromethane-sulfonyl radical, it may also be caused by the steric inability of the alkene-complexed radical to undergo a displacement similar to that proposed for the trichloromethanesulfonyl chloride additions.

The ease of abstraction of halides from the benzenesulfonyl halides increases in the order chloride < bromide < iodide. A comparison of reactivities of these sulfonyl halides with bromotrichloromethane indicates the latter to be more reactive than benzenesulfonyl chloride but somewhat less reactive than benzene-sulfonyl bromide toward displacement.[32]

Sulfuryl chloride reacts with alkenes in a free-radical chain reaction yielding, along with the dichloroalkene, a di-β-chloroalkyl sulfone, (**40**), an addition product involving two molecules of the alkene and one of the sulfuryl chloride.[33] Since no simple 1:1 addition product, namely the β-chloroalkanesulfonyl chloride, is found, the chain sequence proposed for the reaction involves the rearrangement of the adduct radical indicated in reaction 6-92 rather than the

$$\cdot SO_2Cl + RCH{=}CH_2 \rightarrow R\overset{\cdot}{C}HCH_2SO_2Cl \tag{6-91}$$

$$R\overset{\cdot}{C}HCH_2SO_2Cl \rightarrow \underset{\underset{Cl}{|}}{RCHCH_2SO_2}\cdot \tag{6-92}$$

$$\underset{\underset{Cl}{|}}{RCHCH_2SO_2}\cdot + RCH{=}CH_2 \rightarrow \underset{\underset{Cl}{|}}{RCHCH_2SO_2CH_2}\overset{\cdot}{C}HR \tag{6-93}$$

$$\underset{\underset{Cl}{|}}{RCHCH_2SO_2CH_2}\overset{\cdot}{C}HR + SO_2Cl_2 \rightarrow \underset{\underset{Cl}{|}}{RCHCH_2SO_2CH_2}\underset{\underset{Cl}{|}}{CHR} + \cdot SO_2Cl \tag{6-94}$$

40

displacement reaction yielding the expected addition product. The 1,3-shift of the halogen atom in the adduct radical is peculiar to sulfuryl chloride additions since sulfuryl fluoride chloride (**41**) adds to alkenes yielding a β-chloroalkanesulfonyl fluoride (**42**) as the 1:1 addition product. The nature of

$$\cdot SO_2F + RCH{=}CH_2 \rightarrow R\overset{\cdot}{C}HCH_2SO_2F \tag{6-95}$$

$$R\overset{\cdot}{C}HCH_2SO_2F + SO_2FCl \rightarrow \underset{\underset{Cl}{|}}{RCHCH_2SO_2F} + \cdot SO_2F \tag{6-96}$$

41

42

the addition product also illustrates the greater reactivity toward abstraction of the chlorine atom relative to the fluorine atom in sulfuryl halides.

The free-radical additions of bisulfites to alkenes in a chain reaction yield alkane sulfonates. In view of the hybrid nature of the bisulfite-derived radical, it is interesting that, as in the previous reactions, bonding occurs between carbon and the sulfur of the adding sulfuryl radical. Although formation of the carbon-oxygen bond may be more exothermic than formation of the carbon-sulfur bond, other factors, possibly related to the polarizability of the sulfur, influence the direction of addition more than the enthalpic aspects of the reaction.

Organic disulfides participate in free-radical addition reactions that involve a radical displacement on a divalent atom. The addition of di-n-butyl disulfides (**43**) to vinyl acetate yielding 1,2-bis (n-butylthio)ethyl acetate (**44**) is illustrative of this reaction.[34] The displacement reaction 6-98 on sulfur apparently is not prohibitively hindered toward reaction by the vinyl acetate-derived radical as evidenced by the fact that 1:1 addition product is obtained. Most disulfides have comparatively low chain-transfer constants and, although disulfides are

$$n\text{-}C_4H_9S\cdot + CH_2{=}CHOAc \rightarrow n\text{-}C_4H_9SCH_2\overset{\cdot}{C}HOAc \tag{6-97}$$

$$n\text{-}C_4H_9SCH_2\overset{\cdot}{C}HOAc + n\text{-}C_4H_9SSC_4H_9\text{-}n \;\rightarrow$$
$$\textbf{43}$$

$$n\text{-}C_4H_9SCH_2CHOAc + n\text{-}C_4H_9S\cdot \quad \text{(6-98)}$$
$$\underset{\textstyle SC_4H_9\text{-}n}{\overset{\textstyle |}{}}$$
$$\textbf{44}$$

used to moderate the molecular weights of polymers (see Chapter 12), for the most part little use has been made of the reaction for synthetic purposes.

H. ADDITIONS OF PHOSPHORUS-CONTAINING COMPOUNDS

Compounds having a phosphorus-hydrogen bond add to unsaturated compounds in free-radical chain reactions that involve abstraction of the phosphorus-bonded hydrogen by an adduct radical. The radical derived in the displacement step adds to the unsaturated linkage forming a carbon-phosphorus bond. Thus, phosphine adds to alkenes in light, peroxide, and azo-compound induced reactions yielding alkylated phosphines.[35] Substituted phosphines also

$$\cdot PH_2 + RCH{=}CH_2 \;\rightarrow\; R\overset{\cdot}{C}HCH_2PH_2 \qquad \text{(6-99)}$$

$$R\overset{\cdot}{C}HCH_2PH_2 + PH_3 \;\rightarrow\; RCH_2CH_2PH_2 + \cdot PH_2 \qquad \text{(6-100)}$$

react and the product may add to more alkene yielding the di- or trialkyl phosphine. Mono- and di-substituted phosphines can also be added to other unsaturates yielding unsymmetrically substituted phosphines.[36]

The free-radical additions of phosphite esters to unsaturated compounds have been more extensively investigated. These reactions yield alkanephosphonates as the addition products. For example, addition of diethyl phosphite (**45**) to ethylene yields diethyl ethanephosphonate (**46**) as the 1:1 addition product in high yields provided the alkene concentration is maintained at a low concentration in order to prevent telomer formation.[37] Phosphorous acid

$$\overset{\textstyle O}{\underset{\textstyle \|}{}}\cdot P(OC_2H_5)_2 + CH_2{=}CH_2 \;\rightarrow\; \overset{\textstyle O}{\underset{\textstyle \|}{}}\cdot CH_2CH_2P(OC_2H_5)_2 \qquad \text{(6-101)}$$

$$\overset{\textstyle O}{\underset{\textstyle \|}{}}\cdot CH_2CH_2P(OC_2H_5)_2 + \overset{\textstyle O}{\underset{\textstyle \|}{}}HP(OC_2H_5)_2 \;\rightarrow$$
$$\textbf{45}$$

$$\overset{\textstyle O}{\underset{\textstyle \|}{}}CH_3CH_2P(OC_2H_5)_2 + \overset{\textstyle O}{\underset{\textstyle \|}{}}\cdot P(OC_2H_5)_2 \qquad \text{(6-102)}$$
$$\textbf{46}$$

reacts with alkenes in peroxide- and light-induced reactions yielding phosphonic acids as the addition product although much telomer formation also occurs.[38] Reactions of hypophosphorous acid with alkenes yield monoalkanephosphinic

$$\underset{\displaystyle \cdot P(OH)_2}{\overset{\displaystyle O \atop \displaystyle \|}{}} + RCH{=}CH_2 \;\rightarrow\; \underset{}{\overset{\displaystyle O \atop \displaystyle \|}{R\dot{C}HCH_2 P(OH)_2}} \qquad (6\text{-}103)$$

$$\underset{}{\overset{\displaystyle O \quad\quad O \atop \displaystyle \| \quad\quad \|}{R\dot{C}HCH_2 P(OH)_2}} + HP(OH)_2 \;\rightarrow\; RCH_2CH_2 P(OH)_2 + \cdot P(OH)_2 \qquad (6\text{-}104)$$

acids (**47**) which have a phosphorus-hydrogen linkage and are capable of further reaction yielding the dialkanephosphinic acid (**48**). Esters of phosphinic

$$\underset{O}{\overset{H}{H\dot{P}OH}} \left(\text{or } \underset{O}{RCH_2CH_2\dot{P}OH} \right) + RCH{=}CH_2 \;\rightarrow\;$$

$$\underset{O}{\overset{H}{R\dot{C}HCH_2POH}} \left(\text{or } \underset{O}{RCH_2CH_2\dot{P}OH} \overset{RCH_2CH_2}{} \right) \qquad (6\text{-}105)$$

$$\underset{OH}{\overset{O \atop \|}{R\dot{C}HCH_2PH}} \left(\text{or } \underset{OH}{\overset{O \atop \|}{RCH_2CH_2PCH_2CH_2R}} \right) + \underset{O}{H_2POH} \left(\text{or } \underset{OH}{\overset{O \atop \|}{RCH_2CH_2PH}} \right) \rightarrow$$

47

$$\underset{OH}{\overset{O \atop \|}{RCH_2CH_2PH}} \left(\text{or } \underset{\mathbf{48}}{(RCH_2CH_2)_2POH} \right) + \underset{O}{H\dot{P}OH} \left(\text{or } \underset{O}{RCH_2CH_2\dot{P}{-}OH} \right)$$

$$(6\text{-}106)$$

acid add to terminal alkenes by a radical mechanism as illustrated by the addition of ethyl benzenephosphinate (**49**) to alkenes in peroxide-induced reactions.[39]

$$\underset{\underset{\mathbf{49}}{OC_2H_5}}{\overset{O \atop \|}{C_6H_5PH}} + RCH{=}CH_2 \;\xrightarrow{\text{Per.}}\; \underset{OC_2H_5}{\overset{O \atop \|}{C_6H_5PCH_2CH_2R}} \qquad (6\text{-}107)$$

Phosphorus trihalides react with alkenes in two different chain reactions leading to different addition products. In the chain sequence 6-108 and 6-109, transfer occurs on the halogen and a $PX_2\cdot$ is the adding radical.[40a,b] The observation that products such as **50** are formed in a chain reaction,[40b] particularly in the case of PBr_3,[40c] is indicative of the chain sequence 6-110 and 6-111 also being operative. In this case, the adduct radical does not participate

$$PX_2\cdot + RCH{=}CH_2 \rightarrow R\dot{C}HCH_2PX_2 \qquad (6\text{-}108)$$

$$R\dot{C}HCH_2PX_2 + PX_3 \rightarrow \underset{\underset{X}{|}}{RCHCH_2PX_2} + PX_2\cdot \qquad (6\text{-}109)$$

in an abstraction reaction but rather interacts with the phosphorous trihalide yielding a radical that can propagate the chain by reaction with the alkene yielding the adduct radical.

$$R\dot{C}HCH_2X + PX_3 \rightarrow \underset{\underset{\cdot PX_3}{|}}{RCHCH_2X} \qquad (6\text{-}110)$$

$$\underset{\underset{\cdot PX_3}{|}}{RCHCH_2X} + RCH{=}CH_2 \rightarrow \underset{\underset{PX_2}{|}}{RCHCH_2X} + R\dot{C}HCH_2X \qquad (6\text{-}111)$$
$$\textbf{50}$$

I. NITROGEN TETROXIDE AND NITRYL CHLORIDE

Addition of nitrogen tetroxide (**51**) to terminal alkenes yields mixtures of a vicinal dinitroalkanes (**52**) and nitronitrites (**53**).[41] The orientation observed

$$\underset{\textbf{51}}{\overset{\underset{O}{\underset{\parallel}{}}\,\underset{O}{}}{N{-}N}} + RCH{=}CH_2 \rightarrow \underset{\underset{NO_2}{|}}{\underset{\textbf{52}}{RCHCH_2NO_2}} + \underset{\underset{ONO}{|}}{\underset{\textbf{53}}{RCHCH_2NO_2}} \qquad (6\text{-}112)$$

in the nitronitrites is such that the nitro group is always found on the terminal carbon suggesting that the nitro radical $NO_2\cdot$ is the adding species. The

$$N\underset{O\cdot}{\overset{O}{\diagup}} \leftrightarrow \cdot N\underset{O}{\overset{O}{\diagup}} + RCH{=}CH_2 \rightarrow R\dot{C}HCH_2NO_2 \qquad (6\text{-}113)$$

$$R\dot{C}HCH_2NO_2 + \underset{O}{\overset{O}{N{-}N}} \rightarrow \underset{\underset{ONO}{|}}{RCHCH_2NO_2} + \cdot NO_2 \qquad (6\text{-}114)$$

displacement reaction on N_2O_4 occurs predominantly by interactions of the adduct radical with an oxygen of N_2O_4 since the oxygens are less hindered than the nitrogens of this compound. Reactions of nitryl chloride (**54**) with unsymmetrical alkenes yield addition products in a chain reaction in which displacement apparently occurs more readily on the halogen of the adding reagent than on oxygen.[42]

$$ClNO_2 + RCH{=}CH_2 \rightarrow RCHCH_2NO_2 \qquad (6\text{-}115)$$
$$ \underset{\textbf{54}}{} \qquad\qquad\qquad\;\; | \atop Cl$$

via

$$\cdot NO_2 + RCH{=}CH_2 \rightarrow R\dot{C}HCH_2NO_2 \qquad (6\text{-}116)$$

$$R\dot{C}HCH_2NO_2 + ClNO_2 \rightarrow RCHCH_2NO_2 + \cdot NO_2 \qquad (6\text{-}117)$$
$$\qquad\qquad\qquad\qquad\qquad\quad | \atop Cl$$

J. SILICON, GERMANIUM, AND TIN COMPOUNDS

Compounds having at least one hydrogen bonded to silicon add to alkenes in free-radical chain reactions that involve abstraction of the silicon-bonded hydrogen in the transfer step.[43] Many additions of silane (SiH_4), as well as mono-, di-, and trisubstituted silanes to alkenes in peroxide-induced reactions have been recorded. Additions of mono-, di-, and trichlorosilanes, both peroxide-

$$R_3'SiH + RCH{=}CH_2 \rightarrow RCH_2CH_2SiR_3' \qquad (6\text{-}118)$$

via

$$R_3'Si\cdot + RCH{=}CH_2 \rightarrow R\dot{C}HCH_2SiR_3' \qquad (6\text{-}119)$$

$$R\dot{C}HCH_2SiR_3' + R_3'SiH \rightarrow RCH_2CH_2SiR_3' + R_3'Si \qquad (6\text{-}120)$$

$R' = $ H, alkyl or aryl

induced and thermally induced (200–400°C), to a variety of alkenes have been reported to occur in fairly good yields.

Compounds with a germanium-hydrogen bond add to alkenes in free-radical reactions.[44] Addition products are obtained in the reactions of trichlorogermane, triphenylgermane, and various trialkyl germanes with unsaturated compounds. The germanes appear to be more reactive toward free-radical addition than do the corresponding silane derivatives as evidenced by the milder conditions required to effect these additions. Trichlorogermane, for example, reacts with a variety of alkenes at room temperature in an exothermic reaction.[45]

$$Cl_3GeH + RCH{=}CH_2 \rightarrow RCH_2CH_2GeCl_3 \qquad (6\text{-}121)$$

Tin hydrides also add readily to many unsaturated compounds. These additions very likely proceed by a free-chain reaction as evidenced by the influence azo-compounds and light have on increasing the rate of the addition reaction.[46]

K. ADDITIONS OF HYDROGEN HALIDES

The Kharasch additions of hydrogen bromide to alkenes are among the oldest examples of free-radical addition reactions. The details of this reaction are discussed in length in the following chapter. An interesting aspect of these reactions is the stereospecificity observed in addition of this reagent to alkenes and acetylenes.

Hydrogen chloride also reacts with certain alkenes in a free-radical chain process. The chain-transfer constant of this hydrogen halide is, however, comparatively low and considerable amounts of telomer are usually formed in these reactions. No radical reactions of hydrogen fluoride have been reported. In view of the bond dissociation energy of the hydrogen fluorine bond, it is not surprising that this reagent fails to undergo chain-transfer with the adduct radicals. In the case of hydrogen iodide, the transfer constant of the hydrogen halide should be high enough to lead to formation of simple addition products. The failure of the additions to occur is probably due to the reversibility of the iodine atom additions to the alkene.

References

1. R. A. Gregg and F. R. Mayo, *J. Amer. Chem. Soc.*, **70**, 2373 (1948); *Discussions Faraday Soc.*, **2**, 328 (1947).
2. S. Palit and S. K. Das, *Proc. Roy. Soc. London*, **A226**, 82 (1954).
3. S. K. Das, S. R. Chattergee, and S. R. Palit, *ibid.*, **A227**, 252 (1955).
4. S. Basu, J. N. Sen, and S. R. Palit, *ibid.*, **A202**, 485 (1950); C. Walling, *J. Amer. Chem. Soc.*, **70**, 2561 (1948).
5. C. Walling and E. S. Huyser, *Organic Reactions*, Vol. XIII, A. C. Cope, Ed., John Wiley and Sons, New York, 1963, p. 122–131.
6. Shostakovskii, Bogdanove, Zuerov, and Plotnikova, *Inzv. Akad. Nauk. S.S.S.R.*, Otd. Khim. Nauk., **1956**, 1236.
7. M. S. Kharasch, E. V. Jensen, and W. H. Urry, *J. Amer. Chem. Soc.*, **69**, 1100 (1947).
8. M. Weizman, S. Israelashvili, A. Halevy, and F. Bergmann, *ibid.*, **69**, 2569 (1947).
9. R. N. Haszeldine, *J. Chem. Soc.*, **1950**, 3037; **1953**, 3761.
10. E. S. Huyser, *J. Org. Chem.*, **26**, 3261 (1961).
11. E. C. Ladd and L. Y. Kiley, U.S. Patent 2,606,213 (1952).
12. E. S. Huyser and B. Giddings, *J. Org. Chem.*, **27**, 3391 (1962).
13. E. S. Huyser and L. Kim, *ibid.*, **32**, 618 (1967).
14. M. S. Kharasch, W. H. Urry, and B. M. Kuderna, *ibid.*, **14**, 248 (1949).
15. T. M. Patrick, Jr., *ibid.*, **17**, 1009 (1952).

16. T. M. Patrick, Jr., *ibid.*, **17**, 1269 (1952).

17. E. C. Ladd, U.S. Patent, 2,552,980 (1951).

18. W. H. Urry, F. W. Stacey, O. O. Juveland, and C. H. McDonnell, *J. Amer. Chem. Soc.*, **75**, 250 (1953); W. H. Urry, F. W. Stacey, E. S. Huyser, and O. O. Juveland, *ibid.*, **76**, 450 (1954).

19. G. I. Nikishin, V. D. Vorob'ev, A. D. Petrov, *Doklady Akad. Nauk SSSR*, **136**, 360 (1961).

20. W. H. Urry, O. O. Juveland, and F. W. Stacey, *J. Amer. Chem. Soc.*, **74**, 6155 (1952); W. H. Urry and O. O. Juveland, *ibid.*, **80**, 3322 (1958).

21. M. Imato, T. Otsu, T. Ota, H. Takatsugi, and M. Mutsuda, *J. Polymer Sci.*, **22**, 137 (1956).

22. E. S. Huyser, PhD Thesis, University of Chicago, 1954; R. G. Gritter, PhD thesis, University of Chicago, 1955; J. C. Allen, J. I. G. Cadogan, B. W. Harris, and D. H. Hey, *J. Chem. Soc.*, **1962**, 4468; J. C. Allen, J. I. G. Cadogen, and D. H. Hey, *ibid.*, **1965**, 1918.

23. W. H. Urry and E. S. Huyser, *J. Amer. Chem. Soc.*, **75**, 4876 (1953).

24. C. H. McDonnel, PhD Thesis, University of Chicago, 1955.

25. Nagasaka, Nukina, and Odo, *J. Chem. Soc. Japan, Bull. Inst. Chem. Research*, Kyoto University, **33**, 85 (1955).

26. W. E. Hanford, U.S. Patent 2,433,844 (1948).

27. M. S. Kharasch, A. T. Read, and F. R. Mayo, *Chem. and Ind.*, **57**, 752 (1938); S. O. Jones and E. E. Reid, *J. Amer. Chem. Soc.*, **60**, 2452 (1938).

28. C. Walling, *J. Amer. Chem. Soc.*, **70**, 2561 (1948).

29. C. Sivertz, W. Andrews, W. Elliott, and K. Graham, *J. Polymer Sci.*, **19**, 587 (1956).

30. J. I. Cunneen, *J. Chem. Soc.*, **1947**, 134.

31. W. E. Vaughan and F. F. Rust, *J. Org. Chem.*, **7**, 472 (1942).

32. J. H. McNamara, PhD Thesis, Pennsylvania State University (1956), *Diss. Abstr.*, **17**, 226 (1957); E. C. Ladd, U.S. Patent 2,573,580 (1952); G. V. D. Tiers, U.S. Patent 2,846,472 [C.A. **53**, 12175 (1959)].

33. M. S. Kharasch and A. F. Zavist, *J. Amer. Chem. Soc.*, **70**, 3526 (1948); **73**, 964 (1951).

34. K. Yamagishi, K. Araki, T. Suzuki, and T. Hoshino, *Bull. Chem. Soc. Japan*, **33**, 528 (1960).

35. A. R. Stiles, F. F. Rust, and W. E. Vaughan, *J. Amer. Chem. Soc.*, **74**, 3282 (1952).

36. M. M. Rauhat, H. A. Currier, A. M. Semsel, and V. P. Wystrach, *J. Org. Chem.*, **26**, 5138 (1961).

37. A. R. Stiles, W. E. Vaughan, and F. F. Rust, *J. Amer. Chem. Soc.*, **80**, 714 (1958).

38. C. E. Griffen, *J. Org. Chem.*, **25**, 665 (1960).

39. A. R. Stiles and F. F. Rust, U.S. Patent 2,724,718 (1956).

40. (a) M. S. Kharasch, E. V. Jensen, and W. H. Urry, *J. Amer. Chem. Soc.*, **67**, 1864 (1945);
    (b) J. R. Little and P. F. Hartman, *ibid.*, **88**, 96 (1966);
    (c) C. B. Fontal and H. Goldwhite, *J Org Chem.*, **31**, 3804 (1966).
41. N. Levy and C. W. Scaife, *J. Chem. Soc.*, **1946**, 1093, 1096, 1100; N. Levy, C. W. Scaife, and A. E. Wilder-Smith, *ibid.*, **1948**, 52; N. Levy and J. D. Rose, *Quart. Rev.* (London) **1**, 358 (1948).
42. H. Shecter, F. Conrad, A. L. Daulton, and R. S. Kaplan, *J. Amer. Chem. Soc.*, **74**, 3052 (1952).
43. L. H. Sommer, E. W. Pietrusza, and F. C. Whitmore, *ibid.*, **69**, 188 (1947); C. A. Burkhard and R. H. Kreble, *ibid.*, **69**, 2687 (1947); A. J. Barry, L. DePree, J. W. Gilkey, and D. E. Hoak, *ibid.*, **69**, 2916 (1947).
44. M. Lesbre and J. Satge, *Compt. Rend.*, **247**, 471 (1958): H. Gilman and C. W. Gerou, *J. Amer. Chem. Soc.*, **79**, 342 (1959): R. H. Meen and H. Gilman, *J. Org. Chem.*, **22**, 684 (1957).
45. A. D. Petrov, V. F. Mironov, and N. G. Dzhurinskaya, *Proc. Acad. Sci., USSR*, Chem. Sect. **128**, 739 (1959).
46. W. P. Neumann, *Angew. Chem., Intern. Ed.* (Engl.), **2**, 170 (1963).

Chapter 7

Free-Radical Additions: Chemistry of the Unsaturated Compounds

I. Free-Radical Chemistry of Various Unsaturated Linkages

A. INTRODUCTION

The principles of the Kharasch addition reaction outlined in the previous chapter dealt largely with the nature of the adding reagent and its behavior with simple unsaturated linkages. In some cases, however, the nature of the unsaturated species is responsible to a greater extent than the adding reagent for the particular structure and distribution of the reaction products in the addition reaction.

Much of the chemistry of the free-radical additions to species that have a terminal carbon-carbon unsaturated linkage is determined by steric, resonance, and polar factors. The nature of the free-radical additions to unsaturated linkages such as acetylenes, dienes (both conjugated and nonconjugated), and

173

cycloalkenes as well as to unsaturated linkages involving a heteroatom, such as the carbonyl function, can be appreciated in terms of the same general factors outlined for additions to simple alkenes.

In the case of nonterminally unsaturated compounds, the stereochemistry of the addition also becomes a factor. Both the stereospecificity and the lack of specificity in addition reactions are indicative of characteristics of free radicals and their abilities to participate in chain-propagating reactions.

B. ADDITIONS TO ALKYNES

Addition of a free radical to an acetylenic linkage yields a reactive vinyl

$$R\cdot + -C{\equiv}C- \rightarrow \underset{}{\overset{R}{\diagdown}}C{=}C\overset{}{\diagup}\cdot \qquad (7\text{-}1)$$

radical as the adduct radical. Chain transfer by the vinyl radical with the adding reagent occurs readily. The limiting factor in most additions to alkynes is the reaction of the adding radical to the acetylenic linkage, not the chain-transfer reaction. Most of the interesting chemistry of the free-radical additions to alkynes, however, centers around the behavior of this reactive adduct radical.

Reactions of adding reagents that involve transfer of a hydrogen atom to acetylene yield substituted vinyl compounds as the 1:1 addition products. Addition of the adding reagent to the 1:1 addition product is generally more facile than addition to acetylene and, consequently, the product observed is one in which two molecules of the adding reagent are combined with one of acetylene. For example, the peroxide-induced reaction of n-butyl mercaptan (1) with acetylene yields the addition product 3,[1] a product resulting from the further reaction of n-butyl vinyl sulfide (2), the 1:1 addition product of the mercaptan and acetylene.

$$2\ n\text{-}C_4H_9SH + HC{\equiv}CH \xrightarrow{\text{Per.}} n\text{-}C_4H_9SCH_2CH_2SC_4H_9\text{-}n \qquad (7\text{-}2)$$
$$\quad\ \ \mathbf{1} \qquad\qquad\qquad\qquad\qquad\qquad\quad \mathbf{3}$$

via

$$n\text{-}C_4H_9S\cdot + HC{\equiv}CH \rightarrow n\text{-}C_4H_9SCH{=}\dot{C}H \qquad (7\text{-}3)$$

$$n\text{-}C_4H_9SCH{=}\dot{C}H + n\text{-}C_4H_9SH \rightarrow n\text{-}C_4H_9SCH{=}CH_2 + n\text{-}C_4H_9S\cdot \qquad (7\text{-}4)$$
$$\qquad\qquad\qquad\qquad\qquad\qquad\qquad\quad \mathbf{2}$$

$$n\text{-}C_4H_9SCH{=}CH_2 + n\text{-}C_4H_9S\cdot \rightarrow n\text{-}C_4H_9S\dot{C}HCH_2SC_4H_9\text{-}n \qquad (7\text{-}5)$$

$$n\text{-}C_4H_9S\dot{C}HCH_2SC_4H_9\text{-}n + n\text{-}C_4H_9SH \rightarrow$$

$$n\text{-}C_4H_9SCH_2CH_2SC_4H_9\text{-}n + n\text{-}C_4H_9S\cdot \qquad (7\text{-}6)$$
$$\mathbf{3}$$

Additions of compounds that involve transfer of a halogen atom yield a simple 1:1 addition product as observed in the reaction of iodotrifluoromethane to acetylene.[2] The unsaturated linkage of the addition product **4** is likely less reactive than that of acetylene itself because of steric factors. Additions to terminal alkynes yield nonterminally unsaturated 1:1 addition products that

$$ICF_3 + HC{\equiv}CH \ \rightarrow \ F_3CCH{=}CHI \tag{7-7}$$
$$\mathbf{4}$$

generally are less reactive toward addition by the adding radical than the alkyne itself.

If the alkyl portion of the alkyne is long enough, the adduct radical may participate in an intramolecular hydrogen atom abstraction. Thus Heiba and Dessau[3] found that reactions of 1-hexyne (**5**), 1-heptyne (**6**) and 6-methyl-1-heptyne (**7**) with carbon tetrachloride yielded the unsaturated cyclic product **11** along with the expected addition product **8**. The cyclic product resulted from addition to the unsaturated linkage by the radical produced in the intramolecular hydrogen abstraction followed by β-elimination of a chlorine atom as shown in the following sequence. The amount of cyclic product obtained depends on the relative rates of reactions 7-9 and 7-10. In the reactions of **6** and **7** where the hydrogen involved in the intramolecular abstraction process is

$$Cl_3C\cdot + HC{\equiv}C(CH_2)_3CHRR' \ \rightarrow \ Cl_3CCH{=}\overset{\cdot}{C}(CH_2)_3CHRR' \tag{7-8}$$

$$(\mathbf{5}, R = R' = H)$$
$$(\mathbf{6}, R = H, R' = CH_3)$$
$$(\mathbf{7}, R = R' = CH_3)$$

$$Cl_3CCH{=}CCl(CH_2)_3CHRR' + Cl_3C\cdot \tag{7-9}$$
$$\mathbf{8}$$

$$\mathbf{9} \tag{7-10}$$

secondary or tertiary, more cyclic products were obtained than in the reactions of **5** in which the abstracted hydrogen is primary. When bromotrichloromethane was used as the adding reagent, no cyclic product was found presumably because bromine atom abstraction from bromotrichloromethane was much faster than the intramolecular hydrogen abstraction.

$$\underset{9}{\overset{\begin{array}{c}Cl_3C \quad\quad H \\ \diagdown C=C \diagup \\ H \diagup \quad\quad \diagdown CH_2 \\ | \\ R'R\dot{C} \quad CH_2 \\ \diagdown CH_2 \diagup \end{array}}{}} \rightarrow \underset{10}{\overset{\begin{array}{c} H \quad CH_2-CH_2 \\ | \diagup \quad\quad | \\ Cl_3C\dot{C}HC \quad\quad | \\ \diagdown C-----CH_2 \\ \diagup \quad \diagdown \\ R \quad\quad R' \end{array}}{}} \qquad (7\text{-}11)$$

$$\underset{}{\overset{\begin{array}{c} CH_2CH_2 \\ \diagup \quad\quad | \\ Cl_3C\dot{C}HCH \quad\quad | \\ \diagdown C-CH_2 \\ \diagup \quad \diagdown \\ R \quad\quad R' \end{array}}{}} \rightarrow \underset{11}{\overset{\begin{array}{c} CH_2CH_2 \\ \diagup \quad\quad | \\ Cl_2C=CHCH \quad\quad | \\ \diagdown C-CH_2 \\ \diagup \quad \diagdown \\ R \quad\quad R' \end{array}}{}} + Cl\cdot \quad (7\text{-}12)$$

Reaction of n-butyl mercaptan with acetylene in the presence of carbon monoxide (2450 to 2950 atm.) yields, along with the expected 1:2 addition product 12, the aldehyde 13.[1] The aldehyde most likely results from sequence of reactions 7-14 through 7-16, the key reaction being the formation of the acyl

$$n\text{-}C_4H_9SH + HC\equiv CH \xrightarrow[\text{2450 to 2950 atm.}]{CO}$$

$$\underset{12}{n\text{-}C_4H_9SCH_2CH_2SC_4H_9\text{-}n\ (32\%)} + \underset{13}{n\text{-}C_4H_9SCH=CHCHO\ (17\%)} \quad (7\text{-}13)$$

radical by incorporation of the vinyl radical with carbon monoxide. Similar incorporation of carbon monoxide is reported to occur in reactions of other mercaptans and other acetylenes yielding the expected aldehyde. Interestingly,

$$RS\cdot + HC\equiv CH \rightarrow RCH=\dot{C}H \qquad (7\text{-}14)$$

$$RSCH=\dot{C}H + CO \rightarrow RCH=CH\dot{C}O \qquad (7\text{-}15)$$

$$RSCH=CH\dot{C}O + RSH \rightarrow RSCH=CHCHO + RS\cdot \qquad (7\text{-}16)$$

an aldehydic product is reported to be formed in the reaction of ethyl mercaptan with propylene in the presence of carbon monoxide (2900 to 3000 atm.) but aldehydes were not found in the reaction of hydrogen sulfide with ethylene and carbon monoxide (3000 atm.).[4] Reactivity of the vinyl radical obtained from addition to the acetylene apparently is not a necessary feature for interaction with carbon monoxide.

C. ADDITIONS TO DIENES

1. Allenes. Free-radical additions of mercaptans to allene (**14**) yield 1,3-bis(alkanethio)propanes (**16**) in a two-step process which involves formation of an alkyl allyl sulfide (**15**), the 1:1 addition product, followed by addition of the mercaptan to the remaining unsaturated linkage. Under proper conditions, the allyl alkyl sulfide **15** can be isolated in 60 to 75% yields.[5]

$$RSH + CH_2\!=\!C\!=\!CH_2 \rightarrow RSCH_2CH_2CH_2SR \qquad (7\text{-}17)$$
$$\underset{\textbf{14}}{} \underset{\textbf{16}}{}$$

via

$$RS\cdot + CH_2\!=\!C\!=\!CH_2 \rightarrow RSCH_2\overset{\cdot}{C}\!=\!CH_2 \qquad (7\text{-}18)$$

$$RSCH_2\overset{\cdot}{C}\!=\!CH_2 + RSH \rightarrow \underset{\textbf{15}}{RSCH_2CH\!=\!CH_2} + RS\cdot \qquad (7\text{-}19)$$

$$RSCH_2CH\!=\!CH_2 + RS\cdot \rightarrow RSCH_2\overset{\cdot}{C}HCH_2SR \qquad (7\text{-}20)$$

$$RSCH_2\overset{\cdot}{C}HCH_2SR + RSH \rightarrow \underset{\textbf{16}}{RSCH_2CH_2CH_2SR} + RS\cdot \qquad (7\text{-}21)$$

Although additions of mercaptans to allenes appear to involve mainly bonding of the adding thiyl radical with a terminal carbon, some attack at the 2-carbon occurs as evidenced by the formation of 1,2-bis(alkanethio)propanes (**19**) as addition products by the route shown in 7-22 through 7-25. At 17 to 18°C, the

$$RS\cdot + CH_2\!=\!C\!=\!CH_2 \rightarrow \underset{\underset{\textbf{17}}{\underset{|}{SR}}}{\cdot CH_2C\!=\!CH_2} \leftrightarrow \underset{\underset{}{\underset{|}{SR}}}{CH_2\!=\!C\overset{\cdot}{C}H_2} \qquad (7\text{-}22)$$

$$\underset{\underset{}{\underset{|}{SR}}}{CH_2\!=\!C\overset{\cdot}{C}H_2} + RSH \rightarrow \underset{\underset{\textbf{18}}{\underset{|}{SR}}}{CH_2\!=\!CCH_3} + RS\cdot \qquad (7\text{-}23)$$

$$\underset{\underset{}{\underset{|}{SR}}}{CH_2\!=\!CCH_3} + RS\cdot \rightarrow \underset{\underset{}{\underset{|}{SR}}}{RSCH_2\overset{\cdot}{C}CH_3} \qquad (7\text{-}24)$$

$$\underset{\underset{}{\underset{|}{SR}}}{RSCH_2\overset{\cdot}{C}CH_3} + RSH \rightarrow \underset{\underset{\textbf{19}}{\underset{|}{SR}}}{RSCH_2CHCH_3} + RS\cdot \qquad (7\text{-}25)$$

amount of attack at the 2-carbon by the methane thiyl radical, yielding **17** leading to the intermediate **18**, is about 15% and decreases to 6% at −75°C.

Reaction of hydrogen bromide with allene under free-radical conditions yields predominantly 2-bromopropene (**20**). The amount of nonterminal attack (7-26) amounts to about 93% at room temperature, and decreases to about

$$Br\cdot + CH_2\!=\!C\!=\!CH_2 \nearrow \underset{\underset{}{\underset{|}{Br}}}{CH_2\!=\!CCH_2\cdot} \qquad (7\text{-}26)$$
$$\searrow \underset{}{BrCH_2\overset{\cdot}{C}\!=\!CH_2} \qquad (7\text{-}27)$$

$$CH_2\!=\!CCH_2\cdot + HBr \rightarrow CH_2\!=\!CHCH_3 + Br\cdot \qquad (7\text{-}28)$$
$$\underset{Br}{\vert} \qquad\qquad\qquad \underset{\underset{\textbf{20}}{Br}}{\vert}$$

$$BrCH_2\dot{C}\!=\!CH_2 + HBr \rightarrow BrCH_2CH\!=\!CH_2 + Br\cdot \qquad (7\text{-}29)$$

60% at $-70°C$. Addition of iodotrifluoromethane to allene yields only 2-iodo-4,4,4-trifluoro-1-butene (**21**), the product resulting from attack by the adding radical exclusively at the terminal carbon.[6]

$$CF_3\cdot + CH_2\!=\!C\!=\!CH_2 \rightarrow CF_3CH_2\dot{C}\!=\!CH_2 \qquad (7\text{-}30)$$
$$CF_3CH_2\dot{C}\!=\!CH_2 + ICF_3 \rightarrow CF_3CH_2C\!=\!CH_2 + CF_3 \qquad (7\text{-}31)$$
$$\underset{\underset{\textbf{21}}{I}}{\vert}$$

Although the 2-substituted allyl radical produced by the attack at the non-terminal carbon of allene is a resonance-stabilized hybrid, this allylic resonance plays no significant role in lowering the energy of the transition state of this reaction. The geometry of the orbitals in allene is such that bonding by the adding radical at the 2-carbon of allene precludes any overlap of the resulting p-orbital of the 1-carbon with the p-orbitals that comprise the remaining unsaturated linkage. Although once the sigma bond is formed and rotation about the carbon-carbon sigma bond produces the resonance-stabilized allylic radical, the effect of this stabilization cannot influence the transition state of the addition reaction.

$$(7\text{-}32)$$

A plausible explanation for the predominant formation of 2-substituted propenes was presented by Griesbaum and his co-workers.[5] They suggested that the addition always occurs primarily at the terminal carbon producing the 1-substituted adduct radical but that a 1,2-shift of a bromine atom or thiyl radical yielding the 2-substituted allylic radicals (see Chapter 9) follows. If the 1,2-shift is a higher activation energy process than the transfer reaction with the adding reagent, increasing temperature would increase the amount of 2-substituted propene produced in the reactions. The amount of rearrange-

$$\text{X} \cdot + \text{CH}_2\!\!=\!\!\text{C}\!\!=\!\!\text{CH}_2 \;\rightarrow\; \underset{\underset{\text{X}}{|}}{\text{CH}_2\overset{\cdot}{\text{C}}\!\!=\!\!\text{CH}_2} \qquad (7\text{-}33)$$

$$\underset{\underset{\text{X}}{|}}{\text{CH}_2\overset{\cdot}{\text{C}}\!\!=\!\!\text{CH}_2} \;\xrightarrow[1,2\text{-}shift]{\overset{HX}{\nearrow}}\; \begin{array}{l} \underset{\underset{\text{X}}{|}}{\text{CH}_2\text{CH}\!\!=\!\!\text{CH}_2} + \text{X} \cdot \qquad (7\text{-}34) \\[18pt] \underset{\underset{\text{X}}{|}}{\cdot\,\text{CH}_2\text{C}\!\!=\!\!\text{CH}_2} \;\xrightarrow{HX}\; \underset{\underset{\text{X}}{|}}{\text{CH}_3\text{C}\!\!=\!\!\text{CH}_2} + \text{X} \cdot \quad (7\text{-}35) \end{array}$$

ment should depend on the propensity of the adding radical to participate in such a reaction. Bromine atoms undergo 1,2-shifts readily whereas thiyl radicals are less prone to do so. Alkyl groups are not known to migrate and, consequently, only a single product resulting from bonding of the adding radical to the 1-carbon is observed in the addition of iodotrifluoromethane.

2. Conjugated Dienes. Free-radical additions to conjugated dienes generally yield a 1,4-addition product (**23**) as illustrated by the reaction of bromotrichloromethane with 1,3-butadiene (**22**). The hybrid allylic radical **24** produced by the addition reaction 7-37 could react to produce either a 1,2- or 1,4-addition product but in reactions of simple conjugated dienes only the latter is observed.

$$\underset{\textbf{22}}{\text{BrCCl}_3 + \text{CH}_2\!\!=\!\!\text{CHCH}\!\!=\!\!\text{CH}_2} \;\rightarrow\; \underset{\textbf{23}}{\text{Cl}_3\text{CCH}_2\text{CH}\!\!=\!\!\text{CHCH}_2\text{Br}} \qquad (7\text{-}36)$$

via

$$\text{Cl}_3\text{C}\cdot + \text{CH}_2\!\!=\!\!\text{CHCH}\!\!=\!\!\text{CH}_2 \;\rightarrow\; \underset{\textbf{24}}{\overset{\text{Cl}_3\text{CCH}_2\overset{\cdot}{\text{C}}\text{HCH}\!\!=\!\!\text{CH}_2}{\underset{\text{Cl}_3\text{CCH}_2\text{CH}\!\!=\!\!\text{CH}\overset{\cdot}{\text{C}}\text{H}_2}{\updownarrow}}} \qquad (7\text{-}37)$$

$$\underset{\textbf{23}}{\textbf{24} + \text{BrCCl}_3 \;\rightarrow\; \text{Cl}_3\text{CCH}_2\text{CH}\!\!=\!\!\text{CHCH}_2\text{Br} + \text{Cl}_3\text{C}\cdot} \qquad (7\text{-}38)$$

Addition to a conjugated diene is generally quite facile because of the resonance stability of the resulting allylic radical. Although formation of telomeric products is possible, good yields of the 1:1 addition product can be expected if the adding reagent has a high chain-transfer constant.

Reaction of naphthalene-2-thiol with 2,5-dimethyl-2,4-hexadiene (**25**) yields exclusively the 1,2-addition product **26**. In this case the hybrid adduct

$$RSH + (CH_3)_2C=CHCH=C(CH_3)_2 \rightarrow (CH_3)_2CCH_2CH=C(CH_3)_2 \quad (7\text{-}39)$$
$$\underset{\textbf{25}}{} \qquad\qquad\qquad \underset{\substack{|\\SR \\ \textbf{26}}}{}$$

radical **27** prefers to react in such a manner as to form the carbon–hydrogen bond on the secondary rather than on the tertiary carbon, the former reaction likely having a lower activation energy requirement. Reactions of mercaptans

$$RS\cdot + (CH_3)_2C=CHCH=C(CH_3)_2 \rightarrow
\begin{array}{c}
SR \\ | \\ (CH_3)_2\overset{}{C}CHCH=C(CH_3)_2 \\ \updownarrow \\ SR \\ | \\ (CH_3)_2CCH=CH\dot{C}(CH_3)_2 \\ \textbf{27}
\end{array}
\quad (7\text{-}40)$$

$$27 + RSH \rightarrow \underset{\substack{|\\SR}}{(CH_3)_2CCH_2CH=C(CH_3)_2} + RS\cdot \quad (7\text{-}41)$$

and aromatic thiols with 2-methyl-1,3-butadiene (**28**), 2,3-dimethyl 1,3-butadiene (**29**), and 2-chloro-1,3-butadiene (**30**) yield only 1,4-addition products. Reaction of the hybrid adduct radical with the mercaptan to form a primary carbon–hydrogen bond is preferable in each case. Reactions of *cis*- and *trans*-1,3-pentadiene (**31**) with mercaptans yield 1,2- and 1,4-addition

$$\underset{\textbf{28}}{CH_2=C(CH_3)CH=CH_2} \qquad \underset{\textbf{29}}{CH_2=C(CH_3)C(CH_3)=CH_2} \qquad \underset{\textbf{30}}{CH_2=CClCH=CH}$$

$$RS\cdot + CH_2=CHCH=CHCH_3 \rightarrow
\begin{array}{c}
RSCH_2\overset{\bullet}{C}HCH=CHCH_3 \\ \updownarrow \\ RSCH_2CH=CH\dot{C}HCH_3 \\ \textbf{32}
\end{array}
\quad (7\text{-}42)$$
$$\underset{\textbf{31}}{}$$

$$\begin{array}{c}
RSCH_2\overset{\bullet}{C}HCH=CHCH_3 \\ \updownarrow \\ RSCH_2CH=CH\overset{\bullet}{C}HCH_3 \\ \textbf{32}
\end{array}
\begin{array}{l}
\xrightarrow{1,2\text{-add.}} RSCH_2CH_2CH=CHCH_3 + RS\cdot \quad (7\text{-}43) \\ \hspace{5em} \textbf{33} \\[2ex]
\xrightarrow{1,4\text{-add.}} RSCH_2CH=CHCH_2CH_3 + RS\cdot \quad (7\text{-}44) \\ \hspace{5em} \textbf{34}
\end{array}$$

products (**33** and **34**, respectively) since hydrogen atom abstraction by the hybrid adduct radical **32** involves formation of a secondary carbon–hydrogen bond in either case. Allylic radicals generally react in displacement reactions in such a manner that bonding occurs on the least substituted carbon of the hybrid species.[7]

3. Nonconjugated Dienes. Free-radical additions to nonconjugated dienes generally take place in such a manner as to suggest that the two unsaturated centers have no influence on each other. For example, reaction of n-heptanal (**35**) with 4-vinylcyclohexene (**36**) yields the 1:1 addition product **37** resulting from involvement of only the more reactive terminal alkene linkage.[39]

$$n\text{-}C_6H_{13}CHO + \quad\text{(36)} \quad \xrightarrow{\text{Per.}} \quad \text{(37)} \quad\quad (7\text{-}45)$$

35   **36**   **37**

Other nonconjugated dienes that have equivalent unsaturated linkages behave similarly in that additions to each functionality appear to be independent. Addition of carbon tetrachloride to 1,5-hexadiene (**38**)[8] yields both the simple 1:1 addition product **39** and the 1:2 addition product **40**. Reactions of nonconjugated dienes with dithiols yield polymeric products as illustrated by

$$CH_2{=}CH(CH_2)_2CH{=}CH_2 + Cl_3C\cdot \;\rightarrow$$
$$\textbf{38}$$

$$\hspace{4cm} Cl_3CCH_2\dot{C}H(CH_2)_2CH{=}CH_2 \quad (7\text{-}46)$$

$$Cl_3CCH_2\dot{C}H(CH_2)_2CH{=}CH_2 + Cl_4C \;\rightarrow$$

$$\hspace{3cm} Cl_3CCH_2CHCl(CH_2)_2CH{=}CH_2 + Cl_3C\cdot \quad (7\text{-}47)$$
$$\textbf{39}$$

$$Cl_3C\cdot + Cl_3CCH_2CHCl(CH_2)_2CH{=}CH_2 \;\rightarrow$$

$$\hspace{3cm} Cl_3CCH_2CHCl(CH_2)_2\dot{C}HCH_2CCl_3 \quad (7\text{-}48)$$

$$Cl_3CCH_2CHCl(CH_2)_2\dot{C}HCH_2CCl_3 + Cl_4C\cdot \;\rightarrow$$

$$\hspace{2cm} Cl_3CCH_2CHCl(CH_2)_2CHClCH_2CCl_3 + Cl_3C\cdot \quad (7\text{-}49)$$
$$\textbf{40}$$

the light-induced reaction of tetramethylene dithiol (**41**) with **38** producing the polymeric product **42**.[9]

$$HS(CH_2)_4SH + CH_2{=}CH(CH_2)_2CH{=}CH_2 \;\rightarrow$$
$$\textbf{41} \hspace{3cm} \textbf{38}$$

$$\hspace{3cm} {\sim}S(CH_2)_4S(CH_2)_6S(CH_2)_4S(CH_2)_6S{\sim} \quad (7\text{-}50)$$
$$\textbf{42}$$

An interesting reaction of nonconjugated dienes is the formation of cyclic products observed in reactions of certain of these species in their free-radical reactions. The addition of carbon tetrachloride to 1,6-heptadiene (**43**) yields almost exclusively the cyclopentane derivative **44** as the 1:1 addition

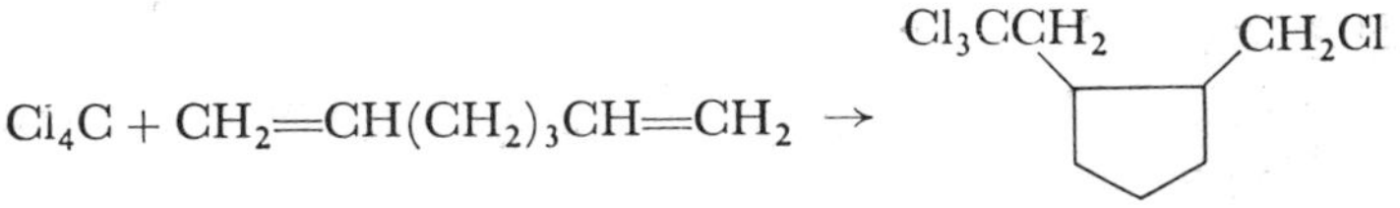

$$Cl_4C + CH_2{=}CH(CH_2)_3CH{=}CH_2 \rightarrow \qquad (7\text{-}51)$$

43   **44**

product.[10, 11] Reaction of heptafluoroiodopropane (**45**) with **43** is reported to yield the cyclopentane derivative **46**[12] although reaction of bromotrichloro-

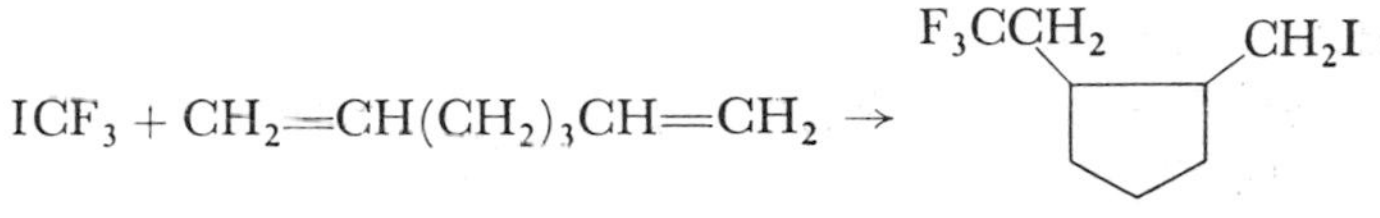

$$ICF_3 + CH_2{=}CH(CH_2)_3CH{=}CH_2 \rightarrow \qquad (7\text{-}52)$$

43   **46**

methane with the same diene yields 85% of the open chain product and 15% of the cyclopentane derivative.[10] Only trace amounts of cyclic product were observed in the reactions of n-butyl mercaptan with 1,6-heptadiene.[11]

The amount of cyclization observed in these reactions depends on the competition between the transfer reaction of the adduct radical **47** with the

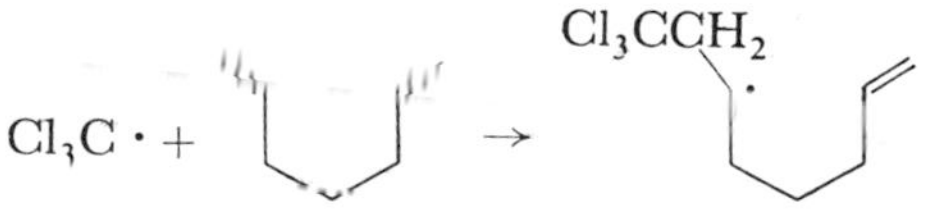

$$Cl_3C\cdot + \qquad \rightarrow \qquad (7\text{-}53)$$

43   **47**

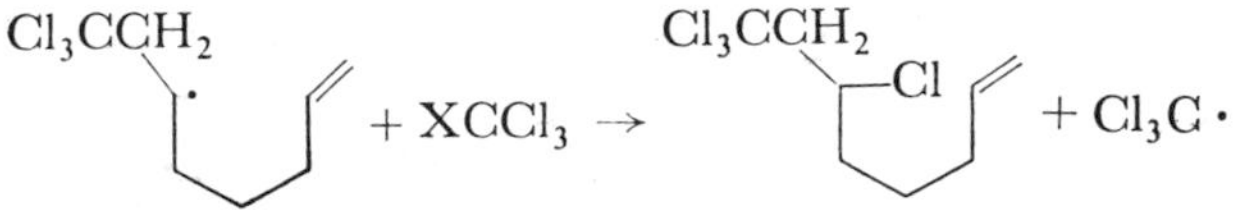

$$+ XCCl_3 \rightarrow \qquad + Cl_3C\cdot \qquad (7\text{-}54)$$

47

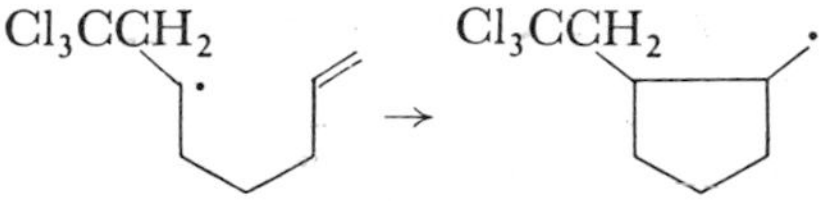

$$\rightarrow \qquad (7\text{-}55)$$

47

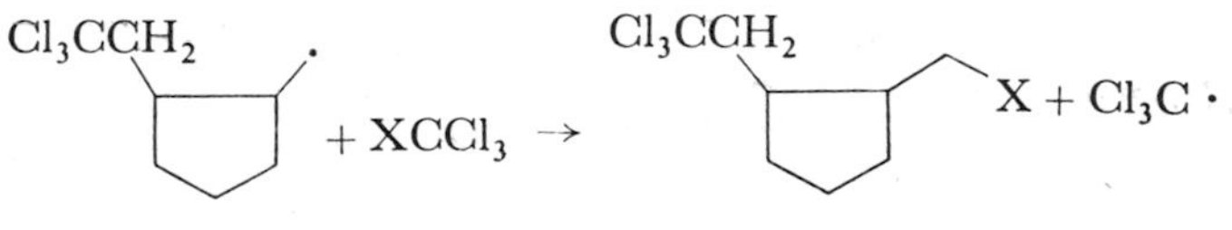

$$+ XCCl_3 \rightarrow \qquad X + Cl_3C\cdot \qquad (7\text{-}56)$$

adding reagent and the intramolecular addition reaction yielding the cyclic adduct radical. The intramolecular addition proceeds in such a manner as to form the unexpected primary alkyl radical. This preference for the intra- molecular addition following the course shown in 7-55 has been observed in other free-radical reactions. Decomposition of 6-heptenoyl peroxide **48** at 77°C in toluene yields methylcyclopentane.[13] The decomposition of the peroxide forms the 5-hexenyl radical (**49**) which undergoes an intramolecular addition yielding the cyclopentylmethyl radical (**50**) reaction of which with the solvent yields methylcyclopentane. Kolbe electrolysis of 6-heptenoic (**51**)

$$(CH_2\!=\!CH(CH_2)_4CO_2)_2 \ \rightarrow \ 2CH_2\!=\!CH(CH_2)_3CH_2\cdot + 2CO_2 \quad (7\text{-}57)$$

$$\underset{\textbf{48}}{} \qquad\qquad\qquad\qquad \underset{\textbf{49}}{}$$

$$(7\text{-}58)$$

acid also results in an intramolecular addition yielding the cyclic radical **50** as evidenced by the formation of the dimeric species **52** and **53** among the

$$CH_2\!=\!CH(CH_2)_4CO_2H \ \xrightarrow[\text{Electrolysis}]{\text{Kolbe}} \ CO_2 + e^{(-)} + H^+ + CH_2\!=\!CH(CH_2)_3\overset{\cdot}{C}H_2$$

$$\underset{\textbf{51}}{} \qquad\qquad\qquad\qquad\qquad\qquad\qquad \underset{\textbf{49}}{} \qquad (7\text{-}59)$$

$$(7\text{-}60)$$

$$(7\text{-}61)$$

$$(7\text{-}62)$$

reaction products.[14] Formation of the 5-hexenyl radical **49** by reactions of 5-hexenyl mercaptan (**54**) with triethylphosphite and of 5-hexenyl bromide (**55**) with trialkyl tin hydride also resulted in formation of the cyclopentylmethyl radical and ultimately methylcyclopentane.[15]

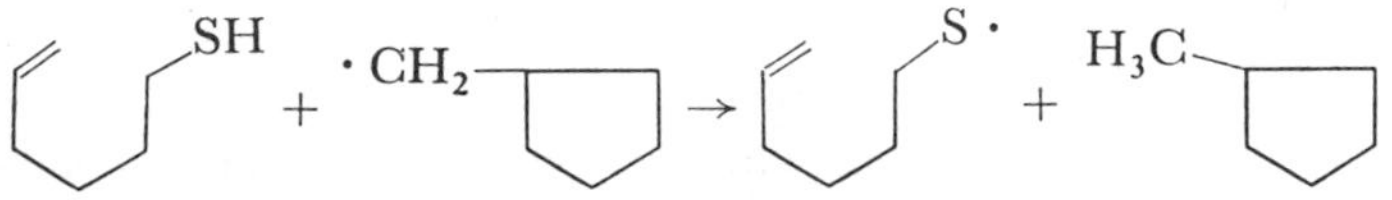

54 **50** (7-63)

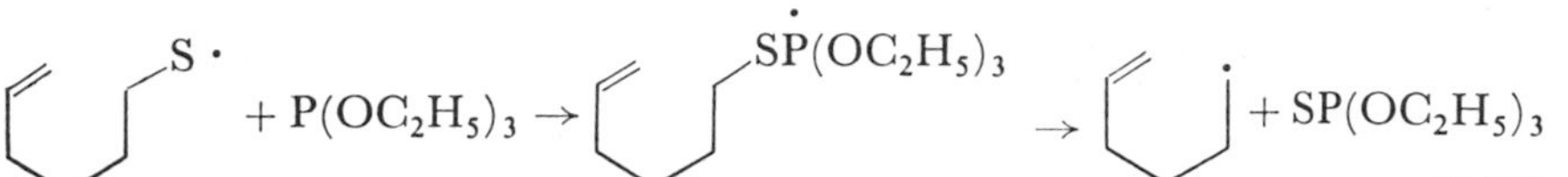

49 (7-64)

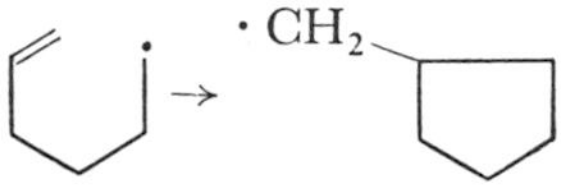

49 **50**

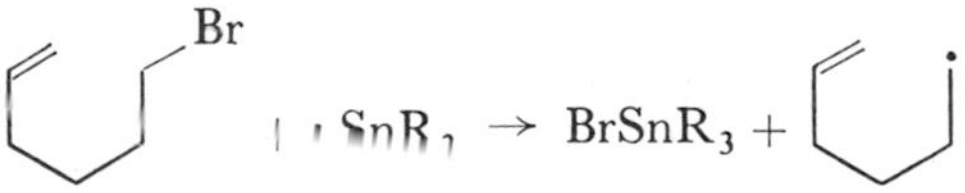

55 **49** (7-65)

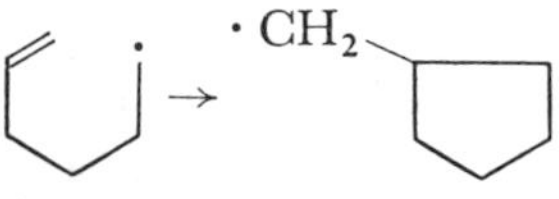

49 **50** (7-66)

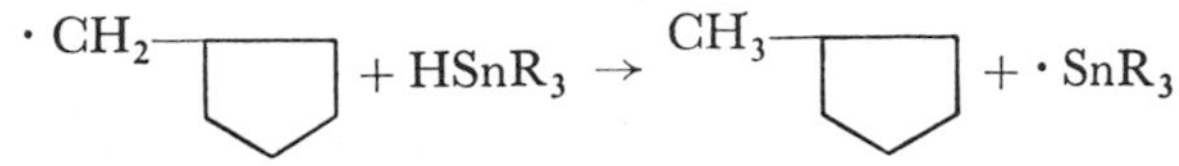

(7-67)

Reaction of carbon tetrachloride with 1,7-octadiene (**56**) produced a mixture of the cyclohexane derivative **57** and the open-chain addition product in about equivalent amounts. This reaction indicates that six-membered rings can be formed in some cases in intramolecular addition reactions. Free-radical addition of mercaptan to **58** is reported to yield both the five- and six-membered ring compounds as addition products with the latter predominating by a factor of

$$\text{56} + \text{Cl}_4\text{C} \rightarrow \overset{\overset{\displaystyle\text{CH}_2\text{Cl}}{\underset{\text{Cl}_3\text{CCH}_2}{}}}{\text{57}} + \text{Cl}_3\text{CCH}_2\text{CHCl}(\text{CH}_2)_4\text{CH}=\text{CH}_2 \quad (7\text{-}68)$$

56   **57**

$$\underset{\underset{\text{CO}_2\text{C}_2\text{H}_5}{\overset{|}{\text{CH}}}}{} + \text{RSH} \rightarrow \underset{\underset{\text{CO}_2\text{C}_2\text{H}_5}{\overset{|}{\text{CH}}}}{\overset{\overset{\displaystyle\text{RSCH}_2 \quad \text{CH}_3}{}}{}} + \underset{\underset{\text{CO}_2\text{C}_2\text{H}_5}{\overset{|}{\text{CH}}}}{\overset{\overset{\displaystyle\text{RSCH}_2 \quad \text{H}}{}}{}} \quad (7\text{-}69)$$

58

three.[16] Quite interesting is the observation that addition to 1,5-hexadiene (**59**) does not yield the cyclic addition product resulting from the intramolecular addition at the terminal carbon although such addition would yield a cyclo-

$$\text{59} + \text{Cl}_3\text{C}\cdot \rightarrow \overset{\overset{\displaystyle\text{Cl}_3\text{CCH}_2-\dot{\text{CH}}}{}}{} \overset{\text{Cl}_4\text{C}}{\nearrow} \overset{}{\searrow} \begin{array}{l} \text{Cl}_3\text{CCH}_2\text{CHCl}(\text{CH}_2)_2\text{CH}=\text{CH}_2 \quad (7\text{-}70) \\ \text{Cl}_3\text{CCH}_2\dot{\text{CH}} \quad (7\text{-}71) \end{array}$$

59

pentane derivative. Small amounts of cyclic product **60** are formed resulting from abnormal addition of the trichloromethyl radical to one of the unsaturated linkages of the diene. This addition produces a radical that can form the cyclic adduct radical leading to the cyclopentane derivative by intramolecular addition that involves bonding at the nonterminal carbon of the remaining unsaturated linkage.[11]

$$\text{59} + \text{Cl}_3\text{C}\cdot \rightarrow \overset{\overset{\displaystyle\text{Cl}_3\text{CCH}-\dot{\text{CH}}_2}{}}{} \rightarrow$$

59

$$\overset{\overset{\displaystyle\text{Cl}_3\text{CCH}-\text{CH}_2}{}}{\underset{\dot{\text{CH}}_2}{}} \overset{\text{Cl}_4\text{C}}{\longrightarrow} \overset{\overset{\displaystyle\text{Cl}_3\text{C}}{}}{\underset{\text{CH}_2\text{Cl}}{}} \quad (7\text{-}72)$$

60

One plausible explanation for the predominant formation of the five-membered ring over the six-membered ring adduct radical is that a complexed radical such as **61** reacts with the adding reagent. The preferred reaction of the complexed radical with the transfer reagent at the 1-carbon may be due both to less steric hindrance at the primary carbon as well as a somewhat higher exothermicity owing to the formation of a new sigma bond on a primary rather than on a secondary carbon.

$$\text{61} \qquad \xrightarrow{\text{RX}}_{\text{1-carbon attack}} \qquad \qquad (7\text{-}73)$$

$$\xrightarrow{\text{RX}}_{\text{2-carbon attack}} \qquad \qquad (7\text{-}74)$$

D. CYCLIC COMPOUNDS

For the most part cyclic compounds that have the unsaturated linkage in the ring are somewhat less reactive toward free-radical addition than terminally unsaturated species. Conformational aspects of cyclic compounds, however, cause many interesting reactions that are not found in the acyclic compounds. Among these are transannular reactions resulting from the proximity of a reactive center across the ring from the radical site. Also of interest are the effects that conformational preferences of the cyclohexane ring system exercise on the free-radical additions to certain substituted cyclohexenes.

1. Transannular Reactions. Free-radical additions of most adding reagents to cyclic conjugated dienes proceed in very much the expected manner as illustrated by the reaction of bromotrichloromethane with 1,3-cyclohexadiene (**62**). Both 1,2- and 1,4-addition products are produced. Nonconjugated cyclic

$$\text{62} + \text{Cl}_3\text{C} \cdot \rightarrow \qquad \leftrightarrow \qquad \qquad (7\text{-}75)$$
$$\text{63}$$

$$\text{63} + \text{BrCCl}_3 \rightarrow \qquad + \qquad \qquad (7\text{-}76)$$

dienes behave somewhat differently in that there is likely no direct involvement of the second unsaturated bond in the initial addition of the adding radical to the first alkene linkage. However, a transannular reaction may occur in a subsequent step. For example, reaction of p-thiocresol with norbornadiene (**64**) yielded a mixture of 1:1 addition products consisting of the unsaturated thio-ether **66** and the saturated nortricyclene derivative **68**.[17] The nortricyclene derivative is formed by the transannular reaction of the adduct radical (**65**) yielding the nortricyl derived radical (**67**) which reacts with the adding reagent

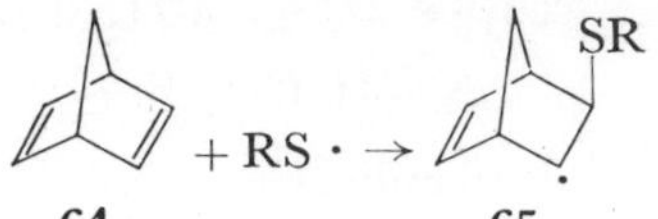

$$(7\text{-}77)$$

64   **65**

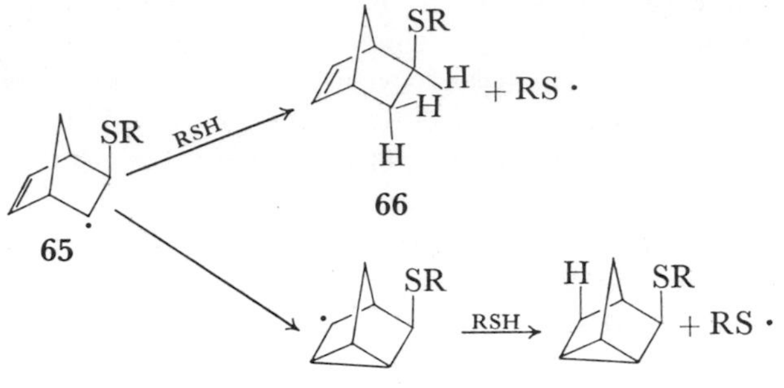

$$(7\text{-}78)$$

66

65

$$(7\text{-}79)$$

67   **68**

$$R = p\text{-}CH_3C_6H_4$$

yielding **68**. The amount of the tricyclene derivative was observed to increase with decreasing concentration of the *p*-thiocresol, an observation consistent with the existence of two discrete reaction paths for the initially formed adduct radical. At higher concentrations of the *p*-thiocresol the bimolecular reaction leading to the unsaturated addition product **66** competes favorably with the unimolecular transannular addition leading to the tricyclene.

Additions to 2-methylene-5-norbornene (**69**) also yield products resulting from a transannular addition. Bromotrichloromethane, carbon tetrachloride, and chloroform react with this diene to yield predominantly the nortricyclene derivatives as the 1:1 addition products.[18] In this case the transannular addition

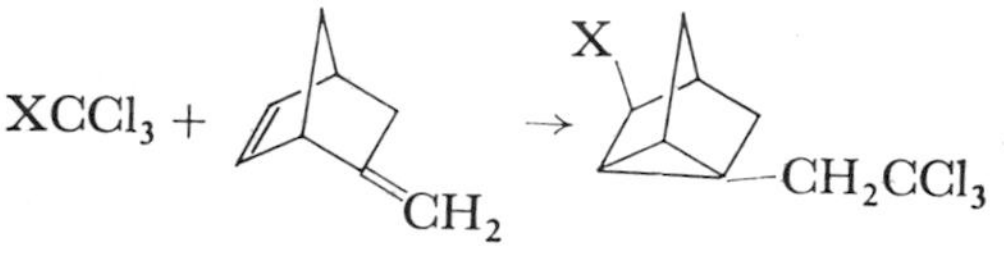

$$(7\text{-}80)$$

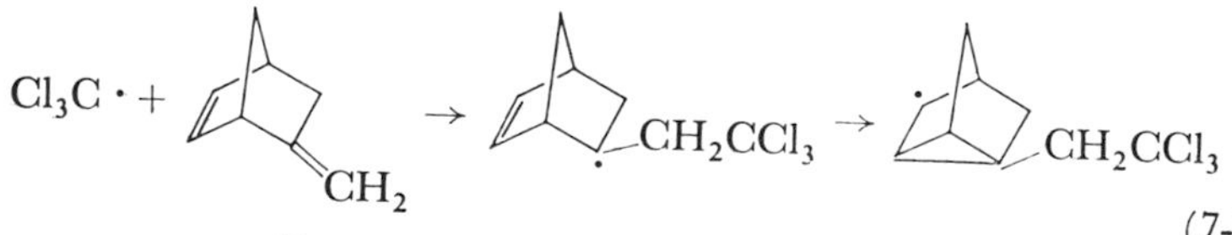

69

via

$$(7\text{-}81)$$

69

$$X = H, Cl \text{ or } Br \qquad (7\text{-}82)$$

of the initially formed adduct radical is appreciably faster than the chain-transfer reaction with the adding reagent. Addition of thiophenol to **69** appears to follow a different course in that the addition products found, including one from a transannular reaction, are exclusively those resulting from preferential addition of the thiyl radical to the endocyclic unsaturated bond.[19] The apparent greater reactivity of the endocyclic double bond toward addition is the result of both the reversibility of the addition of a thiyl radical to alkene-linkages and the greater relief of bond angle strain experienced in the addition to the endocyclic double bond relative to that resulting from addition to the exocyclic unsaturated linkage. This aspect of radical additions to unsaturated linkages is discussed more extensively in Chapter 8.

$$(7\text{-}83)$$

$$(7\text{-}84)$$

$$(7\text{-}85)$$

$$(7\text{-}86)$$

Finding transannular reactions occurring in the norbornenyl radicals cited above may not be too surprising in view of the proximity of the unsaturated linkage to the radical site. It might seem feasible that the unsaturated linkage may render anchimetric assistance in the addition of reactions of the radical. All experiments that have been designed to demonstrate homoconjugation in this type of radical have failed to show any enhancement of the rate of the adduct radical formation that can be attributed to such an effect. Furthermore, higher concentrations of the adding reagent favor formation of the unsaturated addition product, an observation that eliminates from consideration the existence of a single species such as **70** that might react to yield the observed products.

Transannular reactions have been observed in the free-radical reactions of cyclooctane derivatives. Benzoyl peroxide-induced addition of chloroform to 1,5-cyclooctadiene (**71**) yields the bicyclic saturated addition product **72** in 37% yield.[20] The conformational aspects of the eight-membered ring system are such that the radical site can be in the proximity of the unsaturated linkage

across the ring. Reaction of **71** with thioacetic acid, however, yields only the 1,2-addition product.[21] In this case the transfer reaction with the very reactive adding reagent is faster than the transannular addition.

Bromotrichloromethane reacts with cyclooctene (**73**) in a normal manner yielding a mixture of *cis*- and *trans*-1,2-addition products. Reaction of carbon tetrachloride with cyclooctene, on the other hand, yields an anomalous product, namely the 1,4-addition product (**74**). A transannular hydrogen abstraction by

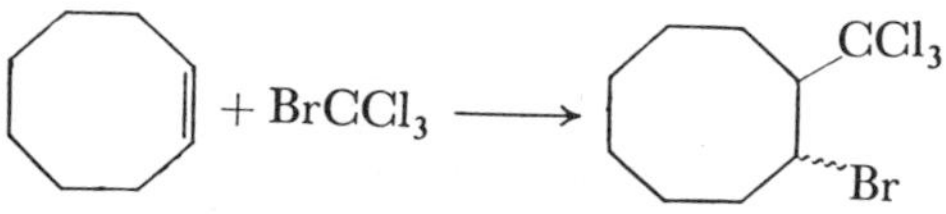

$$\text{(structure)} + \text{BrCCl}_3 \longrightarrow \text{(structure with CCl}_3 \text{ and Br)} \qquad (7\text{-}92)$$

73

the adduct radical, a reaction that leads to the anomalous 1,4-addition product, competes favorably with the chain transfer with the carbon tetrachloride. Reaction of cyclooctene with trichloromethanesulfonyl chloride results in formation of both the 1,2- and 1,4-addition products illustrating that the reactivity of this reagent toward displacement is somewhat greater than carbon tetrachloride but less than bromotrichloromethane.[22, 23]

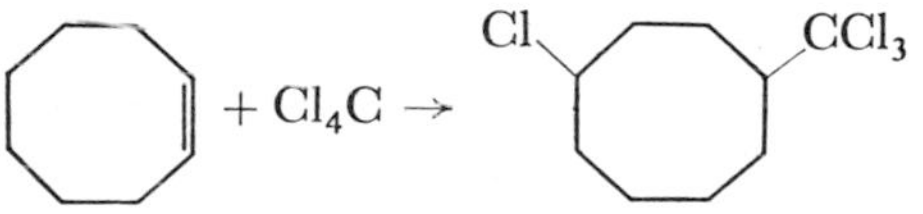

$$\text{(structure)} + \text{Cl}_4\text{C} \rightarrow \text{(structure with Cl and CCl}_3) \qquad (7\text{-}93)$$

73   **74**

via

$$\text{Cl}_3\text{C} \cdot + \text{H} \text{(structure)} \rightarrow \text{(structure)} \rightarrow \text{(structure)} \qquad (7\text{-}94)$$

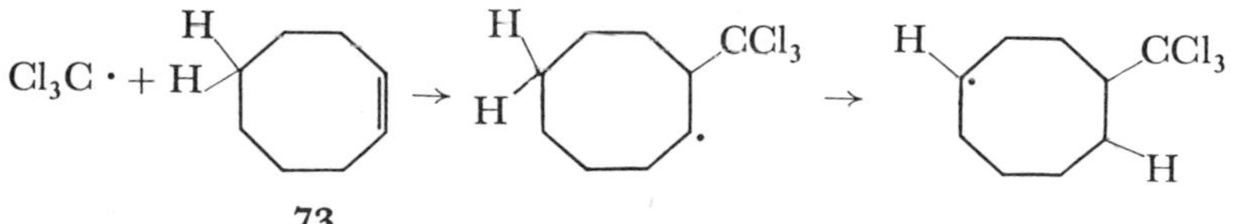

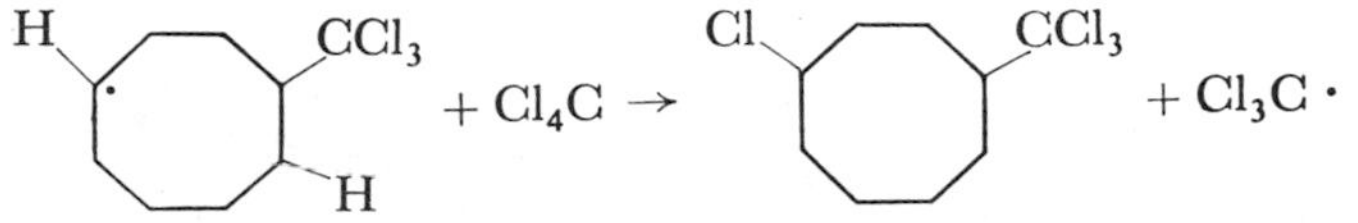

73

$$\text{(structure)} + \text{Cl}_4\text{C} \rightarrow \text{(structure)} + \text{Cl}_3\text{C} \cdot \qquad (7\text{-}95)$$

74

A transannular hydrogen abstraction was observed in the reaction of methylenecyclooctane (**75**) with bromotrichloromethane.[24] Apparently the transannular reaction of the initially formed adduct radical leading to the addition product **76** competes with the transfer process of this radical with bromotrichloromethane.

$$(7\text{-}96)$$

$$(7\text{-}97)$$

2. Conformational Effects. The greater stability of the chair conformer of saturated six-membered rings relative to the twist-boat conformation is responsible for the stereochemical course of certain free-radical additions to cyclohexene derivatives. Addition of thioacetic acid to 4-*t*-butyl-1-methyl-

$$(7\text{-}98)$$

cyclohexene (**77**) in a light-induced reaction yielded a mixture in which the *trans*-addition product **78** predominated over the *cis* product **79**.[25] Free-radical addition of thioacetic acid to 2-chloro-4-*t*-butylcyclohexene (**80**) yielded a mixture of four stereoisomeric addition products.[26] The predominant product (**81**) in this case was also that resulting from *trans*-diaxial bonding of the elements

of the adding reagent to the unsaturated linkage although, at 63°C, about half
of the reaction mixture was composed of the remaining isomers. Both of these

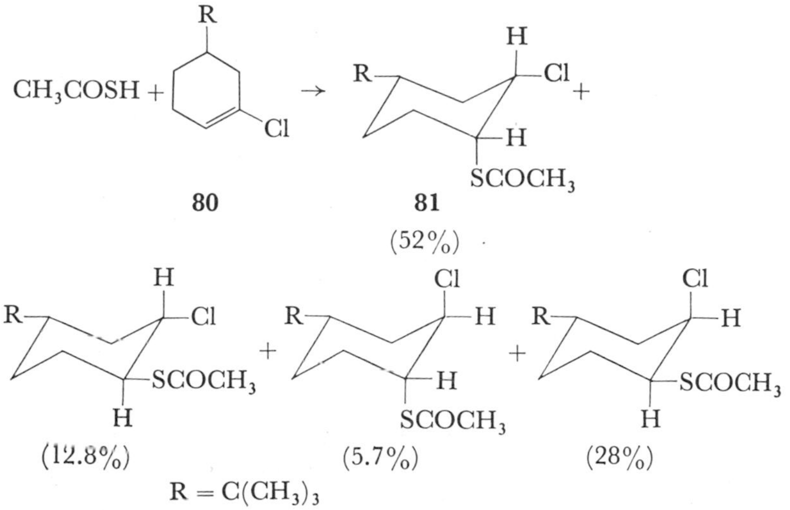

80

81
(52%)

(7-99)

(12.8%)   (5.7%)   (28%)

R = C(CH₃)₃

R = $C(CH_3)_3$

studies indicate a preferential formation of the addition product in which the
elements of the adding reagent not only add *trans* to the unsaturated linkage (see
section on stereochemistry) but also indicate a preference for formation of
isomers with the adding radical in the conformationally less stable axial
position. The preferred formation of the *axial*-product is also evident in the
free-radical addition reactions of methyl mercaptan to 4-*tert*-butyl cyclohexene
(**82**)[27] and *trans*-Δ²-octalin (**83**).[28] The groups responsible for the conforma
tional control of the cyclohexane ring in these reactions are remotely positioned
from the reaction site, thereby excluding the possibility of a direct steric effect
influencing the stereochemistry of the addition.

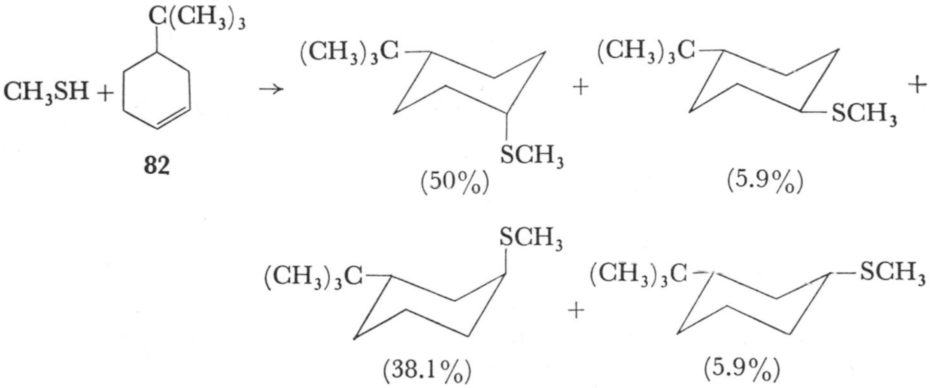

82

(50%)   (5.9%)

(38.1%)   (5.9%)

(7-100)

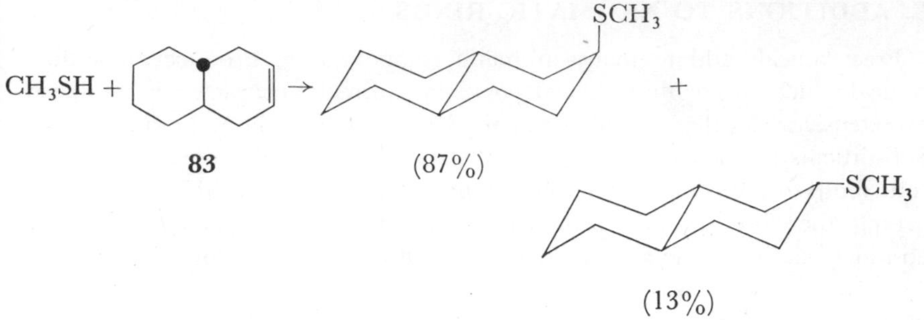

$$CH_3SH + \quad \text{[structure 83]} \quad \rightarrow \quad \text{[product with SCH}_3\text{]} \quad + $$

83   (87%)

[structure with —SCH$_3$]

(13%)

(7-101)

An explanation for the preferred formation of the axially substituted products observed in these additions centers around the conformational factors encountered in formation of the adduct radical. Consider adding a thiyl radical to 4-*t*-butylcyclohexene which exists almost exclusively in the conformation having the *t*-butyl group in an equatorial position. Attack at the 1-carbon from the *cis*-side relative to the equatorial *t*-butyl group yields an adduct radical (**84**) having a chair conformation. On the other hand, approach of the thiyl radical from the side *trans* to the *t*-butyl group yields the adduct radical (**85**) having the twist-boat conformation. Isomerization of **85** yields the adduct radical **86** having the added thiyl group in an equatorial position. Since the twist-boat conformation is considerably less stable than the chair conformation, it is likely that the adduct radical having the chair conformation that leads to the axial product is formed faster.

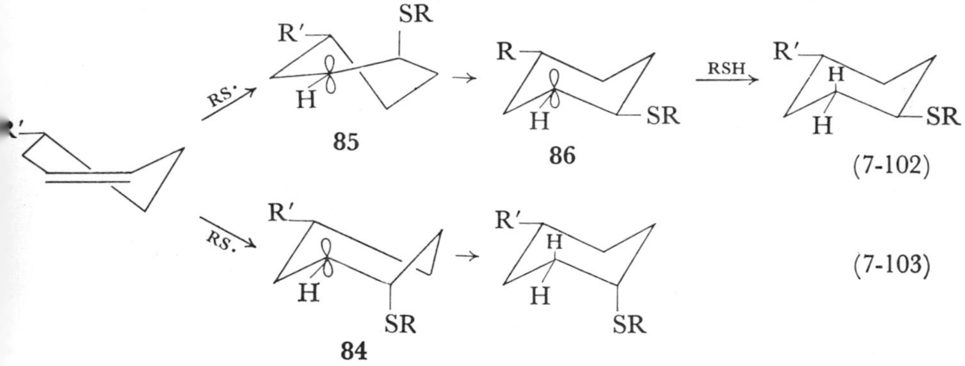

$$\text{[RS· adds to alkene]} \quad \xrightarrow{\text{RS·}} \quad \text{85} \quad \rightarrow \quad \text{86} \quad \xrightarrow{\text{RSH}} \quad \text{[product]} \qquad (7\text{-}102)$$

$$\xrightarrow{\text{RS·}} \quad \text{84} \quad \rightarrow \quad \text{[product]} \qquad (7\text{-}103)$$

Predominant formation of the *trans*-diaxial products in the reactions of **77** and **80** suggests that reaction of the adduct radical might also be such that the preferred attack on the adding reagent is such that an axial bond is formed.

E. ADDITIONS TO AROMATIC RINGS

Free radicals add to many aromatic compounds to produce an adduct radical which, although no longer possessing aromatic character, is stabilized by extensive delocalization of the unpaired electron throughout the ring system. Substituents positioned *ortho* or *para* relative to the site of radical attack also contribute to the resonance stabilization of the adduct radical. It is not surprising, therefore, to find a somewhat greater reactivity at the *ortho* and *para* positions relative to the *meta* position in substituted benzenes toward attack by

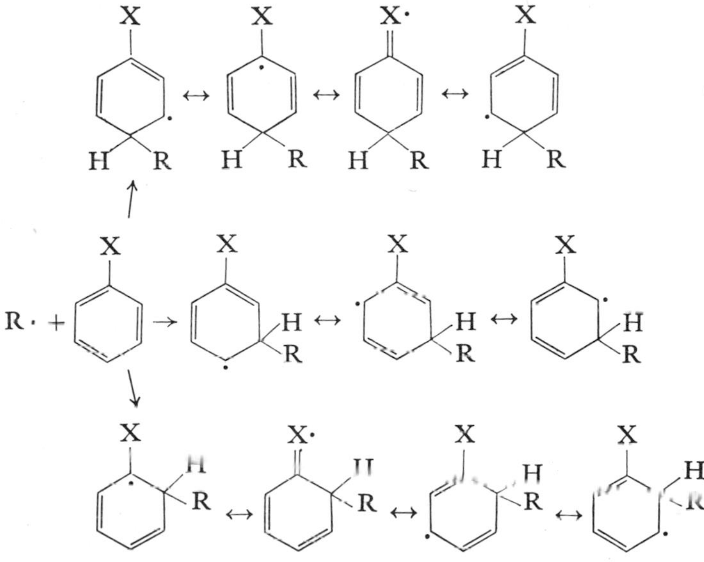

(7-104)

(7-105)

(7-106)

free radicals. Furthermore, both electron-withdrawing groups ($\overset{O}{\overset{\|}{C}}OR$, $\overset{O}{\overset{\|}{C}}R$, NO_2, Cl_2, etc.) and electron-releasing groups (OH, OR, NH_2, R, etc.), in contrast to electrophylic-substitution reactions, exercise similar orientation effects in homolytic aromatic substitutions.

The fate of the adduct radical produced by addition to an aromatic ring depends on the source of the radical. The adding radical may arise from decomposition of a peroxide and, if a sufficient number of radicals is produced in this manner, the adduct radical may participate in a bimolecular reaction as

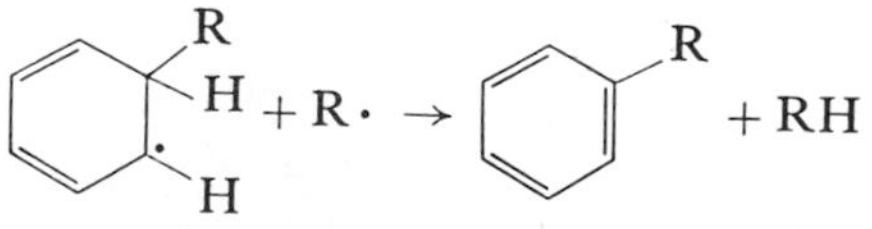

(7-107)

shown in 7-107 yielding the homolytic aromatic substitution product. If an insufficient number of radicals is present, the adduct radicals dimerize. Homolytic arylations of many aromatics using benzoyl peroxide $(R = C_6H_5)$ as the radical source have been extensively investigated. Similarly, methylation of aromatics can be accomplished using acetyl peroxide $(R = CH_3)$ as the source of methyl radicals. The subject of homolytic aromatic substitution, which has been reviewed elsewhere,[29] is outside the scope of this book since the reactions do not appear to be chain processes.

The light-induced addition of chlorine to benzene yielding the isomeric hexachlorobenzenes very likely is a free-radical chain process. The initial attack of benzene by a chlorine atom, the reaction in which the aromaticity of the ring is destroyed, is the limiting factor in the reaction. The formation of a dichloride (either a 1,2- or 1,4-addition product) by reaction of the mono-chloro-adduct could follow in a considerably more facile propagating reaction. The dichlorobenzene is very reactive toward further addition of chlorine yielding first the tetrachlorobenzene and finally the hexachlorobenzene.

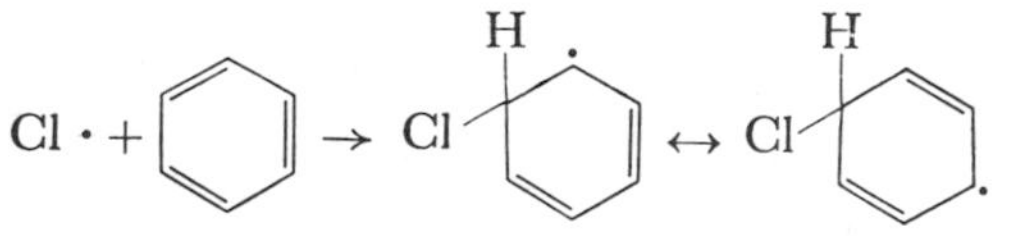

(7-108)

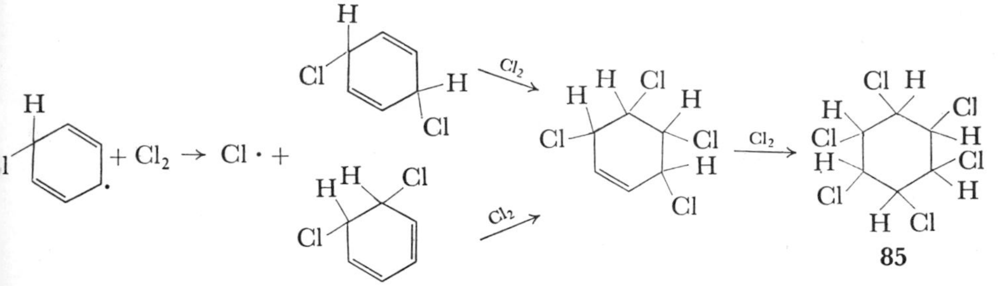

(7-109)

Support for the free-radical chain character of the reaction is the observation that the quantum yield of the reaction (number of moles of products formed per einstein of light absorbed) is about 20.[30]

Bromine adds to benzene in a photolytic reaction but higher temperatures are required for the reaction than for addition of chlorine. This is likely due to the lower reactivity of the bromine atom compared to the chlorine atom in the initial attack of the benzene ring.

Free-radical addition to a benzene derivative yielding a 1:1 addition product occurs in the thermal reactions of iodotrifluoromethane with halobenzenes.

These reactions yield a mixture of *ortho, meta,* and *para* substitution products, fluoroform, and molecular iodine. A mechanism, illustrated for the formation of the *para*-isomer (**87**), that accounts for the formation of these products is one

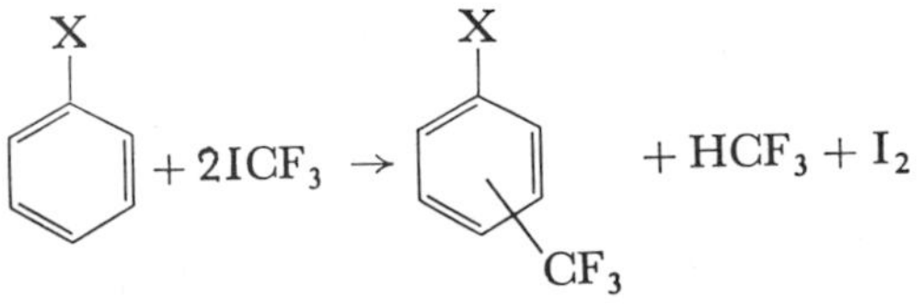

$$\text{(X-phenyl)} + 2ICF_3 \rightarrow \text{(X-phenyl-CF}_3\text{)} + HCF_3 + I_2 \qquad (7\text{-}110)$$

in which the 1:1 addition product (**86**), produced in the chain sequence 7-112 and 7-113, dehydrohalogenates yielding the substitution product. Fluoroform and molecular iodine result from the subsequent reaction of the hydrogen iodide and the polyhalomethane.[31]

$$ICF_3 \xrightarrow[200°]{} I\cdot + \cdot CF_3 \qquad (7\text{-}111)$$

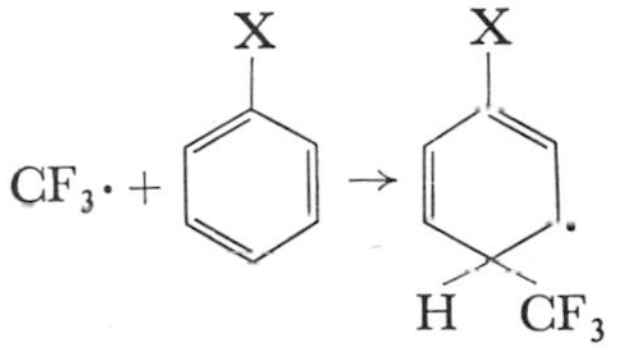

$$CF_3\cdot + \text{(X-phenyl)} \rightarrow \text{(X-ring with H, CF}_3\text{, radical)} \qquad (7\text{-}112)$$

$$\text{(radical, H, CF}_3\text{)} + ICF_3 \rightarrow \text{(X-ring with I, H, CF}_3\text{)} \quad +CF_3\cdot \qquad (7\text{-}113)$$

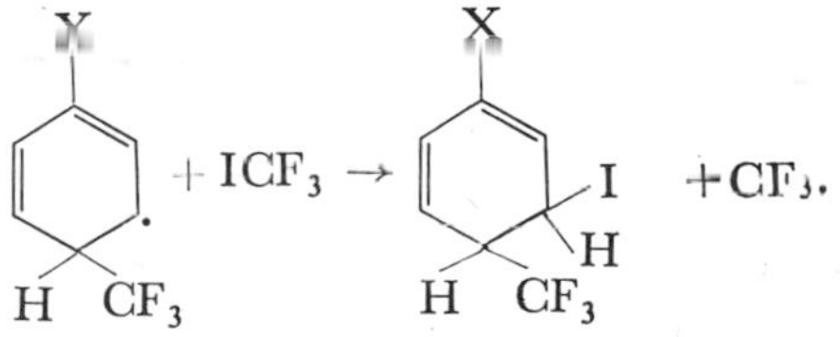

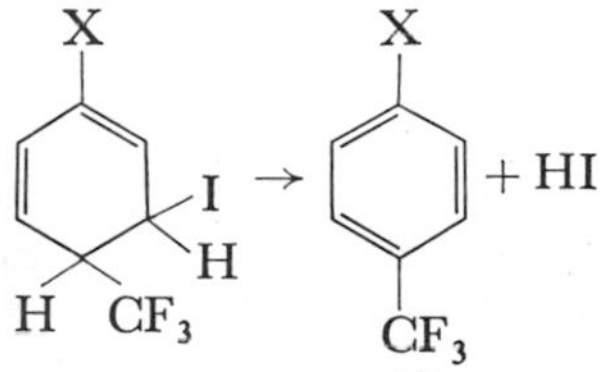

86

$$\text{(X-ring with I, H, CF}_3\text{)} \rightarrow \text{(X-phenyl-CF}_3\text{)} + HI \qquad (7\text{-}114)$$

87

$$HI + ICF_3 \rightarrow I_2 + HCF_3 \qquad (7\text{-}115)$$

Polynuclear aromatics react with bromotrichloromethane and, although never isolated, free-radical addition products may be formed in the manner illustrated for the reaction of bromotrichloromethane with anthracene (**88**).

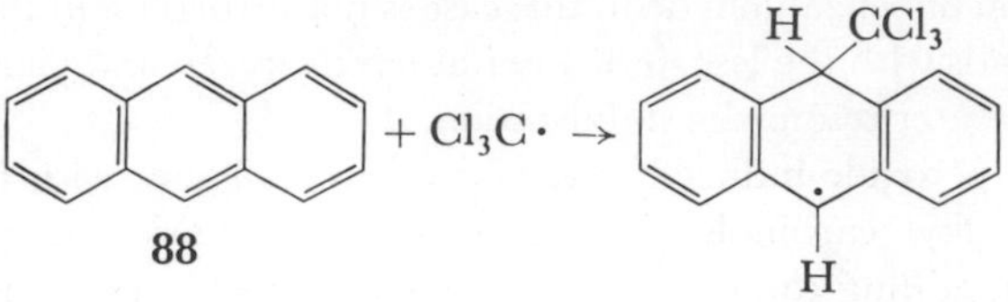

$$(7\text{-}116)$$

88

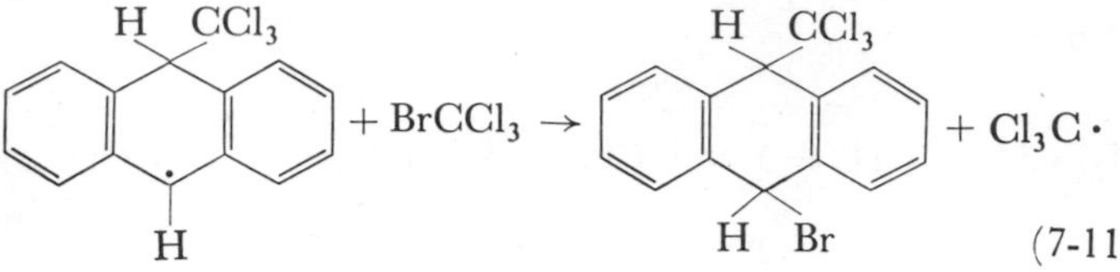

$$(7\text{-}117)$$

The reactivity of polynuclear ring systems toward addition of trichloromethyl radicals is surprisingly great as evidenced by the retarding effect they exercise on the addition of bromotrichloromethane to styrene.[32] The adduct radical, because of resonance stabilization, is relatively unreactive toward transfer with the polyhalomethane.

F. ADDITIONS TO CARBONYL FUNCTIONS

The preceding examples of free-radical additions have involved interaction of the adding reagent with an unsaturated carbon-carbon linkage. Free-radical additions to other unsaturated linkages, particularly the carbonyl function, have also been reported. Of particular interest in these reactions is the observed direction of the addition reaction.

Decomposition of t-butyl peroxide in benzaldehyde yields $meso$-dihydrobenzoin dibenzoate (89) and t-butyl alcohol as reaction products.[33] The ester results from the dimerization reaction of the adduct radicals formed by the addition of a benzoyl radical to benzaldehyde. The direction of addition to the

$$(CH_3)_3COOC(CH_3)_3 \rightarrow 2(CH_3)_3CO\cdot \qquad (7\text{-}118)$$

$$(CH_3)_3CO\cdot + C_6H_5CHO \rightarrow (CH_3)_3COH + C_6H_5\dot{C}O \qquad (7\text{-}119)$$

$$C_6H_5\dot{C}O + O{=}CHC_6H_5 \rightarrow \underset{\underset{O}{\|}}{C_6H_5C}{-}O{-}\dot{C}HC_6H_5 \qquad (7\text{-}120)$$

$$2C_6H_5\underset{\underset{\underset{O}{\|}}{OCC_6H_5}}{\dot{C}H} \rightarrow \underset{\underset{\underset{O}{\|}}{C_6H_5CO}}{C_6H_5CH}{-}\underset{\underset{\underset{O}{\|}}{OCC_6H_5}}{CHC_6H_5} \qquad (7\text{-}121)$$

89

carbon function of benzaldehyde in this case is not surprising in that the attacking radical bonds with the less sterically hindered oxygen and yields the adduct radical with greater resonance stabilization.

The t-butyl peroxide-induced reactions of cycloalkanes with formaldehyde yielding cycloalkyl carbinols (**90**) indicate that addition may occur with bonding of the adding radical to the carbon of the carbonyl linkage.[34] Polar contributions likely play the deciding role in determining the direction of addition.

$$(CH_2)_n CH \cdot + CH_2{=}O \rightarrow \left[(CH_2)_n \overset{\delta+}{CH} \cdots CH_2 \overset{\delta-}{\cdots} O \right] \rightarrow (CH_2)CHCH_2O \cdot \qquad (7\text{-}122)$$

$$(CH_2)_n CHCH_2O \cdot + (CH_2)_n CH_2 \rightarrow (CH_2)_n CHCH_2OH + (CH_2)_n CH \cdot \qquad (7\text{-}123)$$
$$\textbf{90}$$

Addition to a carbonyl linkage likely occurs in the case of the peroxide-induced reaction of biacetyl (**91**) with n-butyraldehyde yielding 2,3-hexadione (**92**).[35] The mechanism suggested for this reaction is one in which an acyl

$$n\text{-}C_3H_7CHO + CH_3\overset{\overset{O}{\|}}{C}\overset{\overset{O}{\|}}{C}CH_3 \rightarrow n\text{-}C_3H_7\overset{\overset{O}{\|}}{C}\overset{\overset{O}{\|}}{C}CH_3 + CH_3CHO \quad (7\text{-}124)$$
$$\qquad\qquad\qquad \textbf{91} \qquad\qquad\qquad \textbf{92}$$

radical adds to the dione in a reversible process yielding an adduct radical **93**. This radical can fragment in such a manner as to yield either the starting dione by β-elimination of the butyryl radical or the hexadione by β-elimination of an acetyl radical. In this case the addition also occurs with bonding at the more hindered carbon atom. Although addition with bonding at the oxygen would yield a resonance stabilized-carbonyl radical, polar contributions to the transition state of the addition reaction of the acyl radical with bonding at the carbon of the carbonyl apparently outweigh both the resonance and steric factors that favor bonding at oxygen.

$$n\text{-}C_3H_7\dot{C}O + \begin{matrix} CH_3 \\ | \\ C{=}O \\ | \\ CCH_3 \\ \diagup\diagdown \\ O \end{matrix} \rightarrow \left[\begin{matrix} CH_3 \\ | \\ n\text{-}C_3H_7\overset{\delta+}{C}\cdots\overset{\delta-}{C}\cdots O \\ \| \quad | \\ O \quad CCH_3 \\ \diagup\diagdown \\ O \end{matrix} \right] \rightarrow \begin{matrix} CH_3 \\ | \\ n\text{-}C_3H_7C{-}C{-}O \cdot \\ \| \quad | \\ O \quad CCH_3 \\ \diagup\diagdown \\ O \end{matrix} \quad (7\text{-}125)$$
$$\qquad\qquad \textbf{91} \qquad\qquad\qquad\qquad\qquad\qquad \textbf{93}$$

$$n\text{-}C_3H_7\overset{\overset{\displaystyle O}{\|}}{C}\overset{\overset{\displaystyle CH_3}{|}}{C}O \cdot \;\; \longrightarrow \;\; n\text{-}C_3H_7\overset{\overset{\displaystyle O}{\|}}{C}\!-\!\overset{\overset{\displaystyle CH_3}{|}}{C}\!=\!O + CH_3\dot{C}O \qquad (7\text{-}126)$$

with substituent $\underset{O}{\overset{\diagup}{C}}CH_3$ labeled **93** and **92** over the product.

II. Stereochemistry of Free-Radical Addition Reactions

In many free-radical addition reactions, a certain degree of stereoselectivity is observed that can be attributed to special steric or conformational features of the unsaturated compound that is undergoing addition. Most often it is not the result of stereospecificity inherent to the addition reaction itself. The most notable exception to this generalization is the free-radical addition of hydrogen bromide to alkenes. All evidence available indicates that this free-radical addition reaction is indeed stereospecific in that the elements of this reagent add *trans* across the unsaturated linkage. This section is concerned with both the stereoselective additions of certain reagents in order to show the influence of steric factors on the course of the transfer reaction and with the stereospecific addition reactions of hydrogen bromide. Conformational aspects and their effect on the stereochemistry of free-radical additions were described earlier in this chapter.

A. STEREOSELECTIVE ADDITIONS

Light-induced additions of bromotrichloromethane to *cis*-2-butene (**94**) and *trans*-2-butene (**95**) yield the same mixture of the *threo* and *erythro* addition

(7-127)

products (**96** and **97**, respectively).[36] The lack of stereospecificity in this addition reaction can be ascribed to reaction of the same adduct radical in the transfer step with bromotrichloromethane. The adduct radicals resulting from addition of the trichloromethyl radical to **94** and **95** apparently undergo conformational equilibration faster than they react in the chain-transfer reaction with bromo-trichloromethane.

In contrast to the reaction with the 2-butenes, the free-radical additions of bromotrichloromethane to both norbornene (**98**) and bicyclo[2.2.2]octene-2 (**99**) appear to be highly stereoselective in that the elements of the polyhalo-

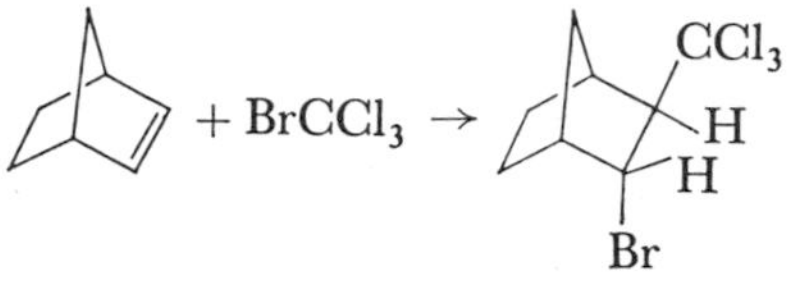

$$
\text{98} + \text{BrCCl}_3 \rightarrow \text{100} \tag{7-128}
$$

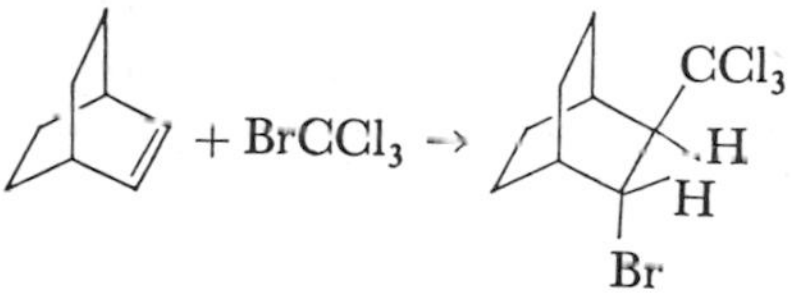

$$
\text{99} + \text{BrCCl}_3 \rightarrow \text{101} \tag{7-129}
$$

ᴍᴇᴛʜᴀɴᴇ add *trans* to the unsaturated linkage of these bicyclic alkenes.[37] A plausible explanation for the apparent stereospecific *trans* additions of bromo-trichloromethane to these bicyclic alkenes stems from a combination of two factors; the steric effect rendered by the trichloromethyl group and the methylene bridge (the ethylene bridge *endo* to the trichloromethyl group in the case of **99**) and the fact that rotation about the carbon-carbon sigma bond formed by the addition to the double bond is prohibited. The preferred entry of the trichloro-methyl group from the *exo*-side in the reaction with norbornene seems reasonable

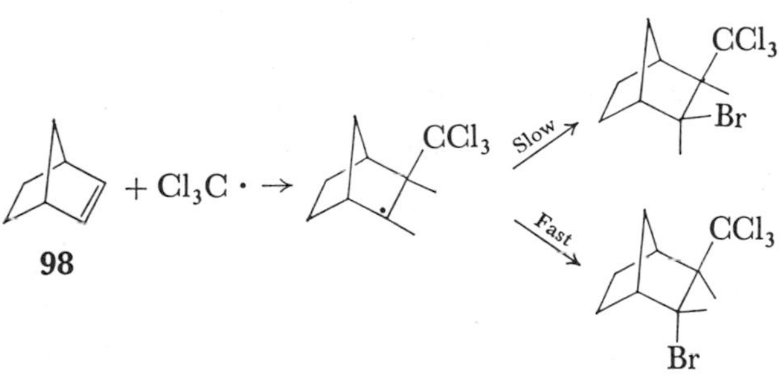

$$
\text{98} + \text{Cl}_3\text{C} \cdot \rightarrow \tag{7-130}
$$

$$
\tag{7-131}
$$

100

in that the methylene bridge exerts less steric hindrance to the addition than the ethylene bridge. The adduct radical formed from *exo*-addition to norbornene is sterically hindered on the *exo*-side due to the combination of both the methylene bridge and trichloromethyl group, and transfer on the adding reagent occurs on the less hindered *endo*-side. In view of the results obtained in the reactions of the 2-butenes, the reaction should be considered to be one showing a high degree of stereoselectivity due to steric effects rather than a stereospecific addition.

Interestingly, free-radical additions of other reagents to norbornene and its derivatives appear to be *cis*-additions. Reaction of ethyl bromoacetate (**102**) to norbornene yields *exo-cis*-2-carboethoxymethyl-3-bromonorbornane (**103**).[39]

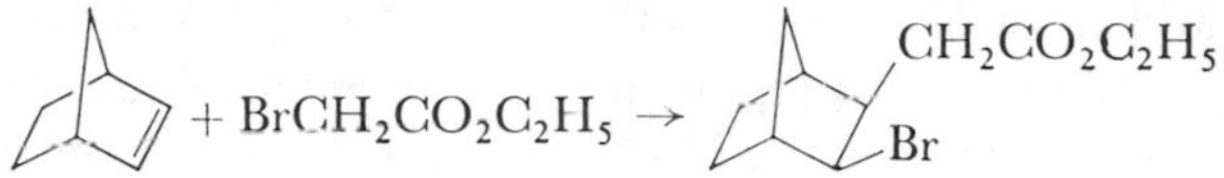
+ BrCH2CO2C2H5 →
CH2CO2C2H5
Br

(7-132)

98 **102** **103**

via

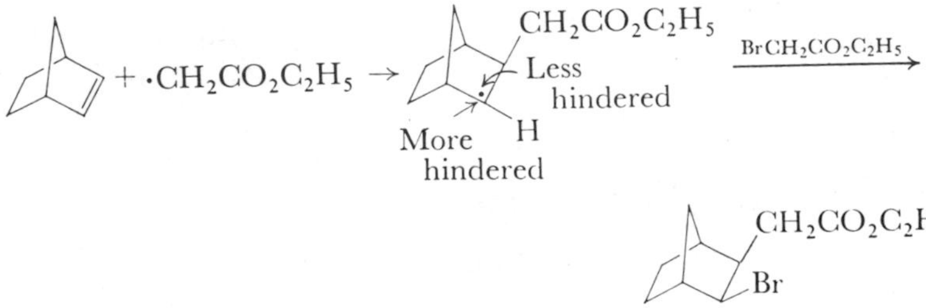
+ ·CH2CO2C2H5 →
CH2CO2C2H5
Less
hindered
More
hindered
H
BrCH2CO2C2H5
CH2CO2C2H5
Br

(7-133)

103

In this case the steric effect of the ethylene bridge in the adduct radical likely outweighs the combined steric effects of the methylene bridge and the carbo-ethoxymethylene group. Similarly, addition of *p*-thiocresol to the norbornene

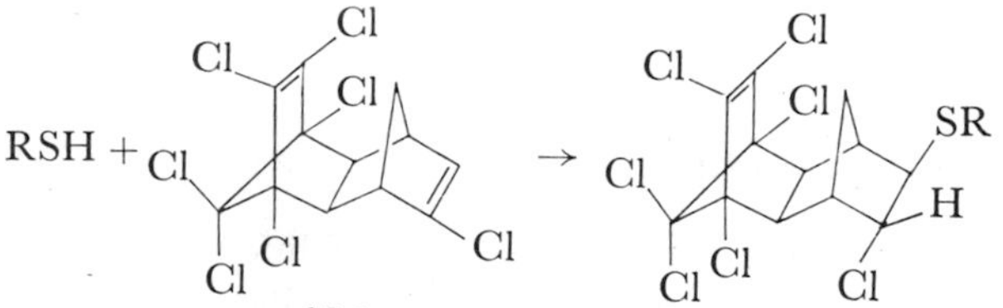
RSH +
Cl
Cl
Cl
Cl
Cl
Cl
→
Cl
Cl
Cl
Cl
Cl
Cl
SR
H

(7-134)

104

$R = p\text{-}CH_3C_6H_4$

derivative aldrin (**104**) results in exclusive formation of the product of an *exo*-attack of the adding thiyl radical but overall *cis*-addition of the *p*-thiocresol.[40]

Free-radical addition of other sulfur-containing compounds to cyclohexene and cyclopentene derivatives appear to yield predominantly (but not exclusively) the product of a *trans*-addition. Additions of thiolacetic acid, hydrogen sulfide, and thiophenol to 1-chlorocyclohexene (**105**) produce *cis*-products resulting

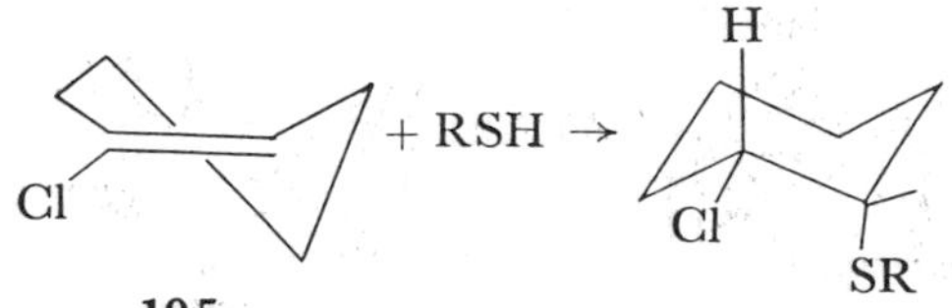

$$+ RSH \rightarrow \qquad (7\text{-}135)$$

105

from *trans*-addition in yields of 66 to 73%, 74.8% and 94.2%, respectively.[41] Similarly, additions of thiolacetic acid to 1-methylcyclohexene and 1-methyl cyclopentene yield predominantly the products of *trans*-additions.[42] However, additions of this same reagent at −78°C to the acyclic *cis*- and *trans*-2-chloro-2-butenes (**106** and **107**, respectively) yield the same mixture of stereoisomers (90% threo and 10% erythro).[43]

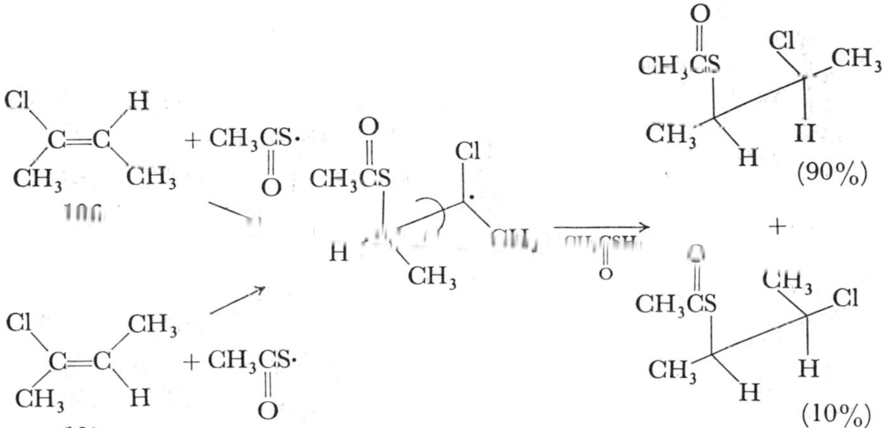

$$(7\text{-}136)$$

B. STEREOSPECIFIC ADDITIONS

In contrast to the above reactions, free-radical additions of hydrogen bromide to unsaturated linkages appear to be truly stereospecific. Additions of hydrogen bromide to 1-methylcyclohexene (**108**) and 1-bromocyclohexene (**109**) yielded almost exclusively the *cis*-addition products by a stereospecific *trans*-addition of the hydrogen halide to the double bond.[44] Stereospecific free-radical additions of hydrogen bromide have been observed in the reactions of hydrogen bromide with 1-chlorocyclohexene (**105**)[45] and 1,2-dimethylcyclohexene (**110**).[46]

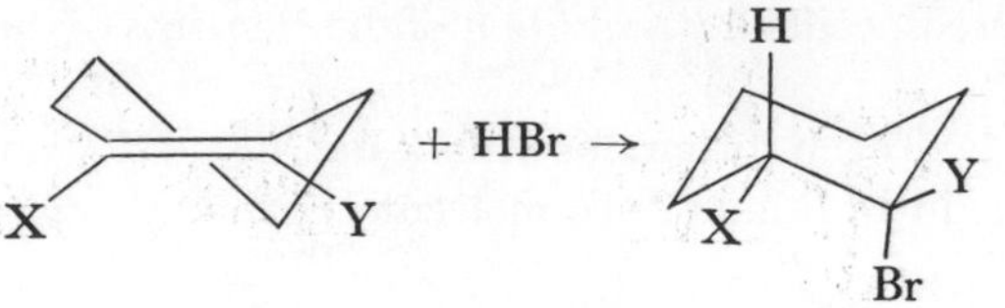

$$(7\text{-}137)$$

(**108**, X = Me, Y = H)
(**109**, X = Br, Y = H)
(**105**, X = Cl, Y = H)
(**110**, X = Y = CH$_3$)

Reactions of hydrogen bromide with both *cis*- and *trans*-2-bromo-2-butene (**111** and **112** respectively) at −80°C gave exclusively the products of a *trans*-addition.[47] Thus, from the *cis*-isomer **111**, only *meso*-2,3-dibromobutane (**113**) was obtained and only the dl-dibromide **114** was formed in the reaction of the

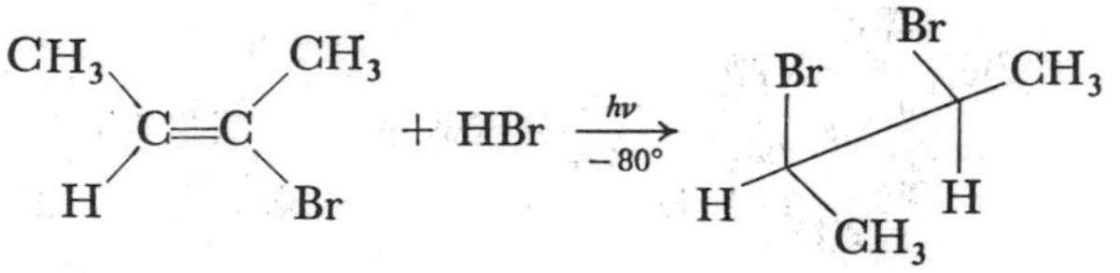

$$(7\text{-}138)$$

111   **113**

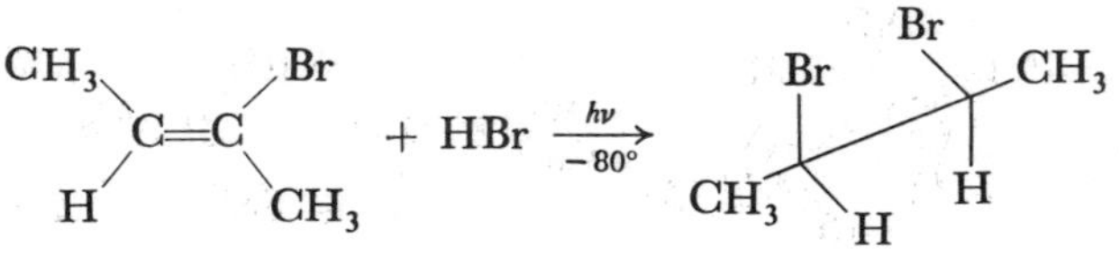

$$(7\text{-}139)$$

112   **114**

trans-isomer **112**. Reactions of *cis* and *trans* 2-butenes with deuterium bromide yield exclusively the *threo*- and *erythro* products (**115** and **116**), respectively, the

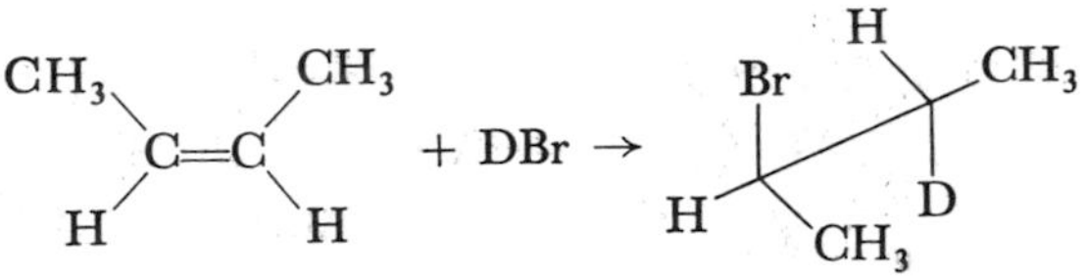

$$(7\text{-}140)$$

115

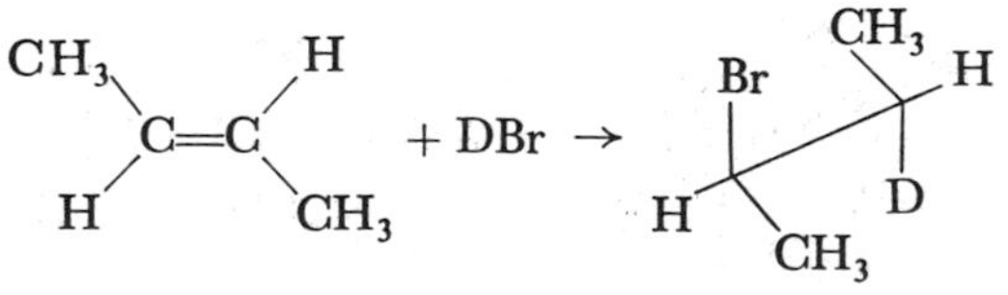

$$(7\text{-}141)$$

116

products of a stereospecific *trans*-addition of the elements of the adding reagent to the alkenes.[48]

Reaction of hydrogen bromide with 2-butyne (**117**) under free-radical conditions results in addition of two molecules of the hydrogen halide yielding

$$CH_3C{\equiv}CCH_3 + 2HBr \rightarrow \qquad\qquad (7\text{-}142)$$

117   **114**

only dl-2,3-dibromobutane (**114**).[49] This product could result either from two successive *cis*-additions or from two successive *trans*-additions of the hydrogen halide. In view of the *trans*-addition observed in the reaction of *trans*-2-bromo-2-butene (**112**) yielding the dl-2,3-dibromobutane, two successive *trans*-additions appear more likely. This would necessitate a *trans*-addition of hydrogen bromide to the acetylene linkage yielding **112**. Support for a stereospecific

$$CH_3C{\equiv}CCH_3 \xrightarrow{\text{HBr}} \qquad \xrightarrow{\text{HBr}} \qquad\qquad (7\text{-}143)$$

117   **112**   **114**

trans-addition of hydrogen bromide to acetylenes comes from the observation that reaction of hydrogen bromide with propyne (**118**) in the liquid phase yields exclusively *cis*-1-bromopropene (**119**).

$$CH_3C{\equiv}CH + HBr \rightarrow \qquad\qquad (7\text{-}144)$$

118   **119**

The stereospecificity observed in these free-radical additions of hydrogen bromide to unsaturated linkages can be ascribed to a combination of two factors. One of these is the very high reactivity of the hydrogen halide toward displacement reactions with free radicals. The other is the ability of bromine atoms to react with the unsaturated linkage forming either a π-complex that has an appreciable degree of stability or a bridged radical (see Chapter 5). The course of the stereospecific addition of hydrogen bromide to an unsaturated alkene linkage may be that shown in 7-145. Interaction of the bromine atom

$$Br\cdot + \underset{T}{\overset{R}{\diagup}}C=C\underset{U}{\overset{S}{\diagdown}} \rightarrow \underset{T}{\overset{R}{\diagdown}}C\overset{Br}{=\!\!=}C\overset{S}{\diagup}_{U} \rightleftharpoons \underset{T}{\overset{R}{\diagdown}}C\overset{Br}{\diagup\diagdown}C\overset{S}{\diagup}_{U}$$

$$\mathbf{120} \qquad\qquad \mathbf{121}$$

$$\text{(7-145)}$$

$$\mathbf{122}$$

$$Br\cdot + \underset{T}{\overset{R}{\diagup}}C=C\underset{S}{\overset{U}{\diagdown}} \rightleftharpoons \underset{T}{\overset{R}{\diagdown}}C\overset{Br}{=\!\!=}C\overset{U}{\diagup}_{S} \rightleftharpoons \underset{T}{\overset{R}{\diagdown}}C\overset{Br}{\diagup\diagdown}C\overset{U}{\diagup}_{S}$$

$$\mathbf{120} \qquad\qquad \mathbf{121}$$

with the π-electron system yields the π-complex **120**, a species that would still retain the stereochemistry of the alkene linkage. The π-complex **120** may react with the hydrogen bromide from the opposite side of the complexed bromine atom, a reaction resulting in *trans*-addition of the elements of hydrogen bromide to the double bond. It is also possible that the π-complex may be converted to the bridge-species **121** which, on reaction with hydrogen bromide, would also yield a *trans*-addition product. The reaction of either **120** or **121** with hydrogen bromide must be considerably faster than the conversion of the bridged radical **121** to the classical adduct radical **122** or the stereochemistry of the original alkene would be lost due to the free rotation about the carbon-carbon sigma bond.

The necessity of a very reactive transfer agent present in appreciable concentration to allow rapid displacement leading to stereospecific addition is evident by the observation that the classical radical apparently is formed in other reactions of alkenes with bromine atoms. *Cis-trans* isomerizations can be accomplished by molecular bromine in small concentrations by a reversible addition of a bromine atom to the alkene linkage indicating the existence of the classical adduct radical. Similar isomerizations are noted in the allylic brominations of alkenes with N-bromosuccinimide, a reaction that likely involves hydrogen bromide as an intermediate in such small amounts that its reaction with the classical radical is precluded.

Although formation of either a π-complex of the adding radical (e.g., $Cl_3C\cdot$, $RS\cdot$) with the unsaturated linkage or a bridged radical (or both) cannot be excluded, the reactions of such species with the adding reagent ($BrCCl_3$ or

RSH) are slower than conversion to the classical adduct radical. This is evidenced both by the isomerization of the alkene as well as the lack of stereospecificity in the addition reactions of these adding reagents.

Addition of deuteromethyl mercaptan (CH_3SD) to cis- and trans-2-butene yielded the same mixture of the threo and erythro isomers (**125** and **126**, respectively). However, in the presence of deuterium bromide, stereospecific trans-addition was observed, an observation that suggests the existence of either the π-complexed thiyl radical **123** or the bridged radical **124** which reacts rapidly enough with the deuterium bromide to yield the stereospecific addition products.[50]

(7-146)

The stereospecific additions of hydrogen bromide to cyclohexene derivatives having remotely positioned substituents can also be explained in terms of reaction of either a π-complexed radical or a bridged species. Addition of hydrogen bromide to 3-bromocyclohexene (**127**), yielded trans-1,3-dibromocyclohexane (**128**)[51] and likely involves formation of the π-complex (or bridged species) by approach of the bromine atom to the alkene linkage from the side trans to the bromine bonded to the 3-carbon for steric reasons. The adduct radical species would react to form either the *trans*-1,2-dibromocyclohexane (**129**) or the *trans*-1,3-dibromocyclohexane (**128**) depending on whether the attack occurs on the 3-carbon or the 2-carbon, respectively. Attack at the 3-carbon leading to the 1,2-dibromocyclohexane would yield an initial product having a twist-boat conformation which would be converted to **129**. Attack at the 2-carbon yields the observed product in the more stable chair conformation.

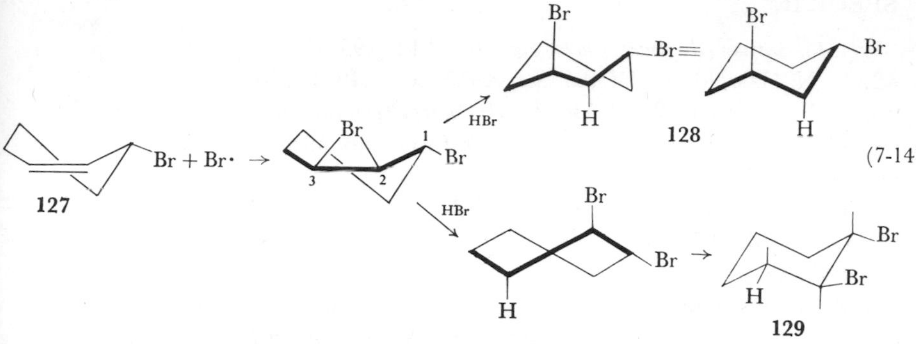

(7-147)

The stereospecific additions of hydrogen bromide to 4-*t*-butyl cyclohexene (**130**) yielded only the *trans*-3-bromo-*t*-butyl cyclohexane (**133**) and *cis*-4-bromo *t*-butyl cyclohexane (**134**), both products having the bromine in the axial position. The *t*-butyl group of **130** is too remotely positioned from the unsaturated linkage to exercise any steric control over the formation of the initial cyclic intermediate and, consequently, both **131** and **132** are formed and react to yield **133** and **134**, respectively.[52]

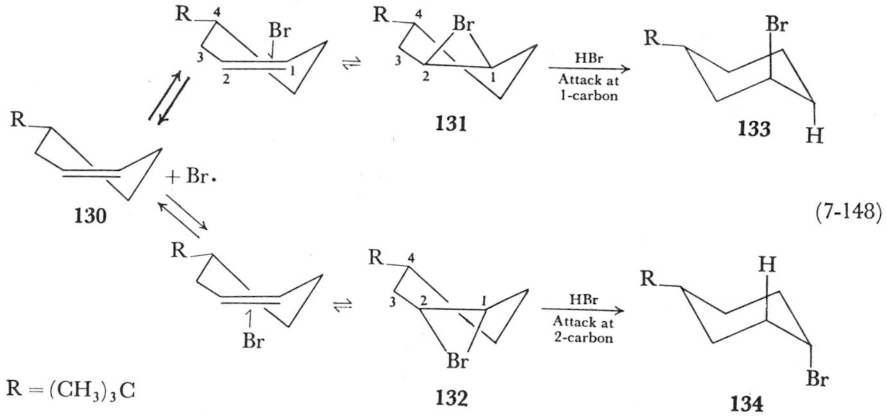

$R = (CH_3)_3C$

(7-148)

References

GENERAL

B. A. Bohm and P. I. Abell, "Stereochemistry of Free-Radical Additions to Olefins," *Chem. Rev.*, **62**, 599 (1962).

SPECIFIC

1. J. C. Saurer, *J. Amer. Chem. Soc.*, **79**, 5314 (1957).
2. R. N. Haszeldine, *J. Chem. Soc.*, **1950**, 3037; **1951**, 588.
3. E. I. Heiba and R. M. Dessau, *J. Amer. Chem. Soc.*, **89**, 3772 (1967).
4. R. E. Foster, A. W. Larchar, R. D. Lipscomb, and B. C. McKusick, *ibid.*, **78**, 5606 (1956).
5. K. Griesbaum, A. A. Oswald, E. R. Quiram, and W. Naegele, *J. Org. Chem.*, **28**, 1952 (1963); K. Griesbaum, A. A. Oswald, and D. N. Hall, *ibid.*, **29**, 2404 (1965).
6. R. N. Haszeldine, K. Leedham, and R. B. Steele, *J. Chem. Soc.*, **1954**, 2020.
7. A. A. Oswald and K. Griesbaum, *Org. Sulfur Compounds*, N. Kharasch, Ed. Vol. 1, Pergamon, Oxford (1961).
8. M. S. Kharasch, E. V. Jensen, and W. H. Urry, *J. Amer. Chem. Soc.*, **69**, 1100 (1947).
9. C. S. Marvel and R. R. Chambers, *J. Amer. Chem. Soc.*, **70**, 993 (1948).
10. M. Julia and M. Maumy, *Bull. Soc. Chim. France*, 434 (1966).
11. C. J. Bredeweg and R. A. Hickner, *Am. Chem. Soc. Meeting Abstr.*, **155** Paper P-80 (1968).
12. N. O. Brace, *J. Amer. Chem. Soc.*, **86**, 523 (1964).
13. R. C. Lamb, P. W. Agers, and M. K. Toney, *ibid.*, **85**, 3483 (1963).
14. R. G. Garwood, C. J. Scott, and B. C. L. Weedon, *Chem. Commun.*, **1**, 14 (1965).
15. C. Walling and M. S. Pearson, *J. Amer. Chem. Soc.*, **86**, 2262 (1964); C. Walling, J. H. Cooley, A. A. Ponaras, and E. J. Racah, *ibid.*, **88**, 5361 (1966).
16. J. I. Cadogen, D. H. Hey, and A. O. S. Hock, *Chem. and Ind.*, **1964**, 753.
17. S. J. Cristol, G. D. Brindell, and J. A. Reeder, *J. Amer. Chem. Soc.*, **80**, 635 (1958).
18. E. S. Huyser and G. Echegaray, *J. Org. Chem.*, **27**, 429 (1962).
19. S. J. Cristol, T. W. Russell, and D. I. Davies, *J. Org. Chem.*, **30**, 207 (1965).
20. R. Dowbenko, *J. Amer. Chem. Soc.*, **86**, 946 (1964).
21. J. M. Locke and E. W. Duck, *Chem. Commun.*, **1**, 151 (1965).
22. J. G. Traynham, T. M. Couvillon, and N. S. Bhacca, *J. Org. Chem.*, **32**, 529 (1967).

23. J. G. Traynham and T. M. Couvillon, *J. Amer. Chem. Soc.*, **87**, 5806 (1965); **89**, 3205 (1967).

24. M. Fisch and G. Ourisson, *Chem. Commun.*, **1**, 407 (1965).

25. F. G. Bordwell, P. S. Landis, and G. S. Whitney, *J. Org. Chem.*, **30**, 3764 (1965).

26. N. A. LeBel and A. DeBoer, *J. Amer. Chem. Soc.*, **84**, 2784 (1967).

27. E. S. Huyser and J. R. Jeffrey, *Tetrahedron*, **21**, 3083 (1965).

28. E. S. Huyser, H. Benson, and H. J. Sinnege, *J. Org. Chem.*, **32**, 622 (1967).

29. G. H. Williams, *Homolytic Aromatic Substitution*, Pergamon Press, London (1960).

30. C. E. Lane and W. A. Noyes, Jr., *J. Amer. Chem. Soc.*, **54**, 161 (1932); W. A. Noyes, Jr., *ibid.*, **55**, 4444 (1933).

31. E. S. Huyser and E. Bedard, *J. Org. Chem.*, **29**, 1588 (1964).

32. F. C. Kooyman and E. Farenhorst, *Trans. Faraday Soc.*, **49**, 58 (1953).

33. F. F. Rust, F. H. Seubold, and W. E. Vaughan, *J. Amer. Chem. Soc.*, **70**, 3258 (1948).

34. G. Fuller and F. F. Rust, *ibid.*, **80**, 6148 (1958).

35. W. H. Urry, M. H. Pai, and C. Y. Chen, *ibid.*, **86**, 5342 (1964).

36. P. S. Skell and R. C. Woodworth, *J. Amer. Chem. Soc.*, **77**, 4638 (1955).

37. M. S. Kharasch and H. N. Friedlander, *J. Org. Chem.*, **14**, 239 (1949).

38. F. S. Fawcett, *Chem. Rev.*, **47**, 219 (1950).

39. J. Weinstock, Abstracts of Papers, 128th American Chemical Society Meeting, (1955) p. 19-0.

40. S. J. Cristol, R. P. Arganbright, G. D. Brindell, and R. M. Heitz, *J. Amer. Chem. Soc.*, **179**, 6039 (1957).

41. H. L. Goering, D. I. Relyea, and D. W. Larsen, *ibid.*, **78**, 348 (1946).

42. F. G. Bordwell and W. A. Hewett, *ibid.*, **79**, 3493 (1957).

43. N. P. Neureiter and F. G. Bordwell, *ibid.*, **82**, 5354 (1960).

44. H. L. Goering, P. I. Abell, and B. F. Aycock, *ibid.*, **74**, 3588 (1952).

45. H. L. Goering and L. L. Sims, *ibid.*, **77**, 3465 (1955).

46. B. M. Vittemberga, PhD Thesis, University of Rhode Island, 1954.

47. H. L. Goering and D. W. Larsen, *J. Amer. Chem. Soc.*, **81**, 5937 (1959).

48. P. S. Skell and R. G. Allen, *ibid.*, **81**, 5583 (1959).

49. C. Walling, M. S. Kharasch, and F. R. Mayo, *J. Amer. Chem. Soc.*, **61**, 1711 (1939).

50. P. S. Skell and R. G. Allen, *ibid.*, **82**, 1511 (1961).

51. M. S. Kharasch, J. S. Sallo, and W. Nudenberg, *J. Org. Chem.*, **21**, 129 (1956).

52. E. S. Huyser, H. Benson, and W. T. Short, unpublished results.

Chapter 8

Elimination Reactions

I. Introduction

Reactions that involve a β-elimination as a key chain-propagating step are examined in this chapter. In some cases the reversible addition of a radical to an unsaturated linkage will be the reaction observed, as in the case of additions

$$\text{Br} \cdot + \underset{/}{\overset{\backslash}{C}} = \underset{\backslash}{\overset{/}{C}} \; \rightleftarrows \; \text{BrC} - \underset{\backslash}{\overset{/}{C}} \cdot \qquad (8\text{-}1)$$

$$\text{RS} \cdot + \underset{/}{\overset{\backslash}{C}} = \underset{\backslash}{\overset{/}{C}} \; \rightleftharpoons \; \text{RSC} - \underset{\backslash}{\overset{/}{C}} \cdot \qquad (8\text{-}2)$$

of bromine atoms and thiyl radicals to alkenes. In other reactions a free radical is generated that fragments readily but may not be prone to undergo the reverse addition reaction as readily. Examples of such reactions are the β-eliminations of alkyl radicals from alkoxy (**1**) and from α-alkoxyalkyl radicals (**2**).

$$\underset{\mathbf{1}}{R_3CO} \cdot \; \underset{\leftarrow}{\longrightarrow} \; R_2C{=}O + R \cdot \qquad (8\text{-}3)$$

$$\underset{\mathbf{2}}{R_2'\dot{C}OR} \; \underset{\leftarrow}{\longrightarrow} \; R_2'C{=}O + R \cdot \qquad (8\text{-}4)$$

The addition-elimination process has the earmarks of an equilibrium reaction. In some instances equilibrium can be attained. In many reactions, however, other chain-propagating reactions of either the adding radical or the adduct radical compete with the addition or elimination reactions, respectively, and equilibrium is never attained. It is significant to point out that temperature and entropy factors are important in the addition-elimination process. Lower temperatures favor the addition reaction, generally an exothermic process involving a significant decrease in entropy. The β-elimination reaction is an endothermic process involving an increase in entropy, and is favored by higher temperatures. Some of the reactions in this chapter involve the β-elimination reaction of a linear radical which results in formation of a second species adding three additional translational degrees of freedom to the system and thereby increasing the entropy significantly. Fragmentations of cyclic radicals, on the other hand, do not result in any additional translational degrees of freedom but several rotational degrees of freedom result. In either case, β-elimination results in an increase in the entropy and should be favored by higher temperatures.

II. Chemistry of β-Elimination Reactions

A. REVERSIBLE ADDITION OF HALOGEN ATOMS

Addition of a bromine atom to an unsaturated carbon–carbon linkage is reversible except at very low temperatures. The reversibility of this reaction has some very interesting effects on many reactions involving bromine atoms. For example, a small amount of bromine causes the isomerization of dimethyl maleate (3) to dimethyl fumarate (5) in a reaction induced by light.[1] Evidence of the free-radical chain character of this reaction comes from the observation that the quantum yield of the reaction (the number of molecules that isomerize per quantum of light) is about 600. The isomerization proceeds by the addition of a bromine atom to 3 yielding the adduct radical 4. The β-elimination of the bromine atom from 4 can yield either 3 or 5. Since 5 can also undergo rapid addition of the bromine atom to yield 4, equilibration of the two unsaturated esters eventually results.

$$3 \qquad\qquad 4 \qquad\qquad 5 \qquad (8\text{-}5)$$

The isomerization of cis-1,2-dibromoethylene (6) using radioactive bromine lends convincing support to the addition-elimination mechanism in that radioactive bromine is found in both the trans-1,2-dibromoethylene (7) and in the unrearranged material. The adduct radical can eliminate either a labelled or a nonlabelled bromine yielding either **6** or **7**. Small amounts of iodine also catalyze the isomerization of various unsaturated compounds from the *cis-* to

(8-6)

the *trans*-isomer. These isomerizations must be performed with small concentrations of the molecular halogen in order to suppress the reaction of the adduct radical with the halogen molecule producing the addition product.[2]

Formation of radicals having a halogen atom bonded to a β-carbon relative to the radical site often results in eliminations of the halogen atom. An excellent example of such a reaction is the addition of bromotrichloromethane to allyl bromide (**8**). This free-radical addition reaction does not yield the expected product but rather a mixture of products which can be explained in terms of the β-elimination of a bromine atom from the adduct radical (**9**) yielding 4,4,4-trichloro-1-butene (**10**). Adding bromotrichloromethane to **10** yields 1,1,1,5,5,5-hexachloro-3-bromopentane (**11**). The eliminated bromine atom can react with another molecule of allyl bromide yielding the adduct radical **12** which, although very likely undergoes elimination (see below), does react with bromotrichloromethane yielding 1,2,3-tribromopropane (**13**).[3] β-Elimina-

$$Cl_3C \cdot + CH_2{=}CHCH_2Br \rightarrow Cl_3CCH_2\dot{C}HCH_2Br \qquad (8\text{-}7)$$
$$\qquad\quad \mathbf{8} \qquad\qquad\qquad\qquad\quad \mathbf{9}$$

$$Cl_3CCH_2\dot{C}HCH_2Br \rightarrow Cl_3CCH_2CH{=}CH_2 + Br \cdot \qquad (8\text{-}8)$$
$$\qquad\quad \mathbf{9} \qquad\qquad\qquad\qquad \mathbf{10}$$

$$Cl_3C \cdot + Cl_3CCH_2CH{=}CH_2 \rightarrow Cl_3CCH_2\dot{C}HCH_2CCl_3 \qquad (8\text{-}9)$$
$$\qquad\qquad \mathbf{10}$$

$$Cl_3CCH_2\dot{C}HCH_2CCl_3 + BrCCl_3 \rightarrow Cl_3CCH_2CHBrCH_2CCl_3 + Cl_3C \cdot \;(8\text{-}10)$$
$$\qquad\qquad\qquad\qquad\qquad\qquad\qquad \mathbf{11}$$

$$\text{Br} \cdot + \text{CH}_2\!\!=\!\!\text{CHCH}_2\text{Br} \;\rightleftarrows\; \text{BrCH}_2\dot{\text{C}}\text{HCH}_2\text{Br} \tag{8-11}$$

8   12

$$\text{BrCH}_2\dot{\text{C}}\text{HCH}_2\text{Br} + \text{BrCCl}_3 \;\rightarrow\; \text{BrCH}_2\text{CHBrCH}_2\text{Br} + \text{Cl}_3\text{C} \cdot \tag{8-12}$$

13

tion of a halogen atom from the radical resulting from addition of a halogen atom to an allyl halide was demonstrated by the incorporation of radioactive iodine in the reaction of allyl iodide with radioactive iodine.[4]

$$\text{I*} \cdot + \text{CH}_2\!\!=\!\!\text{CHCH}_2\text{I} \;\rightleftarrows\; \text{I*CH}_2\dot{\text{C}}\text{HCH}_2\text{I} \;\rightleftharpoons\; \text{I*CH}_2\text{CH}\!\!=\!\!\text{CH}_2 + \text{I*} \cdot \tag{8-13}$$

The equilibration of crotyl bromide (**14**) and methallyl bromide (**15**) by hydrogen bromide at temperatures as low as $-12°C$ involves a similar mechanism.[5] In this case, the same adduct radical is produced and elimination can occur to give either **14** or **15**. A similar mechanism can be used to explain the

$$\text{Br} \cdot + \text{CH}_3\text{CH}\!\!=\!\!\text{CHCH}_2\text{Br} \;\rightleftharpoons\; \text{CH}_3\text{CHBr}\dot{\text{C}}\text{HCH}_2\text{Br} \;\rightleftarrows\;$$

14

$$\text{CH}_3\text{CHBrCH}\!\!=\!\!\text{CH}_2 + \text{Br} \cdot \tag{8-14}$$

15

isomerization of α-bromoacetoacetic esters (**16**) to γ-bromoacetoacetic esters (**17**) in the presence of small amounts of hydrogen bromide and peroxides.[6]

$$\underset{\textbf{16}}{\overset{\overset{\text{O}}{\overset{\|}{}}}{\text{CH}_3\text{CCHBrCO}_2\text{R}}} \;\xrightarrow[\text{[O]}]{\text{HBr}}\; \underset{\textbf{17}}{\overset{\overset{\text{O}}{\overset{\|}{}}}{\text{BrCH}_2\text{CCH}_2\text{CO}_2\text{R}}} \tag{8-15}$$

via

$$\text{Br} \cdot + \underset{\textbf{16}}{\overset{\overset{\text{OH}}{\overset{|}{}}}{\text{CH}_2\!\!=\!\!\text{CCHBrCO}_2\text{R}}} \;\rightleftharpoons\; \underset{\textbf{18}}{\overset{\overset{\text{OH}}{\overset{|}{}}}{\text{BrCH}_2\dot{\text{C}}\text{CHBrCO}_2\text{R}}} \tag{8-16}$$

$$\overset{\overset{\text{OH}}{\overset{|}{}}}{\text{BrCH}_2\dot{\text{C}}\text{CHBrCO}_2\text{R}} \;\rightleftharpoons\; \overset{\overset{\text{OH}}{\overset{|}{}}}{\text{BrCH}_2\text{C}\!\!=\!\!\text{CHCO}_2\text{R}} + \text{Br} \cdot \tag{8-17}$$

$$\downarrow$$

$$\underset{\textbf{17}}{\overset{\overset{\text{O}}{\overset{\|}{}}}{\text{BrCH}_2\text{CCH}_2\text{CO}_2\text{R}}}$$

The β-chloroalkyl radical formed by addition of an alkyl radical to 1,2-dichloroethylene (**18**) undergoes β-elimination of a chlorine atom. The reactivity of the eliminated chlorine atom as a hydrogen abstractor allows for a reaction in which a hydrocarbon, for example, cyclopentane, is condensed

with the 1,2-dichloroethylene in a chain reaction involving addition, elimination, and abstraction as chain-propagating steps.[7] Similar condensation

$$\text{Cl}\cdot\ +\ \text{[cyclopentane with H, H]} \rightarrow \text{HCl} + \text{[cyclopentyl radical with H]} \qquad (8\text{-}18)$$

$$\text{[cyclopentyl radical with H]} + \text{ClCH}=\text{CHCl} \rightarrow \text{[cyclopentane with H, CHClĊHCl]} \qquad (8\text{-}19)$$

18

$$\text{[cyclopentane with H, CHClĊHCl]} \rightarrow \text{[cyclopentane with H, CH=CHCl]} + \text{Cl}\cdot \qquad (8\text{-}20)$$

reactions occur with trichloroethylene (**19**) and tetrachloroethylene (**20**).

$$\text{RH} + \text{ClCH}=\text{CCl}_2 \rightarrow \text{RCH}=\text{CCl}_2 + \text{HCl} \qquad (8\text{-}21)$$
19

$$\text{RH} + \text{Cl}_2\text{C}=\text{CCl}_2 \rightarrow \text{RCCl}=\text{CCl}_2 + \text{HCl} \qquad (8\text{-}22)$$
20

Methallyl chloride (**21**) undergoes a photochemical dimerization yielding
2-methyl-4,4-bischloromethyl-1-pentene (**22**) in a reaction that likely proceeds
by a chain reaction in which there is β-elimination of a chlorine atom from the
adduct radical.[8]

$$\text{Cl}\cdot\ +\ \underset{\underset{\text{CH}_3}{|}}{\text{CH}_2}=\text{CCH}_2\text{Cl} \rightarrow \underset{\underset{\text{CH}_3}{|}}{\text{ClCH}_2\dot{\text{C}}\text{CH}_2\text{Cl}} \qquad (8\text{-}23)$$

21

$$\underset{\underset{\text{CH}_2\text{Cl}}{|}}{\overset{\overset{\text{CH}_3}{|}}{\text{ClCH}_2\text{C}\cdot}}\ +\ \underset{\underset{\text{CH}_3}{|}}{\text{CH}_2}=\text{CCH}_2\text{Cl} \rightarrow \underset{\underset{\text{CH}_2\text{Cl}}{|}}{\overset{\overset{\text{CH}_3}{|}}{\text{ClCH}_2\text{C}}}\text{CH}_2\underset{\underset{\text{CH}_3}{|}}{\dot{\text{C}}}\text{CH}_2\text{Cl} \qquad (8\text{-}24)$$

$$\underset{\underset{\text{CH}_2\text{Cl}}{|}}{\overset{\overset{\text{CH}_3}{|}}{\text{ClCH}_2\text{C}}}\text{CH}_2\underset{\underset{\text{CH}_3}{|}}{\dot{\text{C}}}\text{CH}_2\text{Cl} \rightarrow \underset{\underset{\text{CH}_2\text{Cl}}{|}}{\overset{\overset{\text{CH}_3}{|}}{\text{ClCH}_2\text{C}}}\text{CH}_2\text{C}=\text{CH}_2 + \text{Cl}\cdot \qquad (8\text{-}25)$$

22

Alkyl halides decompose at elevated temperatures to the hydrogen halide and alkene in a mechanism that is the reverse of the anti-Markownikoff addition. Ethylene dichloride (23) is pyrolysed to HCl and vinyl chloride (24) at temperatures in the range of 350 to 500°C in a reaction having kinetic chain lengths for the sequence 8-26 and 8-27 approaching 10^5.[9]

$$Cl\cdot + ClCH_2CH_2Cl \rightarrow HCl + Cl\dot{C}HCH_2Cl \qquad (8\text{-}26)$$
$$23$$

$$Cl\dot{C}HCH_2Cl \rightarrow ClCH{=}CH_2 + Cl\cdot \qquad (8\text{-}27)$$
$$24$$

B. THIYL RADICAL ELIMINATIONS

Compelling evidence for reversibility of thiyl radical additions to unsaturated linkages is found in the isomerization of the *cis*- and *trans*-2-butenes by methyl mercaptan to an equilibrium mixture of the two. This isomerization, as in the case of those with the halogen atoms, most likely involves formation of a common adduct radical that can fragment to produce either alkene.[10]

$$(8\text{-}28)$$

The β-elimination of thiyl radical is evidenced in other reactions in which formation of species having a thiyl radical bonded to a β-carbon relative to the radical site is proposed as an intermediate. For example, *cis* and *trans* crotyl methyl sulfides (25 and 26, respectively) and α-methallyl methyl sulfide (27), undergo addition of a methane thiyl radical yielding a common adduct radical 28. The elimination of a methane thiyl radical results in formation of either 25, 26, or 27. An equilibrium mixture of the three butenyl methyl sulfides was formed when either 26 or 27 were illuminated in methyl mercaptan. In this isomerization both equilibration of the geometric isomers as well as the positional isomers was attained by formation of a common adduct radical.[11] Although reaction of 28 with mercaptan yielded the addition product 29, equilibration was complete by the time 30% of the butenyl methyl sulfide initially present was consumed in the addition reaction.

Another example of a β-elimination of a thiyl radical generated by some reaction other than addition of the thiyl radical to an unsaturated bond is encountered in the peroxide-induced decomposition of 3-ethanethio-2-methyl-propanal (30) to propylene, carbon monoxide, and ethyl mercaptan. The

$$CH_3S\cdot \; + \quad \begin{array}{c} CH_3 \\ \diagdown \\ C=C \\ \diagup \qquad \diagdown \\ H \qquad\qquad H \end{array} \begin{array}{c} CH_2SCH_3 \\ \diagup \end{array} \qquad\qquad \begin{array}{c} CH_3CHCH=CH_2 + CH_3S\cdot \\ | \\ SCH_3 \end{array}$$

25

27

$$\begin{array}{c} H \\ | \\ CH_3S \diagdown \diagup \cdot \\ \diagup \qquad \diagdown CH_2SCH_3 \\ CH_3 \quad H \end{array}$$

28

(8-29)

CH₃SH

$$CH_3S\cdot \; + \quad \begin{array}{c} CH_3 \\ \diagdown \\ C=C \\ \diagup \qquad \diagdown \\ H \qquad CH_2SCH_3 \end{array}$$

26

$$\begin{array}{c} CH_3CHCH_2CH_2SCH_3 \\ | \\ SCH_3 \end{array}$$

29

chain-sequence suggested for this reaction involves the formation of the β-alkanthioalkyl radical (**32**) by decarbonylation of the acyl radical (**31**) resulting from the abstraction of the aldehydic hydrogen by an eliminated thiyl radical.[12]

$$C_2H_5SCH_2CHCHO \xrightarrow{\text{Per.}} C_2H_5SH + CH_2{=}CHCH_3 + CO \quad (8\text{-}30)$$
$$| \atop CH_3$$

30

via
$$C_2H_5SCH_2CH\dot{C}O \rightarrow C_2H_5SCH_2\dot{C}HCH_3 + CO \qquad (8\text{-}31)$$
$$| \atop CH_3$$

31   **32**

$$C_2H_5SCH_2\dot{C}HCH_3 \rightarrow CH_2{=}CHCH_3 + C_2H_5S\cdot \qquad (8\text{-}32)$$

$$C_2H_5S\cdot + C_2H_5SCH_2CHCHO \rightarrow C_2H_5SH + C_2H_5SCH_2CH\dot{C}O \quad (8\text{-}33)$$
$$| \qquad\qquad\qquad\qquad\qquad\qquad | \atop CH_3 \qquad\qquad\qquad\qquad\qquad\qquad CH_3$$

Peroxide-induced reactions of β-hydroxysulfides yield mercaptans and carbonyl compounds in a free-radical chain reaction in which the β-alkanthio-alkyl radical is formed by hydrogen abstraction by an eliminated thiyl radical. 1-Methanethio-2-propanol (**33**) is converted to methyl mercaptan and acetone in a t-butyl peroxide-induced reaction.[13]

$$CH_3CHOHCH_2SCH_3 + CH_3S\cdot \rightarrow CH_3\dot{C}OHCH_2SCH_3 + CH_3SH \quad (8\text{-}34)$$
33

$$CH_3\dot{C}OHCH_2SCH_3 \rightarrow CH_3COH{=}CH_2 + CH_3S\cdot \quad \text{(8-35)}$$
$$\downarrow$$
$$CH_3\overset{}{C}OCH_3$$

The reversibility of the addition of thiyl radicals to unsaturated linkages might well be expected to have some interesting effects on the addition reactions of mercaptans to alkenes. One such effect has already been mentioned, namely equilibration of geometric isomers of alkenes as well as certain positional isomeric compounds. It becomes virtually impossible to observe the reaction products of the addition reaction of the mercaptan to a single geometric or positional isomer of such unsaturated compounds.

The addition-elimination reaction of thiyl radicals becomes important in other aspects of mercaptan additions to unsaturated compounds. Photo-chemically induced additions of methyl mercaptan to isobutylene, propylene, and ethylene in the gas phase show negative overall activation energies meaning that the rate of the reaction *decreases* as the temperature of the reaction is

$$CH_3SH + CH_2{=}CR_2 \xrightarrow{h\nu} CH_3SCH_2CHR_2 \quad \text{(8-36)}$$

$$R = H \text{ or } CH_3$$

increased.[14] The rate of formation of the addition product is the rate of reaction

$$CH_3S\cdot + CH_2{=}CR_2 \rightleftarrows CH_3SCH_2\dot{C}R_2 \quad \text{(8-37)}$$

$$CH_3SCH_2\dot{C}R + CH_3SH \xrightarrow{k_{38}} CH_3SCH_2CHR_2 + CH_3S\cdot \quad \text{(8-38)}$$

of the adduct radical with methyl mercaptan and is determined by the reaction

$$\frac{d[CH_3SCH_2CHR_2]}{dt} = k_{38}[CH_3SCH_2\dot{C}R_2][CH_3SH] \quad \text{(8-39)}$$

rate constant k_{38}, which should be expected to increase with increasing tempera-ture and by the concentrations of the reactants. Although the methyl mercaptan concentration is temperature independent, the adduct radical concentration is not. Its steady-state concentration is determined not only by the rate of forma-tion by addition of thiyl radicals to the alkene and by its rate of reaction with the mercaptan but also by its rate of fragmentation. Apparently the rate constant of the fragmentation reaction increases with increasing temperature faster than do the rate constants for the other reactions. The steady-state concentration of the adduct radical therefore diminishes with increasing temperature and, consequently, so does the rate of formation of the addition product.

The rate of fragmentation of a β-alkanthioalkyl radical is not always faster than the transfer reaction of the adduct radical with the mercaptan. If there is relief of bond-angle strain in the addition of the thiyl radical to the alkene, the elimination process will be one in which angular strain must be introduced. An example of such a system is found in the reactions of norbornene (**34**) with

$$(8\text{-}40)$$

34

mercaptans.[15] The heat of hydrogenation of norbornene is -33.1 kcal/mole compared to -27.1 kcal/mole for cyclohexene indicating a considerable amount of strain introduced by the two sp^2 hybridized carbon atoms in the bicyclic alkene. Addition of a thiyl radical to the π-electron system converts one of these two sp^2-carbons to a sp^3-carbon relieving some of the bond-angle strain. Since elimination of the thiyl radical from the adduct radical would result in formation of the strained norbornene, it is not surprising that the fragmentation of the adduct radical is slow compared to reaction of the adduct radical with mercaptan yielding the addition product.

The prediction can be made that in competition reactions alkenes that yield adduct radicals that do not fragment readily react faster than those that yield adducts that do fragment. Furthermore, if the rate of fragmentation of an adduct radical formed from a strained alkene and its rate of reaction with mercaptan are comparable, then the rate of reaction of the strained alkene with mercaptan will become dependent on the mercaptan concentration. Thus, in competition reactions of norbornene and methylenenorbornane (**35**), which

$$(8\text{-}41)$$

35

yields an adduct radical with no more strain than the parent alkene, the former is not only more reactive but its reactivity relative to the latter is dependent on the mercaptan concentration. At $40°$ and a mole fraction of ethyl mercaptan of 0.87, norbornene is two times more reactive than methylenenorbornane but at a mole fraction of ethyl mercaptan of 0.24, norbornene is 6.7 times more reactive.[16] The derived equation for the relative reactivity ratio of norbornene with respect to methylenenorbornane is given in 8-42 (see Equation 3-173).

$$\frac{\log\,(34_o/34_f)}{\log\,(35_o/35_f)} = \bar{P} = \frac{k_{40}\,k_{-41}}{k_{41}\,k_{tr(41)}[\text{RSH}]} + \frac{k_{40}}{k_{41}} \qquad (8\text{-}42)$$

This equation predicts that the relative reactivity of **34** with respect to **35** is dependent on the concentration of mercaptan and that $\bar{P}$, the measured relative reactivity ratio, should increase with decreasing mercaptan concentration.

C. ELIMINATION REACTIONS OF ALKYL RADICALS

Additions of alkyl and alkyl-derived radicals to alkenes generally proceed readily and fragmentation of the adduct radical is not observed at lower temperatures (up to 200°C). Isomerization of cis-2-butene, for example, does not occur in reaction of this alkene with trichloromethyl radicals.[16] High temperatures are required to reverse the direction of the equilibrium which is normally on the adduct radical side. In some situations the addition process may yield a strained adduct radical and β-elimination relieves the strain allowing the reaction to proceed at lower temperatures. Most examples cited here are of this type.

The opening of a strained ring system by a β-elimination reaction occurs in the free-radical addition of carbon tetrachloride to β-pinene (**36**)[17, 18], which

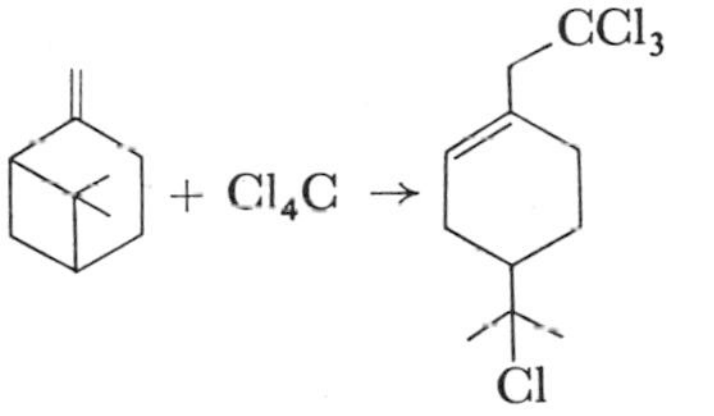

$$(8\text{-}43)$$

36   **37**

yields the cyclohexene derivative (**37**) as the major addition product. The relief

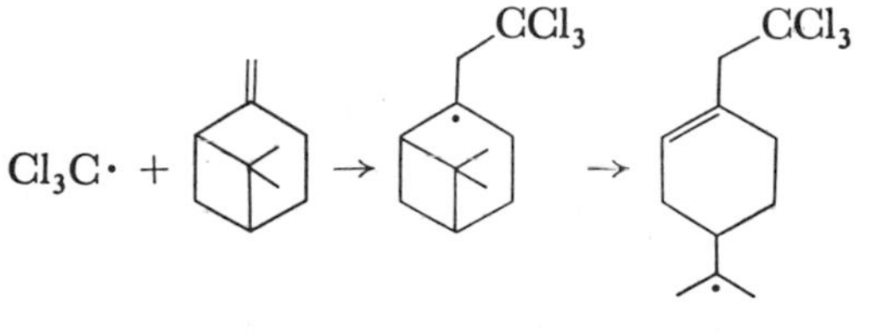

$$(8\text{-}44)$$

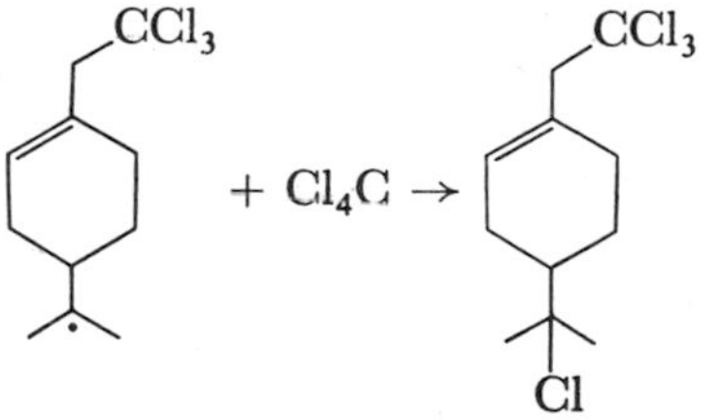

$$(8\text{-}45)$$

37

of the angular strain of the four-membered ring system is the driving force for the β-elimination although the fragmentation results in a small increase in entropy.

The peroxide-induced reaction of phenyl cyclobutyl ketone (**38**) with secondary alcohols also involves a ring opening of a cyclobutylcarbinyl radical.[19] The radical undergoing the β-elimination is generated by the reduction of the carbonyl function with an α-hydroxyalkyl radical (see Chapter 11). The final product of this reaction is phenyl *n*-butyl ketone (**39**), the product of the β-elimination reaction being the enol form of the precursor of this ketone.

$$
\underset{\displaystyle\textbf{38}}{\underset{OH}{CH_3\overset{|}{\underset{\cdot}{C}}C_2H_5} + C_6H_5\overset{O}{\overset{||}{C}}-CH\underset{CH_2}{\overset{CH_2}{\diagdown}}CH_2 \rightarrow CH_3\overset{O}{\overset{||}{C}}C_2H_5 + C_6H_5\overset{OH}{\overset{|}{C}}CH\underset{CH_2}{\overset{CH_2}{\diagdown}}CH_2}
$$

(8-46)

$$
\underset{\;}{C_6H_5\overset{OH}{\overset{|}{C}}-CH\underset{CH_2}{\overset{CH_2}{\diagdown}}CH_2} \rightarrow C_6H_5\overset{OH}{\overset{|}{C}}=CHCH_2CH_2\overset{\cdot}{C}H_2
$$

(8-47)

$$
C_6H_5\overset{OH}{\overset{|}{C}}=CH(CH_2)_2\overset{\cdot}{C}H_2 + CH_3\overset{OH}{\overset{|}{C}}HC_2H_5 \rightarrow
$$

$$
C_6H_5\overset{OH}{\overset{|}{C}}=CH(CH_2)_2CH_3 + CH_3\overset{OH}{\overset{|}{\underset{\cdot}{C}}}C_2H_5
$$

(8-48)

$$
\downarrow
$$

$$
C_6H_5\overset{O}{\overset{||}{C}}C_4H_9\text{-}n
$$

39

Cyclopropylcarbinyl radicals (**40**) undergo ring opening leading to β-vinyl-

$$
\underset{\textbf{40}}{\overset{CH_2}{\underset{CH_2}{\diagup}}\diagdown CHC\overset{\diagup}{\diagdown}} \rightarrow \underset{\textbf{41}}{\overset{\cdot CH_2}{\underset{CH_2}{\diagup}}CH=C\overset{\diagup}{\diagdown}}
$$

(8-49)

alkyl radicals (**41**). Chlorination of methyl cyclopropane (**42**) yields, among other products, 4-chloro-1-butene (**43**).[20]

The addition reactions of vinylcyclopropane (**44**) and 2-cyclopropylpropene (**45**) involve the formation of a cyclopropyl carbinyl system and, consequently, yield noncyclic unsaturated products resulting from fragmentation of this

$$Cl\cdot + CH_3\overset{CH_2}{\underset{CH_2}{\overset{|}{CH}}} \rightarrow HCl + \cdot CH_2\overset{CH_2}{\underset{CH_2}{\overset{|}{CH}}}$$

42

(8-50)

$$Cl\cdot + CH_2{=}CHCH_2CH_2Cl \xleftarrow{Cl_2} CH_2{=}CHCH_2\dot{C}H_2$$

43

radical. Additions of bromotrichloromethane to **44** and **45** yield the unsaturated addition products **46** and **47**, respectively.[21, 22] Addition of carbon tetrachloride

$$Cl_3C\cdot + CH_2{=}\overset{|}{\underset{R}{C}}{-}\overset{CH_2}{\underset{CH_2}{\overset{|}{CH}}} \rightarrow Cl_3CCH_2\overset{|}{\underset{R}{\dot{C}}}{-}\overset{CH_2}{\underset{CH_2}{\overset{|}{CH}}}$$

44, R = H
45, R = CH₃

(8-51)

$$Cl_3C\cdot + Cl_3CCH_2\overset{|}{\underset{R}{C}}{=}CHCH_2CH_2Br \xleftarrow{BrCCl_3} Cl_3CCH_2\overset{|}{\underset{R}{C}}{=}CHCH_2\dot{C}H_2$$

46, R = H
47, R = CH₃

to **45** produced a noncyclic addition product but addition of thiophenol to **45** yielded a product (**49**) in which no rearrangement of the alkene skeleton had occurred. Reaction of **45** with methyl mercaptan yielded a mixture of the

$$C_6H_5SH + CH_2{=}\overset{|}{\underset{CH_3}{C}}\overset{CH_2}{\underset{CH_2}{\overset{|}{CH}}} \rightarrow C_6H_5SCH_2\overset{|}{\underset{CH_3}{CH}}\overset{CH_2}{\underset{CH_2}{\overset{|}{CH}}}$$

48

(8-52)

rearranged product (**49**) and the unrearranged product (**50**), the distribution being dependent on the mercaptan concentration with more of the rearranged product being formed at lower mercaptan concentrations, an observation that can be explained in terms of competing reactions of the adduct radical. The β-elimination reaction 8-55 is independent of the mercaptan concentration and leads to formation of the rearranged product while the unrearranged product results from reaction of the adduct radical with the mercaptan (8-54). Only rearranged products are formed in the additions of polyhalomethanes to these vinylcyclopropanes because the β-elimination reactions of the cyclopropyl carbinyl radicals are faster than their reactions with the polyhalomethanes. The very high reactivity of thiophenol toward hydrogen abstraction

$$CH_3S\cdot + CH_2=CCH \overset{CH_2}{\underset{CH_2}{<}} \quad \rightarrow \quad CH_3SCH_2\overset{\cdot}{C}CH \overset{CH_2}{\underset{CH_2}{<}} \qquad (8\text{-}53)$$

(with CH_3 below each central carbon)

$$CH_3SCH_2\overset{\cdot}{C}CH \overset{CH_2}{\underset{CH_2}{<}} + CH_3SH \quad \rightarrow \quad CH_3SCH_2CHCH \overset{CH_2}{\underset{CH_2}{<}} + CH_3S\cdot$$

$$\underset{\mathbf{50}\ CH_3}{} \qquad (8\text{-}54)$$

$$CH_3SCH_2\overset{\cdot}{C}CH \overset{CH_2}{\underset{CH_2}{<}} \quad \rightarrow \quad CH_3SCH_2\underset{CH_3}{C}=CHCH_2\overset{\cdot}{C}H_2 \qquad (8\text{-}55)$$

$$CH_3SCH_2\underset{CH_3}{C}=CHCH_2\overset{\cdot}{C}H_2 + CH_3SH \rightarrow CH_3SCH_2\underset{CH_3}{C}=CHCH_2CH_3 + CH_3S\cdot$$

$$\underset{\mathbf{49}}{} \qquad (8\text{-}56)$$

is reflected in the fact that the adduct radical does not undergo elimination. The effect of the mercaptan concentration on the distribution of addition products in the methyl mercaptan reactions not only supports this proposed mechanistic scheme but discredits the suggestion that the two products result from the reactions of a single "nonclassical radical" such as **51**. The products

$$\overset{CH_2-CH_2}{\underset{R\overset{\cdot}{C}-CH}{\diagdown\diagup}} \quad\leftrightarrow\quad \overset{\cdot CH_2-CH_2}{\underset{RC=CH}{\diagup}} \quad\leftrightarrow\quad \overset{CH_2-CH_2}{\underset{R\overset{4}{C}\text{-----}\overset{3}{C}H}{\diagup\diagdown}}$$

$$\mathbf{51}$$

observed in these reactions could be explained in terms of reactions at the 4-carbon of such a radical with the adding reagent yielding rearranged products or at the 1-carbon yielding unrearranged products. The product distributions would not, in such a case, be dependent on the adding reagent concentration.

The ethylene oxide ring opens in a β-elimination mechanism.[22] Formation of a radical of the type shown in **52** opens giving the unsaturated alkoxy radical

$$\overset{O}{>C-\overset{}{C}-C<} \quad \rightarrow \quad >C=C-\overset{}{C}-O\cdot \qquad (8\text{-}57)$$

$$\mathbf{52} \qquad\qquad \mathbf{53}$$

53. Cyclohexene oxide (**54**) in a reaction induced with *t*-butyl peroxide yields, among other products, 3-cyclohexenol (**57**)[23] in a chain reaction which involves the β-elimination reaction of **55** yielding the unsaturated alkoxy radical **56**.

The latter radical reacts with the parent compound yielding the product and the chain-carrying radical **55**.

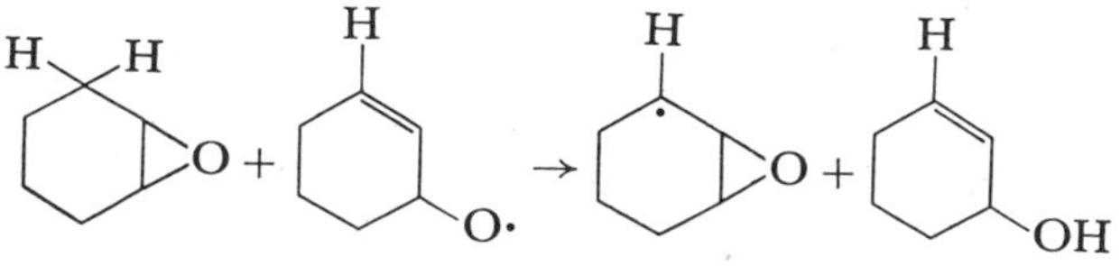

$$\mathbf{54} + \mathbf{56} \rightarrow \mathbf{55} + \mathbf{57} \qquad (8\text{-}58)$$

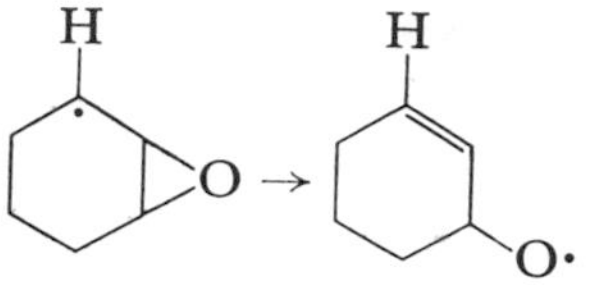

$$\mathbf{55} \rightarrow \mathbf{56} \qquad (8\text{-}59)$$

Cyclooctane reacts with butadiene monoxide (**58**) in a *t*-butyl peroxide induced reaction yielding *trans*-4-cyclooctylbut-2-en-1-ol (**59**) in about 40% yield. The high reactivity of the cycloalkane as an adding agent reflects the

$$C_{10}H_{15}\cdot + CH_2{=}CHCH{-}CH_2 \rightarrow C_8H_{15}CH_2\dot{C}HCH{-}CH_2 \qquad (8\text{-}60)$$

58

$$C_8H_{15}CH_2\dot{C}HCH{-}CH_2 \rightarrow \begin{array}{c} C_8H_{15}CH_2 \\ \end{array} C{=}C \begin{array}{c} H \\ CH_2O\cdot \end{array} \qquad (8\text{-}61)$$

$$\begin{array}{c} C_8H_{15}CH_2 \\ H \end{array} C{=}C \begin{array}{c} H \\ CH_2O\cdot \end{array} + C_8H_{16} \rightarrow \begin{array}{c} C_8H_{15}CH_2 \\ H \end{array} C{=}C \begin{array}{c} H \\ CH_2OH \end{array} + C_8H_{15}\cdot \qquad (8\text{-}62)$$

59

ability of the alkoxy radical to abstract hydrogens from alkanes. The stereo-specific ring opening leading to the *trans* product is due to a lower energy requirement for the two bulky groups which are *trans* rather than *cis* in the transition state of the β-elimination reaction.[22] The reactions of vinylcyclo-propane with polyhalomethanes yield mixtures of the *cis*- and *trans*-isomers. A stereospecifically formed isomer (either *cis* or *trans*) could well have been produced but then isomerized to an equilibrium mixture of the two by halogen atoms that might fortuitously be present in the reaction mixture.

D. FRAGMENTATION OF α-ALKOXYALKYL RADICALS

The simplest method of forming α-alkoxyalkyl radicals which are capable of fragmenting to alkyl radicals and carbonyl compounds is by abstraction of an α-hydrogen from an ether. Benzyl alkyl ethers (60), for example, decompose to benzaldehyde and alkanes when heated with t-butyl peroxide[24] via the chain sequence 8-64 and 8-65. The α-alkoxybenzyl radicals were also observed

$$C_6H_5CH_2OR \xrightarrow{\text{DTBP}} C_6H_5CHO + RH \qquad (8\text{-}63)$$

60

$R = \text{alkyl}$

to dimerize and the amount of fragmentation relative to dimerization increases as the stability of the eliminated alkyl radical increases.

$$C_6H_5CH_2OR + R\cdot \rightarrow C_6H_5\dot{C}HOR + RH \qquad (8\text{-}64)$$

$$C_6H_5\dot{C}HOR \rightarrow C_6H_5CHO + R\cdot \qquad (8\text{-}65)$$

The t-butyl peroxide-induced reaction of n-butyl ether with 1-octene yields, among other products, n-dodecane (62) and 4-dodecanone (63). Both addition products can be accounted for by the addition of the fragments of the β-elimination of an α-n-butoxy-n-butyl radical derived from the ether[25] (see Chapter 6).

$$n\text{-}C_3H_7\dot{C}HOC_4H_9\text{-}n \rightarrow n\text{-}C_3H_7CHO + n\text{-}C_4H_9\cdot \qquad (8\text{-}66)$$

61

$$n\text{-}C_4H_9\cdot + CH_2{=}CHC_6H_{13}\text{-}n \rightarrow n\text{-}C_4H_9CH_2\dot{C}HC_6H_{13}\text{-}n \qquad (8\text{-}67)$$

$$\xrightarrow{\text{R'H}} n\text{-}C_{12}H_{26}$$

62

$$n\text{-}C_3H_7CHO \xrightarrow{\text{R}\cdot} n\text{-}C_3H_7\dot{C}O$$

$$n\text{-}C_3H_7\dot{C}O + CH_2{=}CHC_6H_{13}\text{-}n \rightarrow n\text{-}C_3H_7COCH_2\dot{C}HC_6H_{13}\text{-}n \qquad (8\text{-}68)$$

$$\xrightarrow{\text{R'H}} n\text{-}C_3H_7COC_8H_{17}\text{-}n$$

$R'H = (n\text{-}C_4H_9)_2O \text{ or } n\text{-}C_3H_7CHO$

63

The rearrangement of α-alkoxystyrenes (64) to alkyl phenyl ketones, a reaction first observed by Claisen,[26] has been shown to proceed by a free-radical chain reaction involving fragmentation of an α-alkoxyalkyl radical.[27] Evidence supporting the mechanism for the rearrangement is that (1) the reaction can

$$C_6H_5\overset{\displaystyle OR}{\underset{\displaystyle |}{C}}{=}CH_2 + R\cdot \rightarrow C_6H_5\overset{\displaystyle OR}{\underset{\displaystyle |}{\dot{C}}}CH_2R \qquad (8\text{-}69)$$

64

$$\underset{\text{OR}}{\overset{|}{C_6H_5\overset{|}{C}CH_2R}} \rightarrow \underset{\text{O}}{\overset{\|}{C_6H_5\overset{\|}{C}CH_2R}} + R\cdot \qquad (8\text{-}70)$$

be induced by peroxides, (2) optically active alkyl groups migrate with racemization, (3) the neopentyl group migrates without rearrangement, and (4) at least in the case of α-methoxystyrene, the rearrangement is intermolecular.

Acetals fragment yielding esters and alkanes as reaction products. The t-butyl peroxide-induced reaction of diethyl n-butyral (65) yields ethane and

$$C_2H_5\cdot + \underset{\displaystyle \mathbf{65}}{n\text{-}C_3H_7\overset{\displaystyle OC_2H_5}{\underset{\displaystyle OC_2H_5}{CH}}} \rightarrow C_2H_6 + n\text{-}C_3H_7\overset{\displaystyle OC_2H_5}{\underset{\displaystyle OC_2H_5}{C\cdot}} \qquad (8\text{-}71)$$

$$n\text{-}C_3H_7\overset{\displaystyle OC_2H_5}{\underset{\displaystyle OC_2H_5}{C\cdot}} \rightarrow n\text{-}C_3H_7C\overset{\displaystyle O}{\underset{\displaystyle OC_2H_5}{\diagup}} + C_2H_5\cdot \qquad (8\text{-}72)$$

ethyl butyrate among other products.[28] Cyclic acetals rearrange to esters by a similar reaction path. Benzoate esters are formed from the peroxide-induced reaction of 2-phenyl-1,3-dioxalane (66)[29] and methyl valerate from 2-methoxytetrahydropyran (67).[30]

$$\underset{\displaystyle \mathbf{66}}{C_6H_5\overset{\displaystyle O-CH_2}{\underset{\displaystyle O-CH_2}{CH}}} + C_6H_5\overset{\displaystyle U}{\overset{\|}{C}}OCH_2\dot{C}H_2 \rightarrow C_6H_5\overset{\displaystyle O}{\underset{\displaystyle O-CH_2}{C\cdot}} + C_6H_5\overset{\displaystyle O}{\overset{\|}{C}}OCH_2CH_3 \qquad (8\text{-}73)$$

$$C_6H_5\overset{\displaystyle O-CH_2}{\underset{\displaystyle O-CH_2}{C\cdot}} \rightarrow C_6H_5\overset{\displaystyle O}{\overset{\|}{C}}-OCH_2\dot{C}H_2 \qquad (8\text{-}74)$$

$$\underset{\displaystyle \mathbf{67}}{\overset{\displaystyle}{\bigcirc}} + \cdot CH_2(CH_2)_3CO_2CH_3 \rightarrow \overset{\displaystyle}{\bigcirc} + n\text{-}C_4H_9CO_2CH_3 \qquad (8\text{-}75)$$

$$\overset{\displaystyle}{\bigcirc} \rightarrow \overset{\displaystyle}{\bigcirc} \equiv \cdot CH_2(CH_2)_3CO_2CH_3 \qquad (8\text{-}76)$$

Diethylketene acetal (**68**) rearranges to ethyl butyrate in a *t*-butyl peroxide-induced reaction.[31] The acetal-derived radical undergoing fragmentation in this case is formed by the addition of the eliminated ethyl radical to **68**. Reactions

$$CH_2{=}C{\overset{\displaystyle OC_2H_5}{\underset{\displaystyle OC_2H_5}{}}} \;+\; C_2H_5{\cdot} \;\rightarrow\; C_2H_5CH_2C{\overset{\displaystyle OC_2H_5}{\underset{\displaystyle OC_2H_5}{}}} \qquad (8\text{-}77)$$

68

$$C_2H_5CH_2\overset{\displaystyle .}{C}{\overset{\displaystyle OC_2H_5}{\underset{\displaystyle OC_2H_5}{}}} \;\rightarrow\; n\text{-}C_3H_7\overset{O}{\overset{\|}{C}}OC_2H_5 + C_2H_5{\cdot} \qquad (8\text{-}78)$$

of **68** in the presence of ethers such as 1,4-dioxane or tetrahydrofuran yield a condensation product of the ether and the ketene acetal as illustrated for the reaction of the dioxane and **68**.

$$(8\text{-}79)$$

$$(8\text{-}80)$$

$$(8\text{-}81)$$

Extensive fragmentation occurs in the peroxide-induced reactions of the orthoformates.[32] For example, triethyl orthoformate (**69**) is converted to ethyl carbonate, acetaldehyde, ethyl formate, and ethane in a *t*-butyl peroxide-induced reaction. The ethyl carbonate results from a radical produced by abstraction of the "formyl hydrogen" whereas ethyl formate and acetaldehyde result from two consecutive β-eliminations of the radical produced by the abstraction of an α-hydrogen of one of the ethyl groups.

A mixture of esters is obtained from the reactions of mixed acetals. The ratio of the esters produced in the reactions of mixed acetals is a measure of the relative

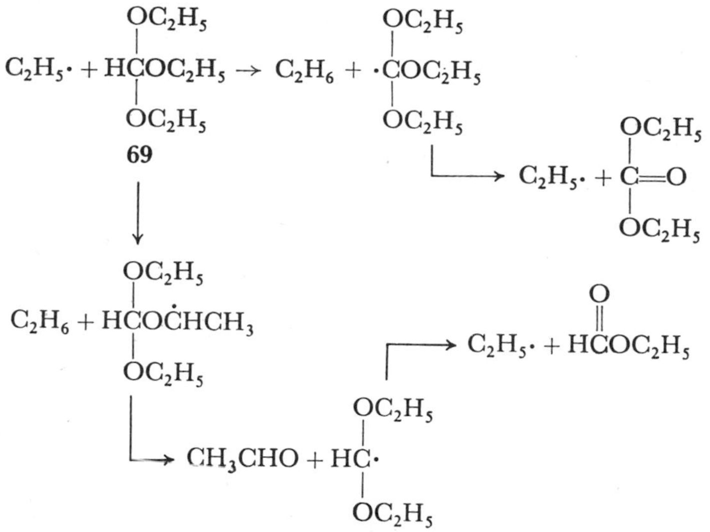

$$C_2H_5\cdot + \overset{\displaystyle OC_2H_5}{\underset{\displaystyle OC_2H_5}{H\overset{|}{\underset{|}{C}}OC_2H_5}} \rightarrow C_2H_6 + \cdot\overset{\displaystyle OC_2H_5}{\underset{\displaystyle OC_2H_5}{\overset{|}{\underset{|}{C}}OC_2H_5}}$$

(8-82)

69

$$\longrightarrow C_2H_5\cdot + \overset{\displaystyle OC_2H_5}{\underset{\displaystyle OC_2H_5}{\overset{|}{\underset{|}{C}}{=}O}}$$

$$C_2H_6 + \overset{\displaystyle OC_2H_5}{\underset{\displaystyle OC_2H_5}{H\overset{|}{\underset{|}{\overset{\displaystyle \cdot}{C}}}O\overset{\displaystyle }{\underset{\displaystyle }{C}}HCH_3}}$$

$$\longrightarrow C_2H_5\cdot + H\overset{\displaystyle O}{\overset{\|}{C}}OC_2H_5$$

$$\longrightarrow CH_3CHO + H\overset{\displaystyle OC_2H_5}{\underset{\displaystyle OC_2H_5}{\overset{|}{\underset{|}{C}}\cdot}}$$

(8-83)

TABLE 8-1 Relative Rates of β-Elimination of Alkyl Radicals from Acetal-Derived Radicals at 130°C

R	k/k' [a]
Isopropyl	5.66
Allyl	>25
n-Butyl	1.00
sec-Butyl	4.10
t-Butyl	18.7
Cyclopropylcarbinyl	3.89
Cyclopentyl	2.53
Cyclohexyl	2.36
Cycloheptyl	9.85
Cyclooctyl	17.2
exo-Norbornyl	2.20
endo-Norbornyl	0.87
Benzyl	22.4

[a] k/k' is reactivity relative to n-butyl (see 8-84 and 8-85).

ease of β-elimination of the two alkyl groups.[33] Table 8-1 lists the relative reactivities (k/k') of several groups toward β-elimination from acetal derived radicals.

$$CH_3C\underset{OC_4H_9\text{-}n}{\overset{OR}{\Bigl\langle}}\;\begin{array}{c}\xrightarrow{\;k\;}\\[6pt]\xrightarrow{\;k'\;}\end{array}\;\begin{array}{l}\underset{O}{\overset{\displaystyle O}{CH_3\overset{\|}{C}OC_4H_9\text{-}n}} + R\cdot \qquad (8\text{-}84)\\[18pt]CH_3\underset{O}{\overset{\displaystyle}{\underset{\|}{C}OR}} + n\text{-}C_4H_9\cdot \qquad (8\text{-}85)\end{array}$$

Resonance stabilization of the radical plays a role in determining its ease of being eliminated. Primary alkyl radicals are eliminated less readily than secondary and secondary alkyl radicals less readily than tertiary. Benzyl and allyl radicals are both eliminated readily. The increased ease of elimination of cycloalkyl groups as the ring size increases is probably due to relief of conformation strain in the larger rings. Both the *endo* and *exo* norbornyl derivatives would be expected to give the same radical ultimately but the stereochemistry of the parent acetal radical is a factor in determining how easily that radical will be formed by the β-elimination reaction. The cyclopropylcarbinyl radical, a primary alkyl radical, is eliminated almost four times faster than the *n*-butyl radical. Although, as pointed out previously in this chapter, there is no evidence for a nonclassical radical such as **51**, the relative ease of elimination of cyclopropylcarbinyl radical seems to indicate that this system is somewhat more stable than a simple primary alkyl radical.

E. FRAGMENTATION OF ALKOXY RADICALS

This section is concerned with the fragmentation reactions of alkoxy radicals, radicals that could be postulated to result from addition of an alkyl radical to a carbonyl function but the adding radical in this case bonds to the carbon of the carbonyl function. A comparison of the modes of fragmentation of alkoxy and α-alkoxyalkyl radicals is interesting because, although the radicals are quite different in structure, they yield the same reaction products.

One of the more direct approaches to the formation of alkoxy radicals is by decomposition of dialkyl peroxides. Thermal decomposition of *t*-butyl peroxide (**70**) yields two *t*-butoxy radicals. Although in many solvents these radicals abstract hydrogens and yield *t*-butyl alcohol and a solvent-derived radical, if the solvent is not prone to hydrogen abstraction, the *t*-butoxy radical fragments yielding a methyl radical and a molecule of acetone. A comparison of the ratios of acetone to *t*-butyl alcohol found in the decomposition of the peroxide

$$\underset{\textbf{70}}{(CH_3)_3COOC(CH_3)_3} \;\rightarrow\; 2(CH_3)_3CO\cdot \qquad (8\text{-}86)$$

$$(CH_3)_3CO\cdot + RH \;\rightarrow\; (CH_3)_3COH + R\cdot \qquad (8\text{-}87)$$

$$(CH_3)_3CO\cdot \;\rightarrow\; CH_3COCH_3 + CH_3\cdot \qquad (8\text{-}88)$$

in various solvents has been used to measure the reactivities of various solvents toward hydrogen abstraction by the t-butoxy radical.[34]

Fragmentation reactions of other alkoxy radicals that are formed by decomposition of the appropriate peroxides show that certain groups are eliminated in preference to others. Decomposition of t-amyl peroxide (71) yields larger amounts of acetone than methyl ethyl ketone indicating that the ethyl

$$
\begin{array}{c}
\overset{\displaystyle CH_3}{\underset{\displaystyle |}{}}\ \overset{\displaystyle CH_3}{\underset{\displaystyle |}{}} \\
C_2H_5\overset{|}{\underset{|}{C}}OO\overset{|}{\underset{|}{C}}C_2H_5 \;\rightarrow\; 2C_2H_5\overset{\displaystyle CH_3}{\underset{\displaystyle CH_3}{\overset{|}{\underset{|}{C}}}}O\cdot \\
\overset{\displaystyle |}{CH_3}\ \overset{\displaystyle |}{CH_3} \\
\mathbf{71}
\end{array}
$$

$$C_2H_5\cdot + CH_3COCH_3 \quad (8\text{-}89)$$
$$CH_3\cdot + C_2H_5COCH_3 \quad (8\text{-}90)$$

radical is eliminated from the alkoxy radical more readily than a methyl radical. The cumyloxy radical (74), which can be obtained from either cumyl peroxide (72) or cumyl hydroperoxide (73), yields only acetophenone indicating that the methyl group is eliminated in preference to the phenyl group.[35]

$$
\begin{array}{c}
\overset{\displaystyle CH_3}{\underset{\displaystyle |}{}}\ \overset{\displaystyle CH_3}{\underset{\displaystyle |}{}} \\
C_6H_5\overset{|}{\underset{|}{C}}OO\overset{|}{\underset{|}{C}}C_6H_5 \\
\overset{\displaystyle |}{CH_3}\ \overset{\displaystyle |}{CH_3} \\
\mathbf{72}
\end{array}
$$

$$
\begin{array}{c}
\overset{\displaystyle CH_3}{\underset{\displaystyle |}{}} \\
C_6H_5\overset{|}{\underset{|}{C}}OOH \\
\overset{\displaystyle |}{CH_3} \\
\mathbf{73}
\end{array}
$$

$$
\begin{array}{c}
\overset{\displaystyle CH_3}{\underset{\displaystyle |}{}} \\
C_6H_5\overset{|}{\underset{|}{C}}O\cdot \;\rightarrow\; C_6H_5COCH_3 + CH_3\cdot \quad (8\text{-}91) \\
\overset{\displaystyle |}{CH_3} \\
\mathbf{74}
\end{array}
$$

Pinane hydroperoxide (75) undergoes rearrangement when heated at 117° yielding the cyclobutane derivative 76.[36] This reaction very likely involves the fragmentation of a cyclic system by the route shown. The primary alkyl radical

$$\mathbf{75} \qquad \mathbf{76} \qquad (8\text{-}92)$$

formed by the β-elimination apparently is more stable than the cyclobutyl radical that would have been formed if the fragmentation followed the alternate route.

Alkyl hypochlorites yield ketones and alkyl chlorides by free-radical chain reactions involving the β-elimination of alkyl radicals from alkoxy radicals. For example, t-amyl hypochlorite (**77**) decomposes when heated yielding ethyl chloride and acetone.[37] Cyclic systems rearrange as evidenced by the formation

$$C_2H_5\cdot + C_2H_5\overset{\displaystyle CH_3}{\underset{\displaystyle CH_3}{C}}OCl \rightarrow C_2H_5Cl + C_2H_5\overset{\displaystyle CH_3}{\underset{\displaystyle CH_3}{C}}O\cdot \qquad (8\text{-}93)$$

77

$$C_2H_5\overset{\displaystyle CH_3}{\underset{\displaystyle CH_3}{C}}O\cdot \rightarrow C_2H_5\cdot + CH_3COCH_3 \qquad (8\text{-}94)$$

of ω-chloromethyl ketones from the appropriate cyclic hypochlorites.[38]

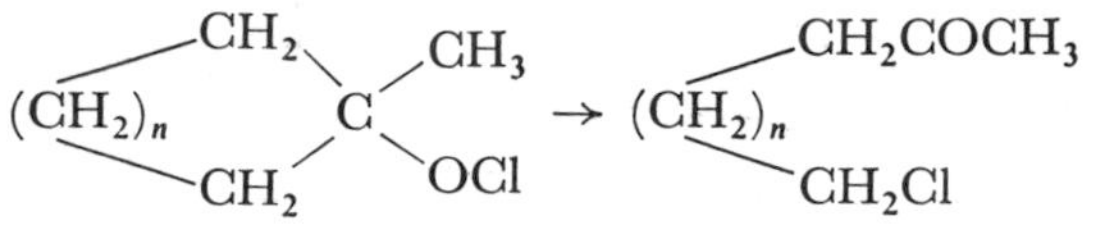

$$(CH_2)_n\underset{CH_2}{\overset{CH_2}{\diagdown}}C\underset{OCl}{\overset{CH_3}{\diagup}} \rightarrow (CH_2)_n\underset{CH_2Cl}{\overset{CH_2COCH_3}{\diagup}} \qquad (8\text{-}95)$$

via

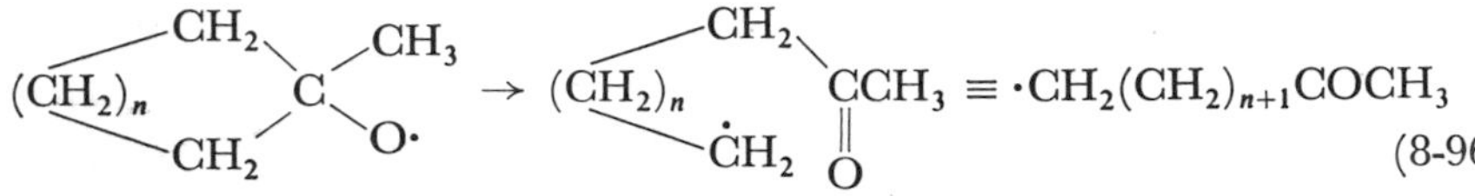

$$(CH_2)_n\underset{CH_2}{\overset{CH_2}{\diagdown}}C\underset{O\cdot}{\overset{CH_3}{\diagup}} \rightarrow (CH_2)_n\underset{\underset{O}{\overset{\|}{C}}H_2}{\overset{CH_2}{\diagdown}}CCH_3 \equiv \cdot CH_2(CH_2)_{n+1}COCH_3 \qquad (8\text{-}96)$$

$$\cdot CH_2(CH_2)_{n+1}COCH_3 + (CH_2)_n\underset{CH_2}{\overset{CH_2}{\diagdown}}C\underset{OCl}{\overset{CH_3}{\diagup}} \rightarrow$$

$$ClCH_2(CH_2)_{n+1}COCH_3 + (CH_2)_n\underset{CH_2}{\overset{CH_2}{\diagdown}}C\underset{O\cdot}{\overset{CH_3}{\diagup}} \qquad (8\text{-}97)$$

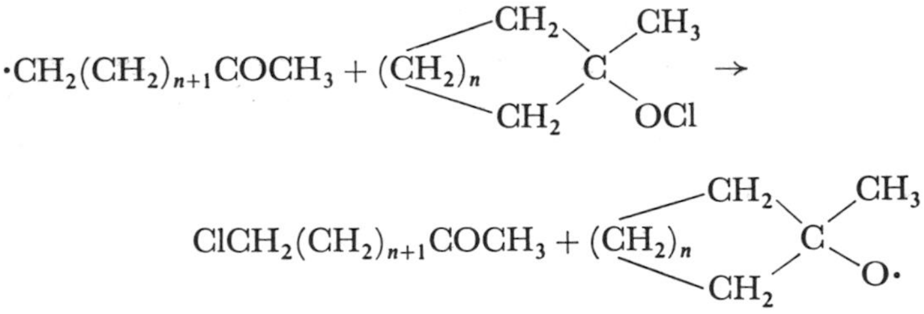

The relative rates of β-elimination of alkyl groups from alkoxy radicals can be determined from the distribution of products obtained from appropriate

$$R'R''C{=}O + R\cdot \xrightarrow{\text{RR'R''COCl}} RCl + RR'R''CO\cdot \quad (8\text{-}98)$$

$$R'{-}\underset{\displaystyle R''}{\overset{\displaystyle R}{\underset{|}{\overset{|}{C}}}}{-}O\cdot \longrightarrow \begin{array}{l} RR''C{=}O + R'\cdot \xrightarrow{\text{RR'R''COCl}} R'Cl + RR'R''CO\cdot \quad (8\text{-}99) \\[2em] RR'C{=}O + R''\cdot \xrightarrow{\text{RR'R''COCl}} R''Cl + RR'R''CO\cdot \end{array}$$

$$(8\text{-}100)$$

alkyl hypochlorites.[39] The relative rates of β-elimination of alkyl radicals from alkoxy radicals and acetal-derived radicals are significantly different. Although some degree of selectivity was noted in the fragmentation of the acetal-derived radicals, it is not nearly so marked as that found for the fragmentation reactions of alkoxy radicals. The difference between elimination of a primary and secondary alkyl radical from the acetal-derived radical is about a factor of 5 but the difference in relative rate of elimination from an alkoxy radical is approximately 50. The difference between a primary and tertiary alkyl group in elimination from the acetal is about 18 but in elimination from the alkoxy radical it is about 300. The very high degree of selectivity observed in the fragmentation reactions of the alkoxy radicals probably is due to polar contributions to the transition state of the reaction in that a donor radical is being formed from an acceptor radical. The contributions that the alkyl radical is

$$R{-}\underset{\displaystyle R}{\overset{\displaystyle R}{\underset{|}{\overset{|}{C}}}}{-}O\cdot \rightarrow \left[\underset{\displaystyle R}{\overset{\displaystyle R}{\underset{|}{\overset{|}{R\overset{\delta+}{\cdots}C\overset{\delta-}{\cdots}O}}}}\right] \rightarrow R\cdot + \underset{\displaystyle R}{\overset{\displaystyle R}{C}}{=}O \quad (8\text{-}101)$$

able to make to the transition of the fragmentation reaction as a donor species determines to a great extent its ease of elimination. It is interesting to note that the relative rates of elimination of the alkyl groups from the alkoxy radicals parallels the stabilities of the corresponding carbonium ions, that is, primary $\ll$ secondary $<$ tertiary.

The comparatively low degree of selectivity found in the fragmentation reactions of the acetals can probably be ascribed to the fact that there are no polar contributions in the transition state of the reaction. Both the fragmenting acetal-derived radical and the eliminated alkyl radical are donor species. It is interesting to note that resonance stabilization of the radicals, although responsible for some of the selectivity, does not impart the high degree of selectivity observed when polar factors become determining factors in the rates of the chain-propagating reactions.

References

1. F. Wackholtz, *Z. physik. Chem.*, **125**, 1 (1927).
2. H. Steinmetz and R. M. Noyes, *J. Amer. Chem. Soc.*, **74**, 4141 (1952).
3. M. S. Kharasch and M. Sage, *J. Org. Chem.*, **14**, 537 (1949).
4. D. J. Sibbett and R. M. Noyes, *J. Amer. Chem. Soc.*, **75**, 763 (1953).
5. W. G. Young and K. Nuzak, *ibid.*, **62**, 311 (1940).
6. M. S. Kharasch, E. Sternfeld, and F. R. Mayo, *ibid.*, **59**, 1655 (1937).
7. L. P. Schmerling and J. P. West, *ibid.*, **71**, 2015 (1949).
8. K. E. Wilzback, F. R. Mayo, and R. Van Meter, *ibid.*, **70**, 4069 (1948).
9. D. H. R. Barton, *J. Chem. Soc.*, **148**, 155 (1949).
10. C. Walling and W. Helmreich, *J. Amer. Chem. Soc.*, **81**, 1144 (1959).
11. E. S. Huyser and R. M. Kellogg, *J. Org. Chem.*, **30**, 2867 (1965).
12. R. E. Foster, A. W. Larchar, R. D. Lipscomb, and B. C. McKusick, *J. Amer. Chem. Soc.*, **78**, 5606 (1956).
13. E. S. Huyser and R. M. Kellogg, *J. Org. Chem.*, **31**, 3366 (1966).
14. C. Sivertz, W. Andrews, W. Elsdon, and K. Graham, *J. Polymer Sci.*, **19**, 587 (1956).
15. E. S. Huyser and R. M. Kellogg, *J. Org. Chem.*, **30**, 3003 (1965).
16. P. S. Skell and R. C. Woodworth, *J. Amer. Chem. Soc.*, **77**, 4638 (1955).
17. D. M. Oldroyd, G. S. Fisher, and L. A. Goldblatt, *ibid.*, **72**, 2407 (1950).
18. G. DuPont, R. Dulon, and G. Clement, *Bull. Soc. Chem. France*, **1950**, 1056, 1115.
19. D. C. Neckers, J. Hardy, and A. P. Schaap, *J. Org. Chem.*, **31**, 622 (1966).
20. H. C. Brown and M. Barkowski, *J. Amer. Chem. Soc.*, **74**, 1894 (1952); J. D. Roberts and R. H. Mazur, *ibid.*, **78**, 2509 (1953).
21. E. S. Huyser and J. D. Taliaferro, *J. Org. Chem.*, **28**, 3442 (1963).
22. E. S. Huyser and L. R. Munson, *ibid.*, **30**, 1436 (1965).
23. E. Sabatino and R. J. Gritter, *ibid.*, **28**, 3437 (1963).
24. R. L. Huang and S. S. Si-Hoe, *Proc. Chem. Soc.* (London), **1957**, 354; H. G. Ang, R. L. Huang, and H. G. Sim, *J. Chem. Soc.*, **1963**, 4841.
25. C. H. McDonnell, PhD Thesis, University of Chicago (1953).
26. L. Claisen, *Ber.*, **29**, 2931 (1896).
27. K. B. Wiberg and B. I. Rowland, *J. Amer. Chem. Soc.*, **77**, 1159 (1955).
28. L. P. Kuhn and C. Wellman, *J. Org. Chem.*, **22**, 774 (1956).
29. E. S. Huyser and Z. Garcia, *ibid.*, **27**, 2716 (1962).
30. E. S. Huyser, *ibid.*, **25**, 1820 (1960).
31. E. S. Huyser, R. M. Kellogg, and D. T. Wang, *ibid.*, **30**, 4377 (1965).
32. E. S. Huyser and D. T. Wang, *ibid.*, **27**, 4696 (1962).
33. E. S. Huyser and D. T. Wang, *ibid.*, **29**, 2720 (1964).
34. E. L. Patmore and R. J. Gritter, *ibid.*, **27**, 4196 (1962).
35. M. S. Kharasch, A. Fono and W. Nudenberg, *J. Org. Chem.*, **16**, 113 (1951).

36. G. S. Schmidt and G. S. Fisher, *J. Amer. Chem. Soc.*, **76**, 5426 (1954).
37. F. D. Chattaway and O. G. Backeberg, *J. Chem. Soc.*, **123**, 2999 (1923).
38. J. W. Wilt and J. W. Hill, *J. Org. Chem.*, **26**, 3523 (1961); T. L. Cairns and B. E. Englund, *ibid.*, **21**, 140 (1956).
39. F. D. Greene, M. L. Savitz, F. D. Osterholtz, H. H. Lau, W. N. Smith, and P. M. Zanet, *J. Org. Chem.*, **28**, 55 (1963).

Chapter 9

Free-Radical Rearrangements

I. Introduction and Scope of Free-Radical Rearrangements

Rearrangements involving a 1,2-shift of a group or an atom are fairly common in reactions of cationic intermediates. Although such rearrangements also occur in radical reactions, they are not as common and far more limited in scope. In this chapter some of the rearrangements that involve a 1,2-shift of either an atom or group in a free radical that is an intermediate in a free-radical reaction are examined.

The rearrangements discussed here are limited to those that are true 1,2-shifts. There are many other free-radical reactions in which extensive rearrangement occurs. Many of these can be explained in terms of other chain-propagating reactions (e.g., β-elimination, addition, and displacement). For example, the rearrangement of a cyclic acetal-derived radical (**1**) to the linear carboxylate

$$RC\cdot \overset{O-CH_2}{\underset{O-CH_2}{<}} \quad \rightarrow \quad RC \overset{O-CH_2}{\underset{O \cdot CH_2}{<}} \qquad (9\text{-}1)$$

$$\textbf{1} \qquad\qquad\qquad \textbf{2}$$

containing radical (**2**) is a β-elimination reaction.[1] Likewise, the rearrangement of the camphane radical **3** to the isocamphane radical **5** most likely results from

235

a β-elimination reaction yielding **4** followed by readdition to yield the rearranged radical.[2] The 1,5-migrations of hydrogen atoms observed in certain radicals are most likely intramolecular hydrogen atom abstractions occurring

$$(9\text{-}2)$$

by a cyclic process such as shown for the rearrangements of aldehyde-derived radical **6** to the acyl radical **7** (reaction 9-3)[3] or the alkoxy radical **8** to the alcohol-derived radical **9** (reaction 9-4).[4]

$$(9\text{-}3)$$

$$(9\text{-}4)$$

An intramolecular 1,2-shift of a group Y in a free radical is limited to those

$$(9\text{-}5)$$

radicals that have very special structural features. The only groups observed to undergo a 1,2-shift are aryl and vinyl groups, some of the halogens, namely chlorine and bromine, and certain acyl functions. There is some evidence that thiyl groups (RS·) may also undergo a 1,2-shift. (See Gen. Ref.)

Generally the product of the 1,2-shift is more stable than the reactant. Migration of Y from a tertiary carbon to a primary carbon may occur readily but the reverse reaction, while not prohibited, is not generally observed.

$$\underset{\underset{CH_3}{|}}{\overset{\overset{CH_3}{|}}{Y-C-CH_2 \cdot}} \rightarrow \underset{CH_3}{\overset{CH_3}{\diagdown}} \cdot C-CH_2Y \qquad (9\text{-}6)$$

These two requirements significantly limit the scope of rearrangements in free radicals but do present some significant advantages in the study of free-radical reactions. Whereas the 1,2-shift of a hydride ion is a common occurrence in carbonium ions (e.g., the rearrangement of primary to a secondary carbonium ion), the 1,2-shift of a hydrogen atom does not occur in alkyl radicals. The

$$CH_3CH_2\overset{(+)}{C}H_2 \rightarrow CH_3\overset{(+)}{C}HCH_3 \qquad (9\text{-}7)$$

$$CH_3CH_2\overset{\cdot}{C}H_2 \nrightarrow CH_3\overset{\cdot}{C}HCH_3 \qquad (9\text{-}8)$$

1,2-shift of an alkyl group from a carbon atom to an adjacent electron-deficient carbon, oxygen, or nitrogen is fairly common in reactions of ionic intermediates (e.g., Meerwine, Baeyer-Villiger, Beckmann, etc. rearrangements). However, since alkyl groups do not migrate in free radicals, rearrangements of alkyl (**10**), alkoxy (**11**), and amino (**12**) radicals such as shown in 9-9 through 9-11 are not observed.

$$\underset{\overset{|}{10}}{\overset{\overset{R}{|}}{-C-C}} \nrightarrow \underset{}{\overset{}{\cdot C-C-}} \qquad (9\text{-}9)$$

$$\underset{\overset{|}{11}}{\overset{\overset{R}{|}}{-C-O\cdot}} \nrightarrow \overset{}{C-OR} \qquad (9\text{-}10)$$

$$\underset{\overset{|}{12}}{\overset{\overset{R}{|}}{-C-\overset{\cdot}{N}}} \nrightarrow \overset{}{C-N}\overset{R}{\diagup} \qquad (9\text{-}11)$$

$$R = alkyl$$

II. Migration of Aryl Group

A. MIGRATIONS TO CARBON

The formation of isobutylbenzene (**16**) in the decarbonylation reaction of β-phenylisovaleraldehyde (**13**) is an example of a free-radical reaction involving the 1,2-shift of a phenyl group.[5] The chain sequence for this reaction is shown in reactions 9-12 through 9-15. The radical undergoing rearrangement is the neophyl radical (**15**) and is formed by the decarbonylation of the acyl radical

$$
\underset{\underset{CH_3}{\overset{CH_3}{|}}}{C_6H_5\overset{|}{C}CH_2\dot{C}O} \rightarrow \underset{\underset{CH_3}{\overset{CH_3}{|}}}{C_6H_5\overset{|}{C}\dot{C}H_2} + CO \tag{9-12}
$$
$$
\text{14} \qquad\qquad \text{15}
$$

$$
\underset{\underset{CH_3}{\overset{\overset{CH_3}{|}}{|}}}{C_6H_5\overset{|}{C}\dot{C}H_2} \rightarrow \underset{CH_3}{\overset{CH_3}{\diagdown}}\dot{C}CH_2C_6H_5 \tag{9-13}
$$
$$
\text{15}
$$

$$
\underset{CH_3}{\overset{CH_3}{\diagdown}}\dot{C}CH_2H_5 + \underset{\underset{CH_3}{\overset{CH_3}{|}}}{C_6H_5\overset{|}{C}CH_2CHO} \rightarrow \underset{\underset{\dot{C}H_3}{\overset{CH_3}{|}}}{CH_3CHCH_2C_6H_5} + \underset{\underset{CH_3}{\overset{CH_3}{|}}}{C_6H_5\overset{|}{C}CH_2\dot{C}O}
$$
$$
\text{13} \qquad\qquad \text{16} \qquad\qquad \text{14} \tag{9-14}
$$

$$
\underset{\underset{CH_3}{\overset{CH_3}{|}}}{C_6H_5\overset{|}{C}\dot{C}H_2} + \underset{\underset{CH_3}{\overset{CH_3}{|}}}{C_6H_5\overset{|}{C}CH_2CHO} \rightarrow \underset{\underset{CH_3}{\overset{CH_3}{|}}}{C_6H_5\overset{|}{C}CH_3} + \underset{\underset{CH_3}{\overset{CH_3}{|}}}{C_6H_5\overset{|}{C}CH_2\dot{C}O} \tag{9-15}
$$
$$
\text{15} \qquad\qquad \text{13} \qquad\qquad \text{17} \qquad\qquad \text{14}
$$

(**14**) produced by abstraction of the aldehydic hydrogen from the parent aldehyde by an arylalkyl radical. *t*-Butyl benzene (**17**) is also formed in this reaction indicating that the rate of rearrangement of the neophyl radical is comparable to that of the rate of hydrogen abstraction of an aldehydic hydrogen. The 1,2-shift of the aryl group in the neophyl radical is one in which a primary alkyl radical is converted to a tertiary alkyl radical.

Other reactions in which the neophyl radical occurs as an intermediate and undergoes rearrangement include the Kolbe electrolysis of β-phenylisovaleric acid (**18**)[6] and the chlorination of *t*-butylbenzene.[7] Migrations of phenyl

$$\underset{18}{\overset{CH_3}{\underset{CH_3}{C_6H_5\overset{|}{\underset{|}{C}}CH_2CO_2H}}} \xrightarrow[-H^{(+)}]{-e^{(-)}} \underset{CH_3}{\overset{CH_3}{C_6H_5\overset{|}{\underset{|}{C}}CH_2CO_2\cdot}} \xrightarrow{-CO_2}$$

$$\underset{15}{\overset{CH_3}{\underset{CH_3}{C_6H_5\overset{|}{\underset{|}{C}}CH_2\cdot}}} \rightarrow \overset{CH_3}{\underset{CH_3}{\diagdown}}CCH_2C_6H_5 \quad (9\text{-}16)$$

$$\underset{CH_3}{\overset{CH_3}{C_6H_5\overset{|}{\underset{|}{C}}CH_3}} \xrightarrow{Cl\cdot} \underset{CH_3}{\overset{CH_3}{C_6H_5\overset{|}{\underset{|}{\overset{.}{C}}}CH_2}} \rightarrow \overset{CH_3}{\underset{CH_3}{\diagdown}}\overset{.}{C}CH_2C_6H_5 \xrightarrow{Cl_2}$$

$$\underset{CH_3}{\overset{CH_3}{Cl\overset{|}{\underset{|}{C}}CH_2C_6H_5}} \quad (9\text{-}17)$$

groups have been observed in decarbonylation reactions of other β-phenyl aldehydes. The extent of rearrangement depends on the increase of resonance stabilization of the rearranged radical over that of the unrearranged radical as illustrated by the extensive rearrangement observed in the decarbonylation of **19**[8] as compared with **20**.[9] The amount of rearranged product formed in these

$$\underset{19}{(C_6H_5)_3CCH_2CHO} \xrightarrow[\substack{24.3\% \text{ yield} \\ 96\% \text{ rearrangement}}]{125°}$$

$$(C_6H_5)_2CHCH_2C_6H_5 + (C_6H_5)_2C{=}CHC_6H_5 \quad (9\text{-}18)$$

$$\underset{20}{C_6H_5CH_2\overset{*}{C}H_2CHO} \xrightarrow[\substack{68\% \text{ yield} \\ 3.3\% \text{ rearrangement}}]{125°}$$

$$C_6H_5CH_2CH_3 + C_6H_5CH{=}CH_2 \quad (9\text{-}19)$$

decarbonylations is also influenced by more subtle effects such as ring size.[10]

$$(RCO_2)_2 \rightarrow 2R\cdot + 2CO_2 \qquad (9\text{-}22)$$

Decompositions of the appropriate diacyl peroxides (see Chapter 10) provide further evidence that the difference in stabilities of the parent and rearranged radicals plays a significant role in determining the extent of rearrangement. Although the extent of rearrangement indicated refers only to the amount deduced from the monomeric hydrocarbons found among the various reaction products resulting from the decomposition,[11] the trend is evident. In each case a primary alkyl radical is formed in the decomposition of the peroxide but the amount of rearrangement increases with increasing stability of the rearranged radical.

The 1,2-shift of a phenyl group occurs in the free-radical addition of n-butyraldehyde to 3,3-diphenyl-1-butene (21), a reaction that yields 7,8-diphenyl-2-octanone (22) as the addition product.[12] Addition of n-butyl

$$n\text{-}C_3H_7CHO + CH_2\!\!=\!\!CHC(C_6H_5)_2CH_3 \;\rightarrow\; n\text{-}C_3H_7COCH_2\overset{\overset{\displaystyle C_6H_5}{\displaystyle |}}{C}HCHCH_3$$

$$\underset{\textbf{21}}{}$$

by

$$\underset{\textbf{22}}{\overset{\displaystyle C_6H_5}{}} \qquad (9\text{-}27)$$

$$n\text{-}C_3H_7\overset{\displaystyle \cdot}{C}O + \textbf{21} \;\rightarrow\; n\text{-}C_3H_7COCH_2\overset{\displaystyle \cdot}{C}HC(C_6H_5)_2CH_3$$

$$n\text{-}C_3H_7\overset{\displaystyle \cdot}{C}O + \textbf{22} \;\xleftarrow{\;n\text{-}C_3H_7CHO\;}\; n\text{-}C_3H_7COCH_2\overset{\overset{\displaystyle C_6H_5}{\displaystyle |}}{C}H\overset{\underset{\displaystyle C_6H_5}{\displaystyle |}}{\overset{\displaystyle \cdot}{C}}CH_3$$

$$\underset{\textbf{(9-28)}}{}$$

mercaptan to **21** yields only an unrearranged addition product **23**. In this case,

$$n\text{-}C_4H_9SH + CH_2\!\!=\!\!CHC(C_6H_5)_2CH_3 \;\rightarrow\; n\text{-}C_4H_9SCH_2CH_2C(C_6H_5)_2CH_3$$

$$\underset{\textbf{21}}{} \qquad\qquad \underset{\textbf{23}}{}$$

by $\qquad\qquad\qquad\qquad\qquad\qquad\qquad\qquad\qquad\qquad\qquad$ (9-29)

$$n\text{-}C_4H_9S\!\cdot + \textbf{21} \;\rightarrow\; n\text{-}C_4H_9SCH_2\overset{\displaystyle \cdot}{C}HC(C_6H_5)_2CH_3 \qquad\qquad (9\text{-}30)$$

$$\xrightarrow{\;n\text{-}C_4H_9SH\;}$$

$$n\text{-}C_4H_9S\!\cdot + \textbf{23}$$

the rate of reaction of the adduct radical with mercaptan apparently is faster than the 1,2-shift of a phenyl group. The difference in the behavior of the adduct radicals in these two addition reactions is indicative of the greater reactivity of mercaptans compared to aldehydes toward hydrogen abstraction. The presence of mercaptan in the decarbonylation reactions involving migration of a phenyl group can markedly lower the amount of rearranged products formed in these reactions. The extent of rearranged product obtained in the decarbonylation reaction shown in 9-21 can be reduced to less than 3% if the reaction is performed in 20% benzyl mercaptan.[10] On the other hand, the extent of rearrangement can be increased to 92% if the aldehyde is diluted to 1M with chlorobenzene. The effect of dilution with an inert solvent is attributed to the fact that the rate of the bimolecular reaction of the unrearranged radical with the aldehyde is decreased if the aldehyde concentration is decreased whereas the rate of the unimolecular 1,2-shift is not affected by a change in aldehyde concentration.

B. MIGRATIONS TO OXYGEN

Decomposition of trityl peroxide (**24**) yields the diphenyl ether of tetra-phenylethylene glycol (**25**). The mechanism for this reaction involves homolytic

$$(C_6H_5)_3COOC(C_6H_5)_3 \rightarrow (C_6H_5)_2C-C(C_6H_5)_2 \qquad (9\text{-}31)$$
$$\underset{\mathbf{24}}{} \qquad \qquad \underset{\mathbf{25}}{\overset{|\quad|}{C_6H_5O \quad OC_6H_5}}$$

cleavage of the peroxide linkage producing two triphenylmethoxy radicals
(**26**). A 1,2-shift of a phenyl group from carbon to oxygen produces the ether
radicals (**27**) which couple to yield the reaction product.[13] The 1,2-shift of the

$$(C_6H_5)_3COOC(C_6H_5)_3 \rightarrow \underset{\mathbf{26}}{2(C_6H_5)_3CO\cdot} \qquad (9\text{-}32)$$

$$\underset{\mathbf{26}}{(C_6H_5)_2\overset{\overset{\displaystyle C_6H_5}{|}}{C}-O\cdot} \rightarrow \underset{\mathbf{27}}{(C_6H_5)_2\dot{C}OC_6H_5} \qquad (9\text{-}33)$$

$$\underset{\mathbf{27}}{2(C_6H_5)_2\dot{C}OC_6H_5} \rightarrow (C_6H_5)_2C-C(C_6H_5)_2 \qquad (9\text{-}34)$$
$$\underset{\mathbf{25}}{\overset{|\quad|}{C_6H_5O \quad OC_6H_5}}$$

phenyl group from the carbon to oxygen in this system occurs readily because
of the greater stability of the rearranged radical **27**. Rearrangement is rapid as
evidenced by the fact that in cumene, a solvent that is quite reactive toward
hydrogen abstraction by alkoxy radicals, extensive rearrangement of the radical
occurs.

C. MIGRATORY APTITUDES

Examination of the relative migratory aptitudes of various aryl groups
indicates little, if any, polar contribution by the migrating group to the transition
state of the reaction. In the case of the triarylmethoxy radical rearrangements,

$$\underset{\overset{\displaystyle |}{C_6H_5}}{\overset{\overset{\displaystyle C_6H_5}{|}}{ArCO\cdot}} \rightarrow \underset{C_6H_5}{\overset{Ar}{>}}COC_6H_5 + \underset{C_6H_5}{\overset{C_6H_5}{>}}\cdot COAr \qquad (9\text{-}35)$$

Ar = p-tolyl, p-nitrophenyl, α-naphthyl, p-diphenyl.

the p-tolyl and phenyl group migrate at about the same rate,[13] the p-nitrophenyl
group about four times more readily than the phenyl,[14] and α-naphthyl and

p-diphenyl both about six times more readily than phenyl.[13] It would appear that the ability of the migrating group to delocalize an unpaired electron is more likely an important factor in the migratory aptitude of these aryl groups.

Studies have been made of the relative migratory aptitudes of aryl groups in the neophyl-type radicals. Table 9-1 lists the amounts of migration of various aryl groups relative to the phenyl group observed in such radicals when obtained from either the decarbonylation reactions of the appropriate aldehydes or the decomposition of the appropriate diacyl peroxides.[15] Although there is

TABLE 9-1 Relative Amounts of Migration of Aryl Groups in $ArC(CH_3)_2\dot{C}H_2$ Radicals[15]

	Radical Source	
Ar	Aldehyde	Peroxide
p-$CH_3OC_6H_4$	0.35	0.36
p-FC_6H_4	0.38	0.40
p-$CH_3C_6H_4$	0.65	0.76
o-ClC_6H_4	0.90	0.9
C_6H_5	1.00	1.00
m-ClC_6H_4	1.55	1.33
m-BrC_6H_4	1.70	1.58
p-BrC_6H_4	1.79	1.90
p-ClC_6H_4	1.82	1.88

excellent agreement between the two systems, there appears to be no direct relationship between the effect that a substituent has on the migratory aptitude of the phenyl group with either the Hammett σ values, which measure electron releasing or withdrawing or ability of the substituents, or Brown's σ^+ values, which measure the ability of the substituent to stabilize a positive charge in the transition state of the reaction. These data also suggest that delocalization of the unpaired electron is the important factor in determining the migratory aptitudes of the aryl groups.

III. Migration of Halogen Atoms

A. ADDITIONS TO HALOGENATED ALKENES

The light-induced addition of hydrogen bromide to 3,3,3-trichloropropene (**28**) yielded 1,1,2-trichloro-3-bromopropane (**29**) as the addition product.[17] Examination of the structure of this addition product shows that a chlorine atom has migrated from one carbon to the adjacent carbon. The mechanism proposed for the formation of this addition product involves the 1,2-shift of a chlorine atom in the adduct radical **30** yielding the rearranged radical **31** which

$$HBr + CH_2{=}CHCCl_3 \rightarrow BrCH_2CHClCHCl_2 \qquad (9\text{-}36)$$
$$28 \phantom{{=}CHCCl_3 \rightarrow BrCH_2}29$$

by

$$Br + CH_2{=}CHCCl_3 \rightarrow BrCH_2\dot{C}HCCl_3 \qquad (9\text{-}37)$$
$$\phantom{Br + CH_2{=}CHCCl_3 \rightarrow BrCH_2}30$$

$$BrCH_2\dot{C}HCCl_3 \rightarrow BrCH_2CHCl\dot{C}Cl_2 \qquad (9\text{-}38)$$
$$\phantom{BrCH_2\dot{C}HCCl_3}30 31$$

$$BrCH_2CHCl\dot{C}Cl_2 + HBr \rightarrow BrCH_2CHClCCl_2H + Br\cdot \qquad (9\text{-}39)$$
$$\phantom{BrCH_2CHCl\dot{C}Cl_2}31 29$$

reacts with hydrogen bromide to produce the observed addition product. It should be noted that this migration is one in which the rearranged radical is more stable than its precursor since delocalization of the unpaired electron in the two chlorines renders more stabilization to the rearranged radical than is found for the secondary alkyl radical. Apparently fluorine atoms do not migrate since addition of hydrogen bromide to 3-fluoro-3,3-dichloropropene (**32**) yields only 1-fluoro-1,2-dichloro-3-bromopropane (**33**) as a rearranged addition product.[18] More unrearranged addition product is formed than rearranged

$$HBr + CH_2{=}CHCFCl_2 \rightarrow$$
$$\phantom{HBr + CH_2{=}CHCFCl}32$$
$$BrCH_2CHClCHClF\ (18\%) + BrCH_2CH_2CCl_2F\ (39\%) \quad (9\text{-}40)$$
$$33$$

product indicating that a fluorine has less stabilizing effect on the rearranged radical than does a chlorine.

Free-radical additions of other reagents to 3,3,3-trichloropropene (**28**) also yield products that result from a 1,2-shift of a chlorine atom in the adduct radical as illustrated by the reactions of this unsaturate with bromotrichloro-methane,[19] bromine,[20] and thiophenol.[21] No 1,2-shift of a chlorine was observed in the addition of bromine to **28** in acetic acid, conditions for ionic addition of

$$BrCCl_3 + CH_2{=}CHCCl_3 \rightarrow Cl_3CCH_2CHClCCl_2Br \qquad (9\text{-}41)$$
$$28$$

$$Br_2 + CH_2{=}CHCCl_3 \rightarrow BrCH_2CHClCCl_2Br \qquad (9\text{-}42)$$
$$28$$

$$C_6H_5SH + CH_2{=}CHCCl_3 \rightarrow C_6H_5SCH_2CHClCHCl_2 \qquad (9\text{-}43)$$
$$28$$

bromine to unsaturated linkages. Similarly, reaction of thiophenol with **28** in the presence of sulfur, a reagent which inhibits the free-radical chain reaction, yielded only unrearranged addition product.

B. REACTIONS OF POLYHALOMETHANES AND DIAZOMETHANE

The light-induced reaction of carbon tetrachloride and diazomethane yields pentaerithitol chloride (**34**) in a chain reaction that can best be explained in terms of a sequence of reactions involving 1,2-shifts of chlorine atoms.[22]

$$Cl_4C + 4CH_2N_2 \xrightarrow{h\nu} \underset{\mathbf{34}}{C(CH_2Cl)_4} \tag{9-44}$$

The mechanism for this reaction not only involves the migration of chlorine atoms but also a radical-propagating displacement of nitrogen from diazomethane. Evidence for the reaction proceeding by the chain sequence shown rather than by a series of carbene insertion reactions is twofold. First, **34** is the only halogenated reaction product formed and products intermediate between carbon tetrachloride and **34** (e.g., Cl_3CCH_2Cl, $Cl_2C(CH_2Cl)_2$, and $ClC(CH_2Cl)_3$) are not found nor do they react with diazomethane to form the product. Second, the quantum yield, that is the number of molecules of product formed per quantum of light absorbed, is over 200.

$$Cl_4C \xrightarrow{h\nu} Cl\cdot + Cl_3C\cdot \text{ (Initiation)} \tag{9-45}$$

Similar reaction products are obtained in the light-induced reactions of diazomethane with bromotrichloromethane and chloroform. However, reaction of diazomethane with carbon tetrabromide yields methylene bromide and vinylidene bromide as reaction products.[23] The 2,2,2-tribromoethyl radical (35), produced by the reaction of the tribromomethyl radical with diazomethane, undergoes β-elimination of a bromine atom yielding vinylidene bromide faster than it rearranges to the 1,1,2-tribromoethyl radical (36). The eliminated bromine atom reacts with diazomethane yielding a bromomethyl

$$Br_3C \cdot + CH_2N_2 \rightarrow Br_3C\dot{C}H_2 \qquad (9\text{-}47)$$
$$\quad \quad \quad \quad \quad \quad 35$$

$$Br_3C\dot{C}H_2 \leftrightarrow Br_2\dot{C}CH_2Br \qquad (9\text{-}48)$$
$$\quad 35 \quad \quad \quad 36$$

$$Br_3C\dot{C}H_2 \rightarrow Br_2C{=}CH_2 + Br \cdot \qquad (9\text{-}49)$$

$$Br \cdot + CH_2N_2 \rightarrow Br\dot{C}H_2 + N_2 \qquad (9\text{-}50)$$

$$Br\dot{C}H_2 + Br_4C \rightarrow BrCH_2Br + Br_3C \cdot \qquad (9\text{-}51)$$

radical which, on reaction with carbon tetrabromide, produces the methylene bromide and a chain-carrying tribromomethyl radical.

C. HALOGENATION OF ALKYL HALIDES

Rearrangements involving the 1,2-shift of bromine atoms occur rapidly enough to compete with eliminations of the bromine atom from free radicals at $-78°C$. The photochemical chlorination of isopropyl bromide at this temperature results in the formation of 1-bromo-2-chloro-propane (37) in 15% yield.[24] The free-radical chain reaction suggested to account for this product is shown in 9-52 through 9-54. Although the rearranged products comprised

$$Cl \cdot + CH_3CHBrCH_3 \rightarrow HCl + CH_3CHBr\dot{C}H_2 \qquad (9\text{-}52)$$

$$CH_3CHBr\dot{C}H_2 \rightarrow CH_3\dot{C}HCH_2Br \qquad (9\text{-}53)$$

$$CH_3\dot{C}HCH_2Br + Cl_2 \rightarrow CH_3CHClCH_2Br + Cl \cdot \qquad (9\text{-}54)$$
$$\quad \quad \quad \quad \quad \quad \quad 37$$

only 15% of the chlorinated product, the results of this reaction illustrate that the 1,2-shift of the bromine atom proceeds at a rate comparable to that of reaction of the 2-bromopropyl radical with chlorine. The rate of the latter reaction is, however, dependent on the chlorine concentration and higher yields of rearranged product would be expected at lower concentrations of the molecular chlorine.

Photochemical chlorination at $-78°C$ of t-butyl bromide (38) with t-butyl hypochlorite yielded 92% of the rearranged product 1-bromo-2-chloro-2-methylpropane (39).[24] The extensive amount of rearrangement in this reaction

$$(CH_3)_3CO\cdot + (CH_3)_3CBr \rightarrow (CH_3)_3COH + (CH_3)_2\underset{\underset{\displaystyle \cdot CH_2}{|}}{CBr} \qquad (9\text{-}55)$$

$$\textbf{38}$$

$$(CH_3)_2\underset{\underset{\displaystyle \cdot CH_2}{|}}{CBr} \rightarrow (CH_3)_2\dot{C}CH_2Br \qquad (9\text{-}56)$$

$$(CH_3)_2\dot{C}CH_2Br + (CH_3)_3COCl \rightarrow (CH_3)_2CClCH_2Br + (CH_3)_3CO\cdot$$
$$\textbf{39}$$
$$(9\text{-}57)$$

is due in part to the large increase in stabilization resulting in the formation of a tertiary alkyl radical relative to the primary alkyl radical. Reaction of the unrearranged radical with t-butyl hypochlorite is slow enough that the rearrangement proceeds faster than the abstraction of chlorine.

IV. Migration of Vinyl Groups

The decarbonylation of 3-methyl-4-pentenal (**40**) yielded a mixture of alkenes consisting of 10.8 parts 1-pentene to 1 part 3-methyl-1-butene (**41**).[25] The formation of 1-pentene requires a rearrangement of the carbon skeleton of the intermediate radical **42** to the linear arrangement of the carbons in

$$CH_2{=}CHCH\underset{\underset{\displaystyle \textbf{40}}{\underset{\displaystyle CH_3}{|}}}{CH_2CHO} \xrightarrow[130°]{DTBP}$$

$$\underset{\text{10.8 parts}}{CH_3CH_2CH_2CH{=}CH_2} + CH_2{=}CHCH\underset{\underset{\displaystyle \textbf{41} \;\; \text{1 part}}{\underset{\displaystyle CH_3}{|}}}{CH_3} + CO \qquad (9\text{-}58)$$

by

$$CH_2{=}CHCH\underset{\underset{\displaystyle CH_3}{|}}{CH_2\dot{C}O} \rightarrow CO + CH_2{=}CHCH\underset{\underset{\displaystyle \textbf{42}}{\underset{\displaystyle CH_3}{|}}}{\dot{C}H_2} \qquad (9\text{-}59)$$

$$CH_2{=}CHCH\underset{\underset{\displaystyle \textbf{42}}{\underset{\displaystyle CH_3}{|}}}{\dot{C}H_2} \rightarrow \underset{\textbf{43}}{CH_3\dot{C}HCH_2CH{=}CH_2} \qquad (9\text{-}60)$$

$$CH_3\dot{C}HCH_2CH{=}CH_2 + CH_2{=}CHCHCH_2CHO \rightarrow$$

$$\underset{\displaystyle CH_3}{|}$$

$$CH_3CH_2CH_2CH{=}CH_2 + CH_2{=}CHCHCH_2\dot{C}O \quad (9\text{-}61)$$

$$\underset{\displaystyle \dot{C}H_3}{|}$$

radical **43**. Although the transformation could be accounted for by the migration of a methyl group, the 1,2-shift of the vinyl group seems more likely since methyl groups have not been observed to migrate in other systems. Decarbonylation of *trans*-3-methyl-4-hexenal (**44**) resulted in an 89% yield of the 2-hexenes (*cis*:*trans* = 1:5.3). Again the rearrangement of the carbon skeleton can best be interpreted in terms of a 1,2-shift of a vinyl group. The observation that the

$$\rightarrow CH_3CH_2CH_2CH{=}CHCH_3 \quad (9\text{-}62)$$
$$(\textit{cis} \text{ and } \textit{trans})$$

via

unrearranged alkene produced in the decarbonylation, namely 4-methyl-2-pentene (**45**), was also a mixture of the *cis*- and *trans*-isomers is of interest since it suggests existence of an intermediate radical in which the π-electron system of the alkene linkage is lost. This point will be discussed further in a subsequent section of this chapter.

The 2-cyclopentenylmethyl radical (**47**), generated by the decomposition of *t*-butyl-2-cyclopentenylperoxy acetate (**46**), rearranges to the 4-cyclohexenyl radical (**48**) as evidenced by the formation of cyclohexene in 47% yield when the perester is decomposed in *p*-cymene.[26] Decomposition of *t*-butylcyclohexene-4-peroxycarboxylate (**49**) in *p*-cymene yielded only cyclohexene (and some cyclohexadiene). It might appear that ring expansion by a 1,2-shift of a vinyl group is feasible but ring contraction by a similar 1,2-shift does not occur. The observations might better be explained in terms of the stabilities of the two radicals in that the primary radical **47** rearranges to the secondary radical **48** whereas the reverse process, as was pointed out before, occurs less readily.

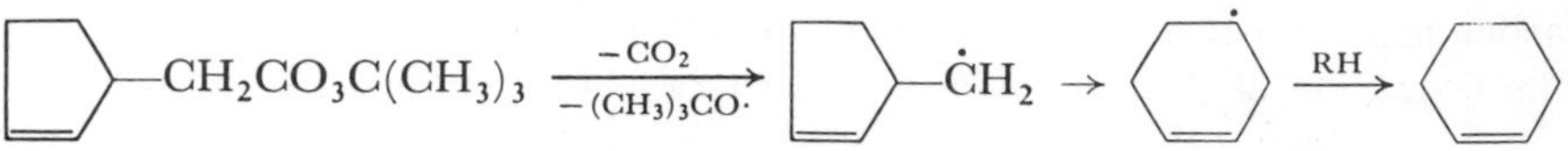

(9-64)

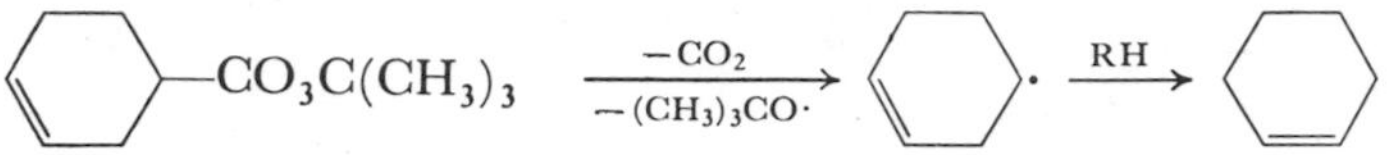

(9-65)

The original observation of a 1,2-shift of a vinyl group is somewhat more subtle than those in the previous examples. Dehydrogenation of 2,4,4-trimethyl-2-pentene (**50**) with iodine in the vapor phase at 500°C yielded p-xylene as a reaction product.[27] Formation of this product requires rearrangement of the carbon skeleton in order to have the necessary six carbons in the linear arrangement for the formation of the aromatic ring. The mechanism suggested in 9-66 through 9-69 finds support in that 2,5-dimethyl-2-hexene (**52**) is also formed in this reaction and may have resulted from reaction of the rearranged radical **51** with hydrogen iodide (9-68).

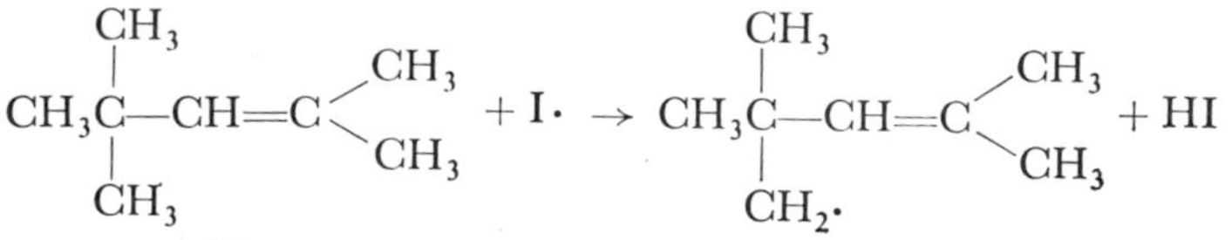

(9-66)

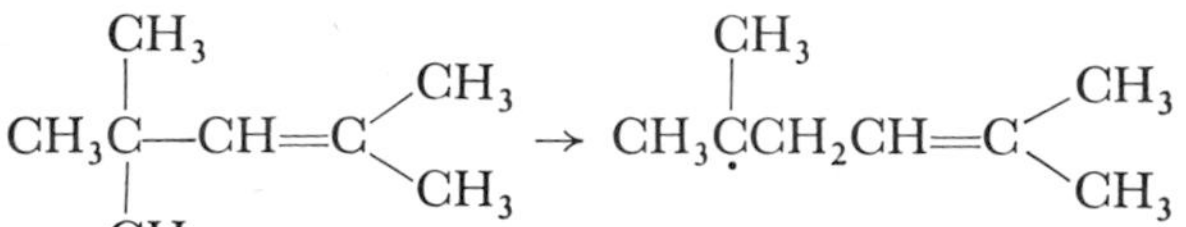

(9-67)

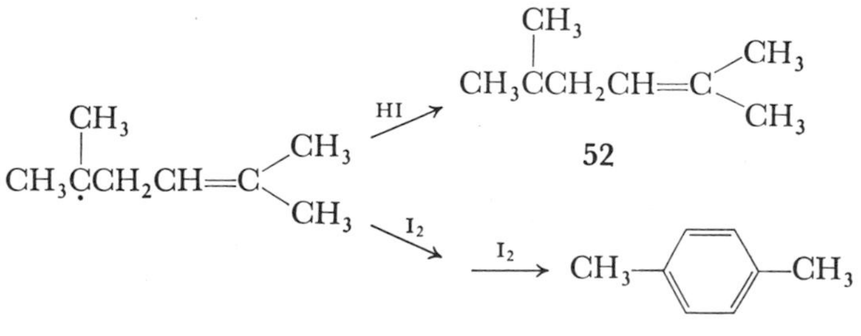

(9-68)

(9-69)

V. Migration of Acyl Groups

Acyl groups migrate in very much the same manner as vinyl groups. Isophorone oxide (**53**) was converted to 2,4,4-trimethylcyclopentanone (**54**) and carbon monoxide in a radical-initiated reaction at temperatures above 200°C.[28] The formation of this rearranged product can be accounted for by a chain

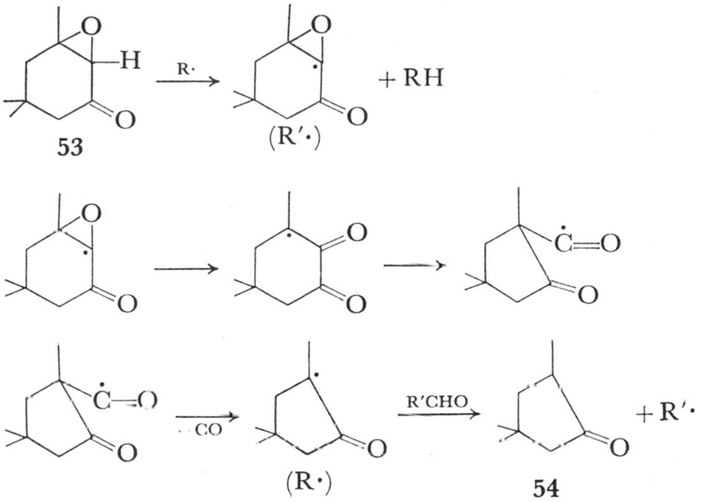

(9-70)

(9-71)

(9-72)

sequence in which a 1,2-shift of an acyl group occurred. Thermal decomposition of *t*-butyl-3-methyl-3-phenylperlevulinate (**55**) at 140°C yielded monomeric and dimeric products in which migration of the acyl group in the radical also had occurred.[29]

$$\underset{\substack{| \\ \text{CH}_3 \\ \text{55}}}{\overset{\substack{\text{C}_6\text{H}_5 \\ |}}{\text{CH}_3\text{COCCH}_2\text{CO}_3\text{C(CH}_3)_3}} \rightarrow \underset{\substack{| \\ \text{CH}_3}}{\overset{\substack{\text{C}_6\text{H}_5 \\ |}}{\text{CH}_3\text{COCCH}_2\cdot}} + \text{CO}_2 + (\text{CH}_3)_3\text{CO}\cdot \quad (9\text{-}73)$$

$$\underset{\substack{| \\ \text{CH}_3}}{\overset{\substack{\text{C}_6\text{H}_5 \\ |}}{\text{CH}_3\text{COCCH}_2\cdot}} \rightarrow \overset{\text{C}_6\text{H}_5}{\underset{\text{CH}_3}{\diagdown}}\text{CCH}_2\text{COCH}_3 \rightarrow \text{Products} \quad (9\text{-}74)$$

Rearrangement of a carboxyl group has been suggested to occur in the Vitamin B_{12} catalyzed rearrangement of methylmalonyl CoA (**56**) to succinyl CoA (**57**).[30] The formation of the primary alkyl radical required, if this is

$$(\text{Co}^{+3}\text{-Cobamide}) + \underset{\underset{56}{\overset{|}{\text{COSCoA}}}}{\overset{\overset{\text{CO}_2\text{H}}{|}}{\text{HCCH}_3}} \quad \rightarrow \quad (\text{Co}^{+2}\text{-Cobamide}) + \underset{\underset{}{\overset{|}{\text{COSCoA}}}}{\overset{\overset{\text{CO}_2\text{H}}{|}}{\text{HCCH}_2 \cdot}} \quad + \text{H}^+$$

$$(9\text{-}75)$$

$$\underset{\overset{|}{\text{COSCoA}}}{\overset{\overset{\text{CO}_2\text{H}}{|}}{\text{HCCH}_2 \cdot}} \quad \rightarrow \quad \overset{\overset{\text{CO}_2\text{H}}{|}}{\text{HCCH}_2\text{COSCoA}} \qquad (9\text{-}76)$$

$$\overset{\overset{\text{CO}_2\text{H}}{|}}{\text{HCCH}_2\text{COSCoA}} + (\text{Co}^{+2}\text{-Cobamide}) + \text{H}^+ \rightarrow$$

$$\underset{\underset{57}{\overset{|}{\text{CH}_2\text{COSCoA}}}}{\overset{\overset{\text{CH}_2\text{CO}_2\text{H}}{|}}{}} + (\text{Co}^{+3}\text{-Cobamide}) \quad (9\text{-}77)$$

indeed the mechanism of this rearrangement, is somewhat surprising but may well be the prerogative of the coenzyme in this enzymatic reaction. Labelling of the 2-carbon of **56** indicated that migration of the thiolester group had occurred in preference to migration of the carboxyl group.

VI. Mechanistic Interpretation

An interesting aspect of these rearrangements is that only certain groups appear to participate in the 1,2-shift. An understanding of the mechanism for the 1,2-shift must come from some characteristic(s) peculiar to those groups that do migrate.

The migrating group in these 1,2-shifts is in a position where it is capable of bonding both with the atom from which it originated and with the atom to which it is migrating. If some partial bonding to both atoms occurs, the question is whether this situation is simply a transition state or if a true intermediate is formed. If it is a transition state, the energy profile for the reaction would be that shown in Figure 9.1. On the other hand, if an intermediate is formed, the energy profile is somewhat more complex as shown in Figure 9.2. In this case, the intermediate can react either to form the rearranged radical or the original radical. Since the rearranged radical is often the more stable of the two, the activation energy requirement for this reaction likely is less than that for returning to the original radical.

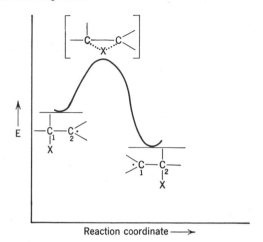

Fig. 9.1. Energy profile for one-step reaction.

Formation of an intermediate radical requires the formation of sigma bonds between the migrating group and both the atom from which it is leaving (migration origin) and the atom to which it is migrating (migration terminus). In the case of vinyl migrations, the intermediate is formed by an addition reaction yielding a cyclopropylcarbinyl radical (**58**) which can undergo a β-elimination reaction readily (Chapter 8). The elimination reaction would be expected to proceed faster in the direction yielding the more stable radical. This mechanism also allows for the *cis-trans* isomerization noted in both the rearranged and nonrearranged radicals in the decarbonylation of **44**,

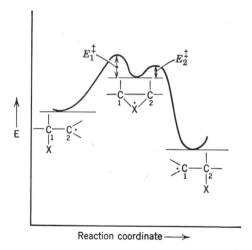

Fig. 9.2. Energy profile for reaction involving an intermediate.

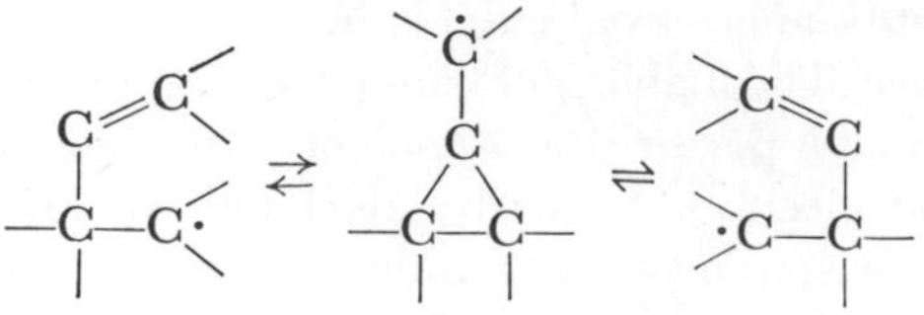

$$\text{(9-78)}$$

58

The susceptibility of aromatic rings toward addition by free radicals suggests that an intermediate cyclic radical such as **60** is formed in aryl migrations.

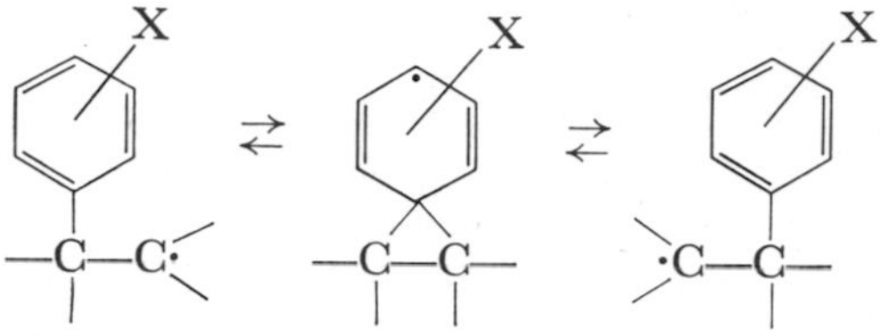

$$\text{(9-79)}$$

59   **60**

A substituent on the aromatic ring may increase the reactivity of the ring toward homolytic attack producing the intermediate radical. Complete formation of the intermediate radical almost certainly does not occur in the formation of β-phenylalkyl radicals. This is evident from the fact that only about 3% rearrangement occurs in the radical obtained from **20**.

In the rearrangements of arylmethoxy radicals, the intermediate radical would be an ethylene oxide-derived radical. The driving force for the observed

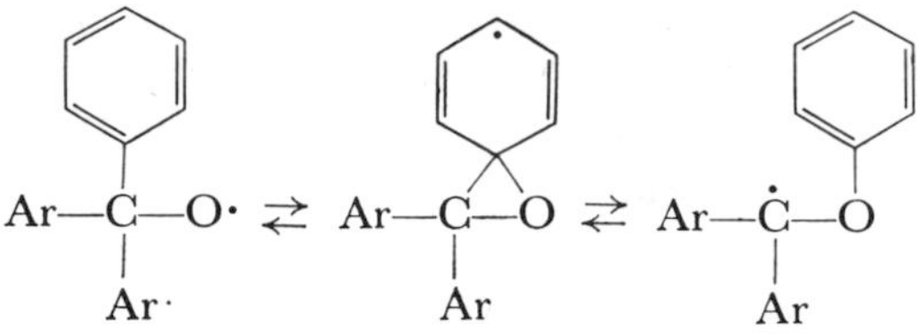

$$\text{Ar—C—O} \cdot \ \rightleftarrows \ \text{Ar—C—O} \ \rightleftarrows \ \text{Ar—C—O} \qquad \text{(9-80)}$$
$$\text{Ar} \cdot \qquad\qquad \text{Ar} \qquad\qquad \text{Ar}$$

ring opening in the direction of the rearranged radical may simply be the exothermicity of the reaction leading to a stabilized radical.

The situation in the case of the halogen migrations is somewhat different in that there is no π-electron system to which the addition of a radical can occur. There are, however, empty d-orbitals in both chlorine and bromine which may accommodate the extra electron. The resulting intermediate would be similar to the bridged radicals involving bromine (see Chapter 5). Although a bridged

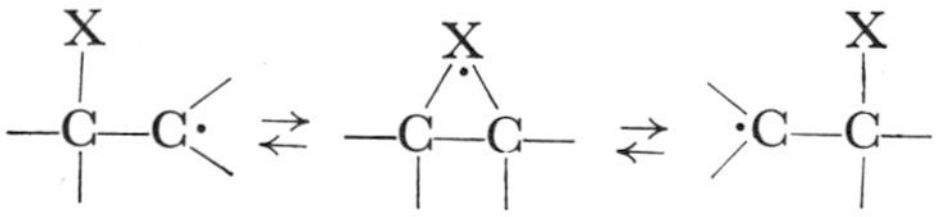

$$\text{(9-81)}$$

radical has not yet been observed for chlorine, this does not exclude the possibility of its existence. The inability of fluorine to migrate in such reactions may
be due to the fact that it possesses no orbitals of energy low enough to accommodate the unpaired electron. Similarly, alkyl radicals and hydrogen atoms,
having neither a π-electron system nor any low-energy orbitals, would not be
expected to form an intermediate similar to those suggested for the radicals
that do have migrating groups.

References

GENERAL

C. Walling, "Free Radical Rearrangements," Chapter 7, *Molecular Rearrange-
ments*, Part I, P. deMayo, Ed., Interscience Publishers, New York, 1963.
R. Kh. Freidlina, "Rearrangements of Radicals in Solution," Chapter 6,
Advances in Free Radical Chemistry, Vol. 1, G. Williams, Ed., Logas Press,
London, 1966.

SPECIFIC

1. E. S. Huyser and Z. Garcia, *J. Org. Chem.*, **27**, 2716 (1962).
2. J. A. Berson, C. J. Olsen, and J. S. Walia, *J. Amer. Chem. Soc.*, **82**, 5000
   (1960); **84**, 3337 (1962).
3. T. J. Wallace and R. J. Gritter, *J. Org. Chem.*, **26**, 5256 (1962).
4. C. Walling and A. Padwa, *J. Amer. Chem. Soc.*, **83**, 2207 (1961); F. D. Greene
   M. L. Bavilz, H. H. Lau, F. D. Osterholtz, and W. N. Smith, *ibid.*, **83**,
   2196 (1961).
5. S. Winstein and F. H. Seubald, *J. Amer. Chem. Soc.*, **69**, 2916 (1947).
6. W. H. Urry, Abstracts, 12th National Organic Symposium of American
   Chemical Society, Colorado, 1951, p. 36.
7. J. D. Backhurst, E. D. Hughes, and Sir C. Ingold, *J. Chem. Soc.*, **1959**,
   2742; H. C. Duffin, E. D. Hughes, and Sir C. Ingold, *ibid.*, **1959**, 2734.
8. D. Y. Curtin and J. C. Kauer, *J. Org. Chem.*, **25**, 880 (1960).
9. L. H. Slaugh, *J. Amer. Chem. Soc.*, **81**, 2262 (1959).
10. J. W. Wilt and H. Philip, *J. Org. Chem.*, **24**, 441 (1959); **25**, 891 (1960).
11. W. Rickatson and T. S. Stevens, *J. Chem. Soc.*, **1963**, 3960.
12. J. Weinstock and S. N. Lewis, *J. Amer. Chem. Soc.*, **79**, 6243 (1957).
13. M. S. Kharasch, A. C. Poshkus, A. Fono, and W. Nudenberg, *J. Org.
    Chem.*, **16**, 1458 (1951).
14. P. D. Bartlett and J. D. Cotman, Jr., *J. Amer. Chem. Soc.*, **72**, 3095 (1950).
15. C. Ruchardt and R. Hecht, *Tetrahedron Letters*, **1961**, 961.
16. See Chapter 5 for discussion of σ and σ^+ values.

17. N. V. Kost, T. T. Sedorova, R. Kh. Freidlina, and A. N. Nesmeyanov, *Doklady Akad. Nauk SSSR*, **132**, 606 (1960).

18. R. Kh. Freidlina, N. V. Kost, M. Ya Khorlina, and A. N. Nesmeyanov, *ibid.*, **137**, 341 (1961).

19. A. N. Nesmeyanov, R. Kh. Freidlina, and L. I. Zakharkin, *ibid.*, **81**, 199 (1951).

20. A. N. Nesmeyanov, R. Kh. Freidlina, and V. N. Kost, *ibid.*, **113**, 828 (1957); *Izvest. Akad. Nauk SSSR, Otdel. Khim. Nauk*, **1958**, 1205; *Tetrahedron*, **1**, 241 (1957).

21. R. Kh. Freidlina, A. B. Terentiev, R. G. Petrova, and A. N. Nesmeyanov, *Doklady Akad. Nauk SSSR*, **152**, 637 (1961).

22. W. H. Urry and J. R. Eiszner, *J. Amer. Chem. Soc.*, **75**, 5822 (1952).

23. W. H. Urry, J. R. Eiszner, and J. W. Wilt, *ibid.*, **79**, 918 (1957).

24. P. S. Skell, R. G. Allen, and N. D. Gilmour, *ibid.*, **83**, 504 (1961).

25. L. K. Montgomery, J. Matt, and J. R. Webster, Abstracts, American Chemical Society Meeting, Spring, 1964, p. 29N.

26. L. H. Slaugh, *J. Amer. Chem. Soc.*, **87**, 1522 (1965).

27. L. H. Slaugh, G. D. Mullineaux, and J. H. Raley, *ibid.*, **85**, 3180 (1963).

28. W. Reusch, C. K. Johnson, and J. A. Manner, *ibid.*, **88**, 2803 (1966).

29. W. Reusch and C. L. Karl, Abstracts, American Chemical Society Meeting, San Francisco, 1968, p. 79.

30. H. Eggeren, P. Overath, P. Lynen, and E. R. Stadtman, *J. Amer. Chem. Soc.*, **82**, 2643 (1960).

Chapter 10
Chemistry of Initiators

I. Introduction

Many free-radical chain reactions were described in the previous chapters but little attention was given, in most cases, to the initiation process of the chain reaction. An appreciation of the initiation processes is important, however, in order to appreciate the potential and limitations of chain reactions.

Some of the chemistry of the more common chemical initiators used in free-radical reactions is discussed in this chapter. Many of these initiators undergo other reactions and their value as initiators of chain reactions may be limited by these other reactions. The format of this chapter consists in examining, first, the chemistry of the unimolecular decompositions of these initiators, reactions that generally produce free radicals that initiate chain processes, followed by discussion of some of the other reactions of the initiators that limit their usefulness in this capacity.

The subject of initiation is limited to chemical initiations. The rudiments of photochemical processes in initiation (as well as high-energy radiation) were mentioned in Chapter 1. Many excellent sources covering radical-producing photochemical reactions are available and the reader is urged to refer to these for further information in this area.[1]

II. Diacyl Peroxides

A. UNIMOLECULAR DECOMPOSITION

The simplest of the diacyl peroxides used as an initiator is acetyl peroxide (**1**) which decomposes at a reasonable rate in the temperature range of 70 to 80°C in most inert solvents (those that do not induce the decomposition of the peroxide). The principle products resulting from its decomposition are carbon dioxide and methane, the latter resulting from hydrogen abstraction from the

$$\underset{\textbf{1}}{CH_3\overset{\overset{O}{\|}}{C}O O \overset{\overset{O}{\|}}{C}CH_3} \xrightarrow{RH} 2CH_4 + 2CO_2 \qquad (10\text{-}1)$$

solvent by the methyl radicals produced in the unimolecular decomposition. Smaller amounts of methyl acetate and ethane are also observed.

$$\underset{}{CH_3\overset{\overset{O}{\|}}{C}O O \overset{\overset{O}{\|}}{C}CH_3} \rightarrow \underset{\textbf{2}}{\left(2CH_3\overset{\overset{O}{\|}}{C}O\cdot \right)} \rightarrow 2CH_3\cdot + CO_2 \qquad (10\text{-}2)$$

$$CH_3\cdot + RH \rightarrow CH_4 + R\cdot \qquad (10\text{-}3)$$

$$2R\cdot \rightarrow R_2 \qquad (10\text{-}4)$$

Coupling of the solvent-derived radicals yields oxidative dimerization products of the solvents. In this manner, alkyl aromatics have been converted to 1,2-diaryl derivates of ethane,[2] carboxylic acids and esters with at least a single α-hydrogen to derivatives of succinic acid and to succinate esters,[3] respectively, and ketones to 1,4-diones.[4] The solvent-derived radical is generally

$$\underset{\textbf{1}}{(CH_3CO_2)_2} + 2ArCHR_2 \rightarrow 2CH_4 + 2CO_2 + ArCR_2CR_2Ar \qquad (10\text{-}5)$$

$$(CH_3CO_2)_2 + R_2CHCO_2H \rightarrow 2CH_4 + 2CO_2 + \begin{array}{c} R_2CCO_2H \\ | \\ R_2CCO_2H \end{array} \qquad (10\text{-}6)$$

$$(CH_3CO_2)_2 + \overset{\overset{\displaystyle O}{\|}}{R}\overset{}{C}CHR_2 \rightarrow 2CH_4 + 2CO_2 + \begin{array}{c} \overset{\displaystyle O}{R_2C} \overset{\nearrow}{C}R \\ | \\ R_2C \overset{\searrow}{C}R \overset{\displaystyle O}{} \end{array} \qquad (10\text{-}7)$$

the one expected in terms of resonance stabilization.

It is significant that acetic acid is not observed as a reaction product in the decomposition of acetyl peroxide in inert solvents. Apparently the acetoxy radicals (2) resulting from the homolytic cleavage of the peroxide linkage decompose to carbon dioxide and methyl radicals more rapidly than they react with the solvent. Attempts to trap the acetoxy radical by interaction with iodine, a species very reactive toward radical attack, have failed to yield any evidence of the existence of the acetoxy radical outside of the solvent cage[5] (see unimolecular decomposition of benzoyl peroxide).

Evidence for the existence of the acetoxy radical in the solvent cage comes from the elegant experiment of Martin and Taylor who showed that when acetyl peroxide labelled with O^{18} in the carbonyl oxygens was allowed to undergo partial observable decomposition, scrambling of the oxygens in the recovered peroxide had occurred.[6] This observation was explained in terms of a reaction in which two acetoxy radicals, species in which the two oxygens become indistinguishable, were formed and then recombined. It would appear,

$$CH_3\overset{\overset{\displaystyle O^*}{\|}}{C}\underset{\underset{\displaystyle O—O}{}}{}\overset{\overset{\displaystyle *O}{\|}}{C}CH_3 \rightleftarrows \left(CH_3\overset{\nearrow O^*}{C\cdots}\overset{*O}{\cdots C}CH_3\right) \rightleftarrows \quad CH_3\overset{\overset{\displaystyle *O—O*}{}}{C}\underset{O}{} \underset{O}{}\overset{}{C}CH_3$$

$$\xrightarrow{\text{Escape from solvent cage}} 2CH_3\cdot + CO_2 \qquad (10\text{-}8)$$

however, that decomposition of acetoxy radicals to carbon dioxide and methyl radicals is very rapid. Methyl acetate, produced in the reaction by recombination of a methyl radical resulting from fragmentation of only one acetoxy radical of the pair present in the solvent cage, also shows that scrambling of the oxygens had occurred. The rate of decomposition of the acetoxy radical must, therefore, be greater than the rate of escape of the radical from the solvent cage.

Unimolecular decomposition of benzoyl peroxide (**3**) yields benzoyloxy radicals (**4**) which are stable enough to react with many substrates at rates comparable to the decarboxylation rate. Decomposition of the peroxide in

$$
\underset{\textbf{3}}{C_6H_5\overset{O}{\overset{\|}{C}}OO\overset{O}{\overset{\|}{C}}C_6H_5} \rightarrow \underset{\textbf{4}}{2C_6H_5\overset{O}{\overset{\|}{C}}O\cdot} \tag{10-9}
$$

$$
C_6H_5\overset{O}{\overset{\|}{C}}O\cdot + I_2 \rightarrow C_6H_5\overset{O}{\overset{\|}{C}}OI + I\cdot \tag{10-10}
$$

$$
C_6H_5\overset{O}{\overset{\|}{C}}OI + H_2O \rightarrow C_6H_5CO_2H + HOI \tag{10-11}
$$

moist carbon tetrachloride containing iodine results in quantitative conversion of benzoyl peroxide to benzoic acid.[7] When formed in less reactive substrates, the unimolecular decarboxylation reaction of the radical competes with its bimolecular interactions with the substrate. In the presence of olefins and polymerizable monomers, benzoate esters were formed as reaction products.[8] Some carbon dioxide was also produced indicating that some decomposition of the benzoyloxy radical had occurred (reaction 10-13) in competition with the addition reaction (e.g., 12 and 36% for p-chloro- and p-bromobenzoyl peroxide,

$$
C_6H_5\overset{O}{\overset{\|}{C}}O + CH_2=CHR \rightarrow C_6H_5\overset{O}{\overset{\|}{C}}OCH_2CHR \tag{10-12}
$$

$$
C_6H_5\overset{O}{\overset{\|}{C}}O\cdot \rightarrow C_6H_5\cdot + CO_2 \tag{10-13}
$$

respectively with styrene). Decompositions in parafins yield benzoic acid resulting from hydrogen abstraction from the alkane by the benzoyloxy radical.[9]

$$
C_6H_5\overset{O}{\overset{\|}{C}}O\cdot + RH \rightarrow C_6H_5\overset{O}{\overset{\|}{C}}OH + R\cdot \tag{10-14}
$$

R = alkyl

However, alkanes are less reactive toward attack by benzoyloxy radicals than most unsaturates are toward addition by the radical as evidenced by the larger amounts of carbon dioxide resulting from decomposition of the benzoyloxy radical being observed in reactions of the peroxide in cyclohexane (48%) compared to cyclohexene (15%).[9]

Decomposition of benzoyl peroxide in aromatics generally results in formation of biphenyls and some higher polyphenyls. These products result from addition reactions of phenyl radicals, produced by fragmentation of benzoyloxy radicals, to the aromatic compound. The resulting adduct radicals can interact with other radicals present in the system yielding either a biphenyl (10-16) or a coupling product (10-17). The latter should be quite reactive to further addition which probably best accounts for the ultimate formation of polyphenyls.[10]

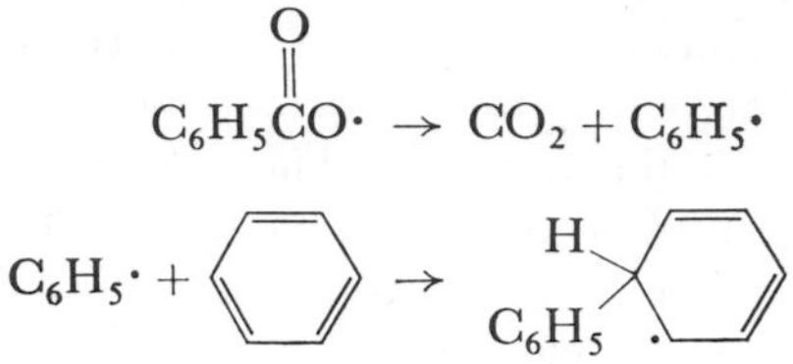

$$C_6H_5\overset{\overset{\displaystyle O}{\displaystyle \|}}{C}O\cdot\ \rightarrow\ CO_2 + C_6H_5\cdot$$

(10-15)

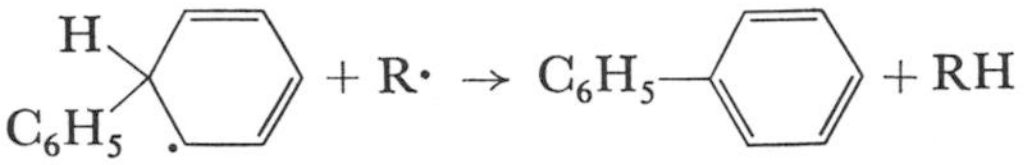

(10-16)

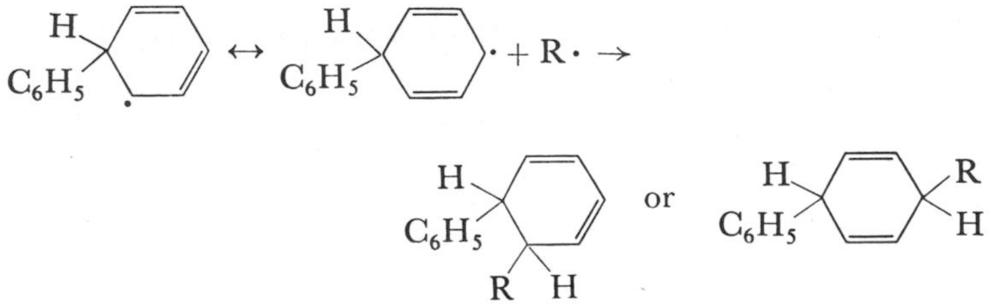

(10-17)

The unimolecular decomposition rates of *meta*- and *para*-substituted benzoyl peroxides indicates that electron-withdrawing groups decrease whereas electron-releasing groups increase the rate of homolytic cleavage of the peroxide.[11] The effect, however, is not large ($\rho < 1$).

Unimolecular decomposition reactions of other aroyl peroxides and acyl peroxides are similar to those of benzoyl and acetyl peroxide, although yields of radicals may vary. Table 10-6 lists some of the more readily available diacyl peroxides and the half-lives of their unimolecular decomposition reactions at temperatures convenient for their use as initiators of free-radical reactions.

B. INDUCED DECOMPOSITIONS

The rates of decomposition of diacyl peroxides are markedly enhanced in certain solvents. This rate enhancement in some cases is the result of a direct interaction of solvent with the peroxide in polar reactions (e.g., phenols[12]). In other cases solvent-derived radicals react with the peroxide in chain-propagating reactions.[13] The consequence of the latter is the establishment of a free-radical chain reaction in which the solvent is generally oxidized. This section covers the solvent-induced decompositions of acyl peroxides by chain reactions and by polar interactions with amines as well as a brief examination of a unimolecular decomposition path yielding products other than free radicals.

1. Induced Decomposition in Ethers. The half-life of the unimolecular decomposition of benzoyl peroxide in most inert solvents at 80°C is about 5 hours. At the same temperature the half-life of this peroxide is less than 5 minutes in ethyl ether and about 15 minutes in n-butyl ether.[13] This marked increase in rate can be accounted for by a chain-propagating reaction of an ether-derived radical with the peroxide. The nature of this reaction was elucidated by Cass who found the major products of the reaction of ethyl ether and benzoyl peroxide at 37°C to be benzoic acid and 1-ethoxyethylbenzoate (5) plus smaller amounts of carbon dioxide.[14] The chain sequence 10-18 and 10-19 accounts for these products.

$$(C_6H_5CO_2)_2 \xrightarrow{\;k_9\;} 2C_6H_5CO_2\cdot \qquad (10\text{-}9)$$
$$\quad\; 3 \qquad\qquad\qquad\quad 4$$

$$C_6H_5CO_2\cdot + C_2H_5OCH_2CH_3 \xrightarrow{\;k_{18}\;} C_6H_5CO_2H + C_2H_5O\dot{C}HCH_3 \quad (10\text{-}18)$$

$$C_2H_5O\dot{C}HCH_3 + (C_6H_5CO_2)_2 \xrightarrow{\;k_{19}\;} C_2H_5OCHCH_3 + C_6H_5CO_2\cdot \quad (10\text{-}19)$$
$$\qquad\qquad\qquad\qquad\qquad\qquad\qquad |$$
$$\qquad\qquad\qquad\qquad\qquad\qquad OCOC_6H_5$$
$$\qquad\qquad\qquad\qquad\qquad\qquad\quad 5$$

$$C_6H_5CO_2\cdot + C_2H_5O\dot{C}HCH_3 \xrightarrow{\;k_{\text{III}}\;} C_2H_5OCHCH_3 \qquad (10\text{-}20)$$
$$\qquad\qquad\qquad\qquad\qquad\qquad\qquad |$$
$$\qquad\qquad\qquad\qquad\qquad\qquad OCOC_6H_5$$
$$\qquad\qquad\qquad\qquad\qquad\qquad\quad 5$$

The rate of decomposition of benzoyl peroxide in ethyl ether was observed to be first order in peroxide. This observation would be consistent with the reaction sequence shown in which the rate of decomposition of the peroxide is

$$\frac{-d[\text{Per.}]}{dt} = k_9[\text{Per.}] + k_{19}[C_2H_5O\dot{C}HCH_3][\text{Per.}] \qquad (10\text{-}21)$$

described by 10-21 and termination is a cross-coupling reaction such as shown in 10-20. The derived rate law for the reaction would be given by 10-22. Evidence for participation of the benzoyl peroxide in a chain reaction comes

$$\text{Rate} = \left[\frac{3}{4}k_9 \pm \frac{k_9}{2}\sqrt{\frac{1}{4} + 2\frac{k_{18}k_{19}}{k_9k_{20}}[(C_2H_5)_2O]} \right][\text{Per.}] \qquad (10\text{-}22)$$

from the observations that the rate of its decomposition in ethers can be reduced by addition of inhibitors or polymerizable olefins that would react with either of the chain-carrying radicals. Reaction of benzoyl peroxide with 1,4-dioxane (6) and with 1,2-diethoxyethane (7) yield similar products.

$$\text{(6)} + (C_6H_5CO_2)_2 \rightarrow \underset{\text{—OCOC}_6H_5}{\text{(dioxane)}} + C_6H_5CO_2H \quad (10\text{-}23)$$

6

$$C_2H_5OCH_2CH_2OC_2H_5 + (C_6H_5CO_2)_2 \rightarrow$$

7

$$\overset{OCOC_6H_5}{\underset{|}{C_2H_5OCHCH_2OC_2H_5}} + C_6H_5CO_2H \quad (10\text{-}24)$$

The formation of the acylal in a free-radical chain reaction can be accounted for in two ways. One route involves attack by the α-alkoxyalkyl radical at a carbonyl-oxygen of the peroxide producing the intermediate adduct radical which, on fragmentation, yields the products of the reaction. Another route to

$$R\cdot + \overset{C_6H_5}{\underset{|}{O=COOCOC_6H_5}} \rightarrow$$

$$\overset{C_6H_5}{\underset{|}{ROC\text{—}O\text{—}OCOC_6H_5}} \rightarrow \overset{C_6H_5}{\underset{|}{ROC=O}} + \cdot \overset{O}{\underset{||}{OCC_6H_5}} \quad (10\text{-}25)$$

the same products would be a direct displacement of a benzoyloxy radical from the peroxide linkage by the α-alkoxyalkyl radical as shown in 10-26. Labelling

$$R\cdot + \overset{O^*}{\underset{|}{O\text{—}OCC_6H_5}} \rightarrow R\text{—}O + \cdot\overset{O}{\underset{||}{OCC_6H_5}} \quad (10\text{-}26)$$
$$\underset{*O}{\overset{CC_6H_5}{\diagup}} \qquad \underset{*O}{\overset{C}{\diagup}\diagdown C_6H_5}$$

$$\xrightarrow{\text{LiAlH}_4} C_6H_5CH_2\overset{*}{O}H$$

the carbonyl oxygen of the peroxide with O^{18} gave an acylal that on reduction with LiAlH$_4$ retained about 80% of the original label[15] indicating that most of the attack by the α-alkoxyalkyl radical occurred on the peroxide oxygen as shown in 10-26. Although this may appear to be somewhat less favorable from a steric standpoint than attack at the carbonyl-oxygen, the polar contributions to the transition state of the displacement reaction may outweigh the steric problems.

$$
\left[
\begin{array}{c}
\overset{\delta+}{R}\cdots O\cdots\overset{\delta-}{O}\overset{\overset{\displaystyle O}{\|}}{C}C_6H_5 \\[2mm]
\underset{O}{\overset{\diagup}{\diagdown}}\overset{C}{\diagdown}C_6H_5
\end{array}
\right]
$$

R = α-alkoxyalkyl radical

2. Induced Decompositions in Alcohols.

Oxidations of primary and secondary alcohols to aldehydes and ketones respectively by diacyl peroxides involve induced decomposition of the peroxide. The half-life of benzoyl peroxide in ethyl and isopropyl alcohols at 80°C is about 2 to 3 minutes[13] in contrast to half-lives of the peroxide up to 4 to 6 hours at the same temperature in inert solvents. The decomposition rate of acetyl peroxide in ethyl alcohol is at least twenty times that of the peroxide in solvents such as benzene, toluene, and carbon tetrachloride.[16] Decomposition of acetyl peroxide in isopropyl alcohol yields acetone and only about half of the expected amounts of carbon dioxide and methane as reduction products of the peroxide. Acetic acid and isopropyl acetate, in contrast to decompositions of this peroxide in inert solvent, are observed as reaction products.[17] The latter results most likely from reactions of the alcohol with acetic acid produced in the free-radical reaction. The formation of acetic acid is indicative of some mode of decomposition of the peroxide other than the unimolecular homolytic cleavage that generally yields only methane and CO_2 as reaction products. The mechanism outlined in 10-27 through 10-29 accounts for both the enhanced rate of reaction of the peroxide and for the reaction products which, for the most part, are formed in the chain sequence 10-27 and 10-28.

$$(CH_3CO_2)_2 \rightarrow 2CH_3\cdot + CO_2$$

$$CH_3\cdot + (CH_3)_2CHOH \rightarrow CH_4 + (CH_3)_2\dot{C}OH \tag{10-27}$$

$$(CH_3)_2\dot{C}OH + (CH_3CO_2)_2 \rightarrow (CH_3)_2C{=}O + CH_3CO_2H + CO_2 + CH_3\cdot \tag{10-28}$$

$$CH_3\cdot + (CH_3)_2\dot{C}OH \rightarrow CH_4 + (CH_3)_2C{=}O \tag{10-29}$$

A displacement on one of the peroxy-oxygens by an α-hydroxyalkyl radical, analogous to the displacement reaction for the α-alkoxyalkyl radical in the induced decompositions of benzoyl peroxide in ethers, would produce α-

$$(CH_3)_2\overset{\underset{\textstyle OH}{|}}{\underset{\underset{\textstyle O}{\|}}{\underset{\textstyle CCH_3}{|}}}C\cdot + O-O\overset{\overset{\textstyle O}{\|}}{C}CH_3 \rightarrow (CH_3)_2\overset{\underset{\textstyle OH}{|}}{C}-O + \cdot O\overset{\overset{\textstyle O}{\|}}{C}CH_3 \qquad (10\text{-}30)$$

acetoxyisopropyl alcohol (**8**) and the acetoxy radical. Decomposition of the latter would produce carbon dioxide and the chain-carrying methyl radical. The hemiketal **8** is unstable and decomposes to acetone and acetic acid.

$$(CH_3)_2\underset{\underset{\textstyle OH}{|}}{\overset{\overset{\textstyle O}{\|}}{C}}OCCH_3 \rightarrow (CH_3)_2C{=}O + CH_3CO_2H \qquad (10\text{-}31)$$

8

Another route to the products of the chain-propagating reaction 10-28 is a hydrogen atom transfer reaction from the α-hydroxyalkyl radical to the peroxide

$$R_2\dot{C}OH + O-O\overset{\overset{\textstyle O}{\|}}{C}CH_3 \rightarrow \left[R_2C\cdots O\cdots H\cdots O\cdots O\overset{\overset{\textstyle O}{\|}}{C}CH_3 \right] \rightarrow$$

$$R_2C{=}O + HO\overset{\overset{\textstyle O}{\|}}{C}CH_3 + (\cdot O\overset{\overset{\textstyle O}{\|}}{C}CH_3) \qquad (10\text{-}32)$$
$$\rightarrow CH_3\cdot + CO_2$$

linkage. This mode of interaction of α-hydroxyalkyl radicals appears to be important in the induced decompositions of *t*-butyl peroxide by alcohols and may be operative in the induced decompositions of diacyl peroxides as well.
3. Induced Decompositions in Amines. The behavior of acyl peroxides with amines differs from that in alcohols and ethers in that the reactions are spontaneous at lower temperatures and apparently do not require any unimolecular decomposition of the peroxide to start a chain process. Although many of these reactions are not necessarily free-radical processes, some do, however, shown definite free-radical character. The reaction of benzoyl peroxide with dimethylaniline (**9**) proceeds readily at temperatures as low as 0°C, and apparently involves formation of free radicals since the presence of these components in styrene causes the free-radical polymerization of the latter.

The products of the reaction of dimethylaniline and benzoyl peroxide in chloroform are mainly benzoic acid, resulting from reduction of the peroxide, and formaldehyde and monomethylaniline resulting from oxidation of the amine. Studies of this reaction indicate that the primary reaction possibly is a nucleophilic displacement by the amine on the peroxide yielding the cationic intermediate **10** and a benzoate ion.[18] The observation that p-methyldimethyl-

$$C_6H_5\overset{\cdot\cdot}{N}(CH_3)_2 + BzOOBz \rightarrow \left(\begin{array}{c} CH_3 \\ | \\ C_6H_5N-OBz \\ | \\ CH_3 \end{array}\right)^{+} + BzO^{-} \quad (10\text{-}33)$$

$$\underset{\mathbf{9}}{} \qquad\qquad \underset{\mathbf{10}}{}$$

aniline reacts faster and p-cyanodimethylaniline very much slower than dimethylaniline supports the proposed nucleophilic reaction 10-33 (electron-releasing methyl group enhances the nucleophilicity of the nitrogen whereas the electron-withdrawing p-cyano group decreases it).[19] The cationic intermediate can decompose by two routes, one leading to free radicals (10-34) and one to the imine intermediate **11** which on hydrolysis would yield the dealkylated amine and formaldehyde.

$$\left(\begin{array}{c} CH_3 \\ | \\ C_6H_5N-OBz \\ | \\ CH_3 \end{array}\right)^{+} \rightarrow C_6H_5\overset{+}{N}\overset{CH_3}{\underset{CH_3}{\diagup}} + \cdot OBz$$

$$ (10\text{-}34)$$

$$\underset{\mathbf{10}}{}$$

$$\left(\begin{array}{c} CH_3 \\ | \\ C_6H_5N-OBz \\ | \\ CH_3 \end{array}\right)^{+} \rightarrow \left[\begin{array}{c} CH_3 \\ | (+) \\ C_6H_5N-O \\ H_2C \diagdown \; C-C_6H_5 \\ H \diagup\!\!\!\diagdown O \end{array}\right] \rightarrow$$

$$C_6H_5\overset{CH_3}{\underset{CH_2}{\overset{\diagup}{N^{(+)}}}} + \overset{O}{\underset{HO}{\diagdown}}CC_6H_5 \quad (10\text{-}35)$$

$$\underset{\mathbf{11}}{}$$

Although primary and secondary amines react rapidly with benzoyl peroxide, there is no evidence that the reactions either produce free radicals or involve free-radical intermediates. The products of the reactions can generally be

explained in terms of ionic reactions resulting from prior formation of intermediates such as **10** formed by nucleophilic displacement on the peroxide by
the amine. However, reactions of a secondary amine with an acyl peroxide that
may well proceed by a free-radical chain reaction are the oxidations of 1,4-
dihydro-2,6-dimethyl-3,5-dicarboethoxypyridine (**12**) by both acetyl and
benzoyl peroxides. In the case of acetyl peroxide oxidation of **12**, the products
of the reaction are the pyridine (**13**), acetic acid, methane, and carbon dioxide.[20]
These products could be expected if the chain sequence 10-37 and 10-38 is

operative. The reaction takes place at room temperature suggesting that a
bimolecular reaction producing a pair of chain-carrying radicals may be
involved in the initiation step (see Chapter 11).

4. Ester Formation. The formation of esters as decomposition products of
certain acyl peroxides can account for a major part of their decomposition.

One simple way of explaining formation of esters is by a solvent cage recombination of an acyloxy radical with an alkyl radical, the latter resulting from decarboxylation of one of the acyloxy radicals produced by the homolytic cleavage of the peroxide linkage. The complete scrambling of the oxygens in

$$\text{RCOOCR} \rightleftarrows \left(\text{RCO·} + \text{·OCR} \right) \rightarrow \left(\text{RCO·} + \text{R·} \right) + CO_2 \quad (10\text{-}40)$$

Caged radical pair   Caged radical pair

$$\left(\text{RCO·} + \text{R·} \right) \xrightarrow{\text{Cage combination}} \text{RCOR} \quad (10\text{-}41)$$
$$\longrightarrow 2R· + CO_2 \quad (10\text{-}42)$$

the methyl acetate formed in the decomposition of carbonyl-oxygen labelled acetyl peroxide supports this route for the formation of ester.[6] Other paths leading to the formation of esters must, however, be considered. Decomposition of optically active methylethylacetyl peroxide (**14**) produced optically active *sec*-butyl methylethylacetate (**15**) which on hydrolysis yielded 2-butanol which

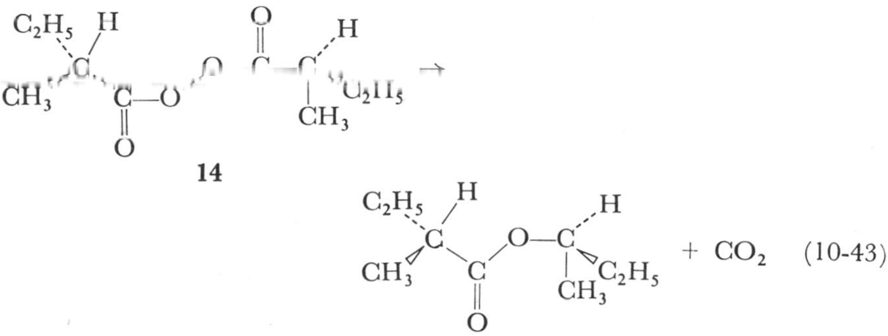

14

$+ CO_2 \quad (10\text{-}43)$

15

showed no change in configuration at the asymmetric center.[21] It is possible, of course, that recombination occurs more rapidly than racemization of the alkyl radical in the solvent cage. Greene and his co-workers showed that the esters found in acyl peroxide decomposition reactions may result from decarboxylation of the "carboxyl inversion product" produced by the peroxide. Such reactions proceed with retention of configuration of the alkyl group as evidenced by the isolation of **17** from **16** and the decarboxylation of **17** to *trans*-4-*t*-butylcyclohexyl *trans*-4-*t*-butyl-cyclohexylcarboxylate (**18**).[22] Rearrangement to the "carboxyl inversion product" which decomposes to the ester may be a general route of decomposition of secondary aliphatic acyl peroxides.

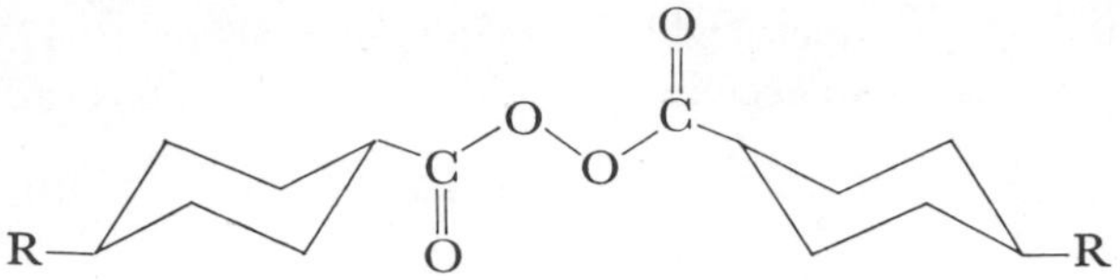

16

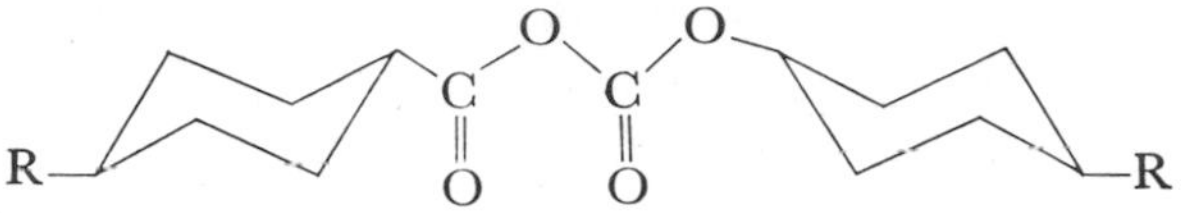

(10-44)

17

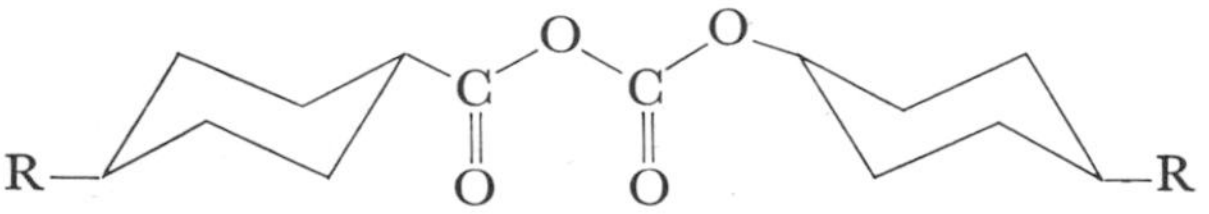

17

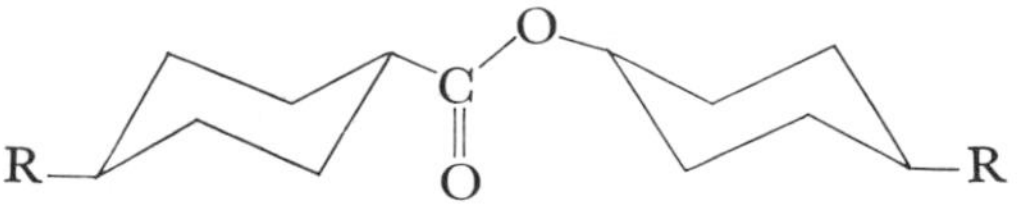

(10-45)

$R = C(CH_3)_3$ 18

III. Dialkyl Peroxides

A. UNIMOLECULAR DECOMPOSITION

1. Products. *t*-Butyl peroxide (**19**) is the most readily available of the dialkyl peroxides and, therefore, has received the most study. It is a convenient source of free radicals, decomposing at higher temperatures than the acyl peroxides discussed in the previous section (see Table 10-6). Decomposition in the gas phase is a unimolecular reaction yielding two *t*-butoxy radicals as shown in reaction 10-46. The products of the gas phase decomposition reaction are ethane and acetone.[23] Decomposition of *t*-butyl peroxide in solution often

$$(CH_3)_3COOC(CH_3)_3 \rightarrow 2(CH_3)_3CO \cdot$$

(10-46)

19

$$(CH_3)_3CO \cdot \rightarrow CH_3COCH_3 + CH_3 \cdot \qquad (10\text{-}47)$$

$$2CH_3 \cdot \rightarrow C_2H_6 \qquad (10\text{-}48)$$

yields t-butyl alcohol indicating that hydrogen atom abstraction by the t-butoxy radical from the solvent (10-49) competes with fragmentation reaction 10-47. The ratio of t-butyl alcohol relative to acetone observed in these reactions can

$$(CH_3)_3CO \cdot + RH \rightarrow (CH_3)_3COH + R \cdot \qquad (10\text{-}49)$$

serve as an indication of the reactivities of various solvents toward hydrogen abstraction by t-butoxy radicals.[24] The t-butyl alcohol:acetone ratio is higher in solvents that are reactive toward hydrogen abstraction provided the solvent is present in a sufficiently high concentration.

t-Amyl peroxide (**20**) behaves similarly in that homolytic fission of the peroxide linkage yields alkoxy radicals. It is interesting, however, that generally little t-amyl alcohol is found as a reaction product in even fairly reactive solvents. The fragmentation of t-amyloxy radicals to acetone and ethyl radicals (reaction 10-51) apparently is more facile than the fragmentation of t-butoxy radicals (reaction 10-47) probably because of the greater stability of the ethyl radical relative to the methyl radical.

$$
\begin{array}{c}
\overset{\displaystyle CH_3}{\underset{\displaystyle CH_3}{|}}\ \overset{\displaystyle CH_3}{\underset{\displaystyle CH_3}{|}} \\
C_2H_5COOCC_2H_5 \rightarrow 2C_2H_5CO \cdot \\
\mathbf{20}
\end{array}
\qquad (10\text{-}50)
$$

$$
\begin{array}{c}
\overset{\displaystyle CH_3}{|} \\
C_2H_5CO \cdot \rightarrow CH_3COCH_3 + C_2H_5 \cdot \\
\underset{\displaystyle CH_3}{|}
\end{array}
\qquad (10\text{-}51)
$$

Cumyl peroxide (**21**) decomposes at about the same rate as t-butyl peroxide. Fragmentation of cumyloxy radicals yields acetophenone indicating a preference for β-elimination of a methyl radical rather than a phenyl radical from this alkoxy radical.

$$
\begin{array}{c}
\overset{\displaystyle CH_3}{\underset{\displaystyle CH_3}{|}}\ \overset{\displaystyle CH_3}{\underset{\displaystyle CH_3}{|}} \\
C_6H_5COOCC_6H_5 \rightarrow 2C_6H_5CO \cdot \\
\mathbf{21}
\end{array}
\qquad (10\text{-}52)
$$

$$
\begin{array}{cc}
\mathrm{CH_3} & \mathrm{O} \\
| & \| \\
\mathrm{C_6H_5CO\cdot} \rightarrow & \mathrm{C_6H_5CCH_3 + CH_3\cdot} \\
| & \\
\mathrm{CH_3} &
\end{array}
\qquad (10\text{-}53)
$$

2. Kinetics. The decomposition rate of t-butyl peroxide in the gas phase has been the object of many investigations.[25] Although rate data are readily obtained for the gas phase reaction because of an increase in the number of moles of gas in proceeding from reactants to products, determination of the decomposition rate of the peroxide in the liquid phase (either neat or in solution) presents a problem. t-Butyl peroxide does not react readily at ambient temperatures with most reducing agents (e.g., iodide ion) allowing for facile determination of peroxide by methods employed for acyl peroxides. Kinetic data for decomposition of the peroxide, however, have been determined both by infrared and gas chromatographic analysis of unreacted peroxide.

The gas phase kinetic studies indicate that the peroxide decomposes in a unimolecular reaction having an activation energy of 38.1 kcal/mole. If the assumption is made that the recombination reaction of two t-butoxy radicals is a process requiring essentially no activation energy, then, by the principle of microscopic reversibility, the transition state of the decomposition reaction must involve essentially complete rupture of the oxygen-oxygen bond. Furthermore, it can be concluded that the bond dissociation energy of the peroxide linkage in t-butyl peroxide is 38.1 kcal/mole, the enthalpy of activation for the homolytic cleavage. This use of kinetic data is illustrative of one of the more valuable means of determining bond-dissociation energies.

$$(CH_3)_3COOC(CH_3)_3 \ \rightleftarrows \ [(CH_3)_3CO\cdot\cdot OC(CH_3)_3] \ \rightleftarrows \ 2(CH_3)_3CO\cdot (10\text{-}54)$$

Early reports[26] indicated that the rates of decomposition of t-butyl peroxide in solution were essentially the same as those for the gas phase decomposition. Subsequent work has shown that the peroxide does undergo induced decompositions (see next section) and, furthermore, the rate of the unimolecular decomposition is influenced to some degree by the nature of the solvent.[27] Table 10-1 lists some rate data observed for the first-order decomposition of t-butyl peroxide in solvents that do not induce the decomposition. Although the decomposition rates of the peroxide in these solvents are not too different from that in the gas phase, the activation parameters indicate that the solvent does influence the decomposition reaction significantly. Most marked are the effects of acetonitrile and acetic acid, solvents reported to solvate effectively t-butoxy radicals[28] (see Chapter 5). These activation parameters can be explained in terms of solvation of the transition state of the decomposition reaction thereby lowering the enthalpy requirement for the transition state and

TABLE 10-1   Kinetic Data for Decomposition of *t*-Butyl Peroxide in Solution

| Solvent | $k \times 10^5$ sec^{-1} | | Activation Parameters[a] | |
	125°C	135°C	$\Delta H^{\ddagger}$ (*kcal/mole*)	$\Delta S^{\ddagger}$ (*eu*)
Cyclohexane	1.52	4.64	41	21
Triethylamine	1.69	5.55	41	21
Dimethylaniline	1.89	5.84	38	14
Cyclohexene	1.38	4.41	37	12
Tetrahydrofuran	1.84	5.76	37	12
t-Amyl alcohol	2.34	6.80	36	9
Nitrobenzene	2.39	7.20	36	9
Ethyl benzoate	1.92	5.90	35	8
Benzene	1.99	6.19	35	8
t-Butyl alcohol	2.49	7.32	34	6
Acetic acid	2.98	(6.29)[b]	33	5
Acetonitrile	3.47	(5.63)[b]	31	-1.5
Gas phase	1.1	3.6	38	14

[a] Activation parameters calculated from rate constants determined at two additional temperatures as well (see Ref. 27).

[b] Rate constants at 130°C.

concurrently lowering the entropy. One factor balances the other resulting in essentially no change in the rate of decomposition in solvents that do solvate the transition state of the reaction.

The rate of decomposition of *t*-amyl peroxide is very similar to that of *t*-butyl peroxide in the gas phase and in most inert solvents. The decomposition reaction in the gas phase is influenced by the pressure of *t*-amyl peroxide suggesting that an induced decomposition of the peroxide is also operative.[29] This induced decomposition possibly involves a mechanism similar to that proposed for the induced decomposition of *t*-butyl peroxide in the liquid phase (see next section).

Extensive kinetic studies of other alkyl peroxides have not been made because of difficulty in analyzing the peroxides. The information available concerning *t*-butyl and *t*-amyl peroxides would indicate that kinetic data on other peroxides would prove quite informative.

B. INDUCED DECOMPOSITIONS

1. Alkyl Attack. Thermal decomposition of undiluted *t*-butyl peroxide is decidedly more rapid than in the gas phase or when diluted with an inert solvent. An induced decomposition of the peroxide involving an intramolecular displacement on the peroxide linkage by the peroxide-derived radical **22** has been proposed to account for this enhanced rate of decomposition.[30] Isolation of isobutylene oxide (23) as a reaction product led to the proposal of this mechanism.

$$(CH_3)_2C-O-OC(CH_3)_3 \rightarrow (CH_3)_2C-O + \cdot OC(CH_3)_3 \quad (10\text{-}55)$$
$$\overset{|}{CH_2}\cdot \qquad\qquad\qquad \overset{\diagup}{CH_2}$$

22   **23**

$$(CH_3)_3CO\cdot + (CH_3)_3COOC(CH_3)_3 \rightarrow (CH_3)_3COH + (CH_3)_2\overset{|}{C}OOC(CH_3)_3$$
$$\overset{|}{CH_2}\cdot \quad (10\text{-}56)$$

or $$(CH_3)_3CO\cdot \rightarrow CH_3COCH_3 + CH_3\cdot \qquad (10\text{-}47)$$

$$CH_3\cdot + (CH_3)_3COOC(CH_3)_3 \rightarrow CH_4 + (CH_3)_2\overset{|}{C}OOC(CH_3)_3\cdot$$
$$\overset{|}{CH_2}\cdot \qquad (10\text{-}57)$$

A similar induced decomposition of *t*-butyl peroxide is operative in the hydrogen chloride catalyzed decompositions of the peroxide.[31] Isobutylene oxide is also formed by the intramolecular displacement of the peroxide-derived radical **22** which, in this case, is formed by reaction of a chlorine atom with *t*-butyl peroxide. Reaction of a methyl radical with hydrogen chloride serves as the source of chlorine atoms. Other products are also found but the formation

$$Cl\cdot + (CH_3)_3COOC(CH_3)_3 \rightarrow HCl + (CH_3)_2\overset{|}{C}OOC(CH_3)_3$$
$$\overset{|}{CH_2}\cdot \qquad (10\text{-}58)$$
$$\textbf{22}$$

$$(CH_3)_2\overset{|}{C}OOC(CH_3)_3 \rightarrow (CH_3)_2C-O + \cdot OC(CH_3)_3 \quad (10\text{-}55)$$
$$\overset{|}{CH_2}\cdot \qquad\qquad\qquad \overset{\diagup}{CH_2}$$
$$\textbf{22} \qquad\qquad\qquad \textbf{23}$$

$$(CH_3)_3CO\cdot \rightarrow CH_3COCH_3 + CH_3\cdot \qquad (10\text{-}47)$$

$$CH_3\cdot + HCl \rightarrow CH_4 + Cl\cdot \qquad (10\text{-}59)$$

of isobutylene oxide can be accounted for by this route as can the enhanced rate of reaction of the peroxide. The enhanced rates of reaction of *t*-butyl peroxide in reactions with vinyl chloride and tetrachloroethylene[32] likely involve a similar sequence of reactions, the hydrogen chloride in this case resulting from some product or intermediate radical produced in the reactions.

Reaction of t-butyl peroxide with the styrene-derived radical produced in styrene polymerization has been suggested to involve abstraction of one of the alkyl hydrogens.[33] In this case, no marked enhancement of the rate of decomposition of the peroxide was observed.

2. Oxygen Attack. The rate of decomposition of t-butyl peroxide in primary and secondary alcohols is markedly faster than is the rate of the gas phase decomposition.[34] An induced decomposition of the peroxide by an alcohol-derived α-hydroxyalkyl radical (**24**) is probably involved in these reactions. The chain sequence 10-60 and 10-61 accounts for the formation of ketones and aldehydes resulting from reaction of secondary and primary alcohols, respectively, with t-butyl peroxide. It also accounts for the enhanced rate of decomposition of the peroxide since the overall rate of decomposition of the peroxide

$$(CH_3)_3COOC(CH_3)_3 \xrightarrow{k_{46}} 2(CH_3)_3CO\cdot \qquad (10\text{-}46)$$

$$(CH_3)_3CO\cdot + R_2CHOH \xrightarrow{k_{60}} (CH_3)_3COH + R_2\dot{C}OH \qquad (10\text{-}60)$$
$$\mathbf{24}$$

$$R_2\dot{C}OH + (CH_3)_3COOC(CH_3)_3 \xrightarrow{k_{61}}$$
$$\mathbf{24}$$

$$R_2C{=}O + (CH_3)_3COH + (CH_3)_3CO\cdot \qquad (10\text{-}61)$$

involves the two reactions in which it is consumed. (The details of a kinetic analysis of the reaction of t-butyl peroxide are given in Chapter 3).

$$\frac{-d[Per.]}{dt} = k_{46}[Per.] + k_{61}[R_2\dot{C}OH][Per.] \qquad (10\text{-}62)$$

Of particular interest is the chain-propagating reaction 10-61, a reaction in which an α-hydroxyalkyl radical reduces the peroxide. In view of the work of Cass, Martin, and Denny with the ether-derived α-alkoxylalkyl radicals and acyl peroxides, a direct displacement by the α-hydroxyalkyl on the peroxide linkage (10-63) might be proposed. The hemiketal or hemiacetal (**25**) produced would decompose to the observed products. In contrast to the diacyl peroxides,

$$
R_2\underset{\underset{OH}{|}}{C}\cdot \; + \; (CH_3)_3C\overset{\overset{\displaystyle C(CH_3)_3}{\diagup}}{\underset{\diagdown}{O-O}} \;\longrightarrow\; R_2\underset{\underset{OH}{|}}{C}OC(CH_3)_3 + (CH_3)_3CO\cdot
$$
$$\mathbf{25} \qquad (10\text{-}63)$$

$$R_2\underset{\underset{OH}{|}}{C}OC(CH_3)_3 \;\longrightarrow\; R_2C{=}O + (CH_3)_3COH \qquad (10\text{-}64)$$
$$\mathbf{25}$$

α-alkoxyalkyl radicals do not interact readily with t-butyl peroxide. There is neither an induced decomposition of the peroxide nor are acetals or ketals found as products in the reactions of ethers with t-butyl peroxide. The oxygens of the peroxide linkage are protected by the t-butyl group from any attacking radical that could displace a t-butoxy radical, a situation similar to that encountered in the very slow S_N2 reactions of neopentyl halides.

$$R_2C\cdot \underset{RO}{\overset{}{|}} + \underset{(CH_3)_3C}{\overset{}{\diagup}} O{-}O \overset{C(CH_3)_3}{\diagup} \overset{\longrightarrow}{\not\longrightarrow} R_2\underset{RO}{\overset{}{\underset{|}{C}}}{-}OC(CH_3)_3 + (CH_3)_3CO\cdot$$

$$(10\text{-}65)$$

The interaction of the α-hydroxyalkyl radical with the peroxide can be accounted for in terms of a hydrogen atom transfer reaction. This reaction not only would be sterically more favorable than a displacement by a larger radical but is also energetically quite exothermic and involves a favorable polar effect in the transition state. It also accounts for the fact that α-alkoxyalkyl radicals

$$R_2\dot{C}{-}OH + \underset{(CH_3)_3C}{\overset{}{\diagup}} O{-}O \overset{C(CH_3)_3}{\diagup} \rightarrow \left[R_2\overset{\delta+}{C}{\cdots}O{\cdots}H{\cdots}O{\cdots}\overset{\delta-}{O}C(CH_3)_3 \atop \underset{(CH_3)_3C}{\overset{}{|}} \right] \rightarrow$$

$$R_2C{=}O + HOC(CH_3)_3 + \cdot OC(CH_3)_3 \quad (10\text{-}66)$$

do not react with the peroxide whereas α-hydroxyalkyl radicals do. There is no apparent reason why the hydrogen atom transfer reaction may not be operative in the induced decompositions of acyl peroxides by primary and secondary alcohols as well.

Primary and secondary amines induce the decomposition of t-butyl peroxide whereas tertiary amines do not. Interaction of primary or secondary amines with the peroxide in a chain reaction involving the hydrogen-atom transfer reaction (10-68) accounts for the induced decomposition of the peroxide.[35] The

$$R_2CHNHR' + (CH_3)_3CO\cdot \ \rightarrow\ R_2\dot{C}NHR' + (CH_3)_3COH \quad (10\text{-}67)$$

$$R_2\underset{\overset{|}{R'}}{\dot{C}NH} + (CH_3)_3COOC(CH_3)_3 \ \rightarrow$$

$$R_2C{=}NR' + (CH_3)_3COH + (CH_3)_3CO\cdot \quad (10\text{-}68)$$

$R' =$ alkyl, aryl, or hydrogen

product of the oxidation reaction is an imine and, in the case of primary amines, the imine produced may undergo a condensation reaction with unreacted primary amine. Decomposition of t-butyl peroxide in benzhydryl amine (**26**) yields ammonia and the condensation product benzhydrylidenebenzhydryl amine (**28**) resulting from interaction of benzophenone imine (**27**), the product of the free-radical chain reaction, with unreacted amine.[36]

$$\underset{\mathbf{26}}{(C_6H_5)_2CHNH_2} + (CH_3)_3CO\cdot \rightarrow (C_6H_5)_2\dot{C}NH_2 + (CH_3)_3COH \qquad (10\text{-}69)$$

$$(C_6H_5)_2\dot{C}NH_2 + (CH_3)_3COOC(CH_3)_3 \rightarrow \underset{\mathbf{27}}{(C_6H_5)_2C{=}NH} + (CH_3)_3OH + (CH_3)_3CO\cdot \qquad (10\text{-}70)$$

$$\underset{\mathbf{27}}{(C_6H_5)_2C{=}NH} + \underset{\mathbf{26}}{(C_6H_5)_2CHNH_2} \rightarrow \underset{\mathbf{28}}{(C_6H_5)_2C{=}NCH(C_6H_5)_2} + NH_3 \qquad (10\text{-}71)$$

A marked induced decomposition of *t*-butyl peroxide occurs in the presence of the dihydropyridine **12** which is oxidized to the pyridine **13** in the process.

$(CH_3)_3CO\cdot +$ 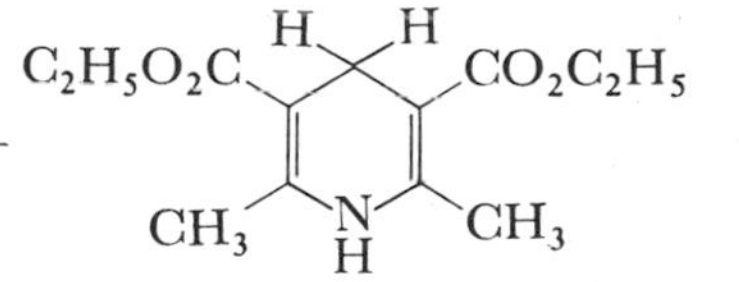 $\rightarrow$

12

$(CH_3)_3COH +$ 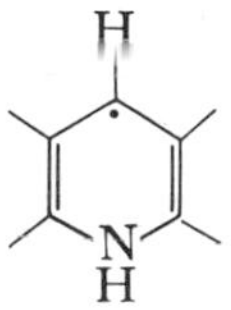 (10-72)

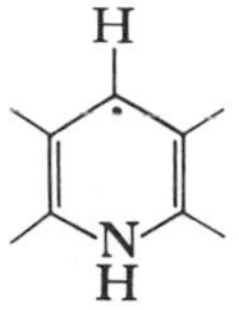 $+ (CH_3)_3COOC(CH_3)_3 \rightarrow$

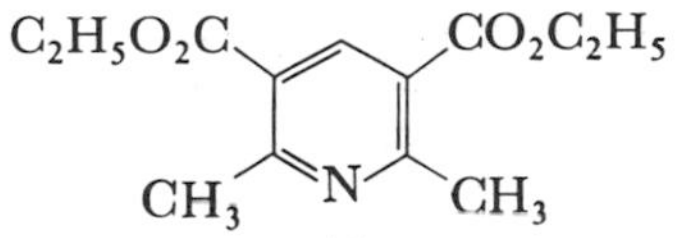 $+ (CH_3)_3COH + (CH_3)_3CO\cdot$ (10-73)

13

Both propagating steps in the chain sequence occur readily because of the reactivity of allylic hydrogens toward hydrogen abstraction and because of the energy gained in the formation of the aromatic ring in the hydrogen-atom transfer reaction. Both chain-propagating reactions also have favorable polar effects.

Tertiary amines do not induce the decomposition of *t*-butyl peroxide. Reaction of the *N*-methyl Hantzsch compound (**29**) with *t*-butyl peroxide yielded the oxidative dimerization product **30**. The radical produced by the hydrogen abstraction process has no nitrogen-bonded hydrogen which can be transferred to the peroxide linkage.[47]

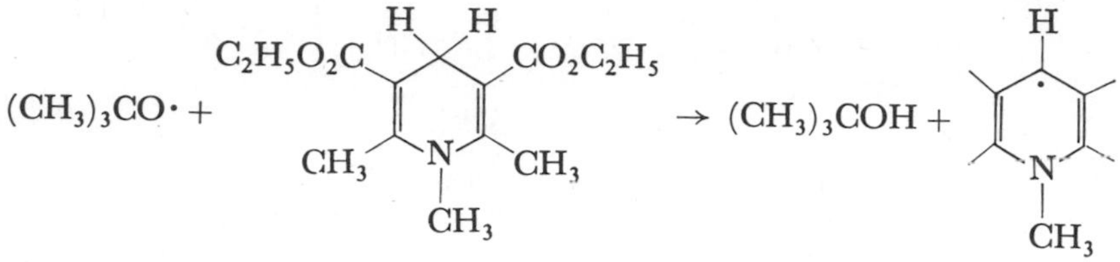

29

(10-74)

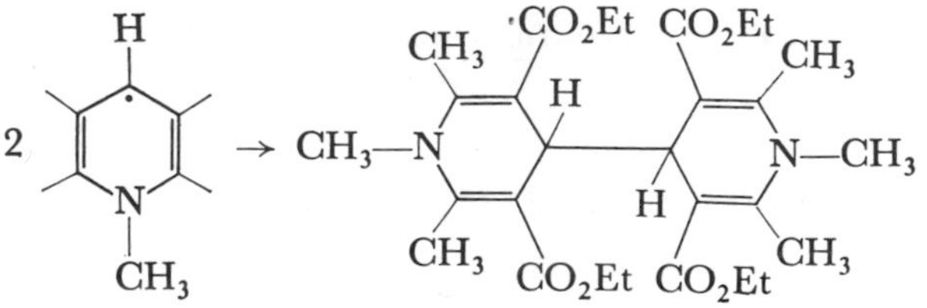

30

(10-75)

IV. Peresters

A. UNIMOLECULAR DECOMPOSITION

Peresters have functionalities of both acyl and alkyl peroxides and the unimolecular decompositions of peresters show characteristics of both acyl and alkyl peroxides. For example, the unimolecular decomposition rates of the more common peresters, namely *t*-butyl peracetate (**31**) and *t*-butyl perbenzoate (**32**), are intermediate between that of *t*-butyl peroxide and the corresponding acyl peroxides (see Table 10-6).

$$CH_3\overset{\overset{O}{\|}}{C}OOC(CH_3)_3 \qquad\qquad C_6H_5\overset{\overset{O}{\|}}{C}OOC(CH_3)_3$$

31 **32**

A characteristic feature of *t*-butyl peresters having the general structure

$RCOOC(CH_3)_3$ is that their unimolecular decomposition rates parallel the stability of the radical R·.[37] This observation suggests that the decomposition reaction involves not only cleavage of the oxygen-oxygen linkage but also simultaneous carbon-carbon bond cleavage as shown for the transition state of reaction 10-76. Decompositions of *t*-butyl phenylperacetate (**33**), *t*-butyl

$$RCOOC(CH_3)_3 \rightarrow \left(R \cdot \cdot C \overset{O}{\underset{\|}{}} O \cdot \cdot OC(CH_3)_3 \right) \rightarrow$$

$$R \cdot + CO_2 + \cdot OC(CH_3)_3 \quad (10\text{-}76)$$

trichloroperacetate (**34**), and *t*-butyl trimethylperacetate (**35**) are all first-order reactions yielding the *t*-butoxy radical, carbon dioxide and the benzyl, trichloromethyl, and *t*-butyl radicals, respectively. The activation parameters ($\Delta H^{\ddagger}$ and $\Delta S^{\ddagger}$) for these decompositions given in Table 10-2 with those of

TABLE 10-2   Activation Parameters for Unimolecular *t*-Butyl Perester ($RCOOC(CH_3)_3$) Decompositions[38]

Perester	R·	$\Delta H^{\ddagger}$	$\Delta S^{\ddagger}$
t-Butyl peracetate (**31**)	$CH_3 \cdot$	34.3	10.4
t-Butyl phenyperacetate (**33**)	$C_6H_5\dot{C}H_2$	28.1	2.2
t-Butyl trichloroperacetate (**34**)	$Cl_3C \cdot$	36.3	9.4
t-Butyl trimethylperacetate (**35**)	$(CH_3)_3C \cdot$	30.0	11.1

t-butyl peracetate show that the enthalpy requirements for decomposition parallel the stabilities of the radicals produced in the decomposition. The somewhat lower $\Delta S^{\ddagger}$ observed in the decomposition of peresters compared to *t*-butyl peroxide possibly reflects the more rigid nature of the transition state required for the concerted decompositions. In such a decomposition, which could be considered a four-centered reaction, a π-electron system is created and the stereochemical requirements leading to formation of the carbonyl linkage must be met. It is interesting to note that the perester leading to the benzyl radical, namely **33**, possibly involves the most extensive carbon-carbon bond cleavage in the transition state and also has the lowest entropy of activation.

Substituents in the *meta-* and *para*-positions of *t*-butyl perbenzoate effect the unimolecular decomposition rate suggesting that an inductive polar effect may be operative in these decompositions.[39] Electron-releasing groups increase the decomposition rate whereas withdrawing groups retard it. The effect has been interpreted as being the result of the substituent increasing or decreasing the ionic character of the peroxide linkage. Assuming that the *t*-butyl group always electron-releasing, more ionic character would be induced in the oxygen-oxygen linkage of the benzoate perester having electron-withdrawing groups whereas electron-releasing groups would have the opposite effect.

Decompositions of *meta-* and *para*-substituted *t*-butyl phenylperacetates (**36**) have been reported to be homolytic in character. The decomposition rate data correlate better in a $\rho\sigma^+$ relationship than in a $\rho\sigma$ relationship indicating that more than an inductive effect of the substituent is operative in these decompositions. A polar effect in which the benzyl radical produced exerts its donor qualities and the *t*-butoxy radical its acceptor qualities in the decomposition transition state accommodates the rate data.[40]

t-Butyl peracetate can apparently decompose unimolecularly in certain inert solvents that do not induce the decomposition (see next section) yielding acetic acid. These reactions most likely do not proceed by oxygen-oxygen cleavage followed by reaction of the resulting acetoxy radical with some hydrogen donating solvent since appreciable amounts of acetic acid ($\sim$30% based on peroxide consumed) are observed in the decomposition of *t*-butyl peracetate in carbon tetrachloride at 100 to 120°C.[41] An intramolecular reaction in which the hydrogen is supplied by the *t*-butoxy moiety may be involved. The activation parameters for acetic acid formation ($\Delta H^{\ddagger} = 28.2$ kcal/mole and $\Delta S^{\ddagger} = -15.62$ eu) support the suggestion that such a process may be operative. Isobutylene oxide was not, however, observed as a reaction product but it may have polymerized under the conditions of the reaction.

$$CH_3\overset{O}{\overset{\|}{C}}OOC(CH_3)_3 \rightarrow \left[CH_3\overset{O}{\overset{\|}{C}} \overset{H}{\underset{O——O}{\overset{\overset{H_2}{C}}{\rightleftharpoons}}} C(CH_3)_2 \right] \rightarrow CH_3\overset{O}{\overset{}{C}}{\underset{OH}{\diagdown}} +$$

31

$$\overset{CH_2}{\underset{O}{|}}C(CH_3)_2 \quad (10\text{-}78)$$

B. INDUCED DECOMPOSITIONS

Peresters, having structural features of both acyl and alkyl peroxides, might be expected to undergo induced decompositions in a manner similar to these peroxides. The data in Table 10-3 show that the decomposition rate of t-butyl

TABLE 10-3   Half-Life of t-Butyl Peracetate in Various Solvents at 95°C[41]

Solvent	Half-Life (min.)
Cyclohexane	1517
Toluene	1393
Collidine	1381
Acetonitril	1213
t-Butyl alcohol	1174
Carbon tetrachloride	1031
Tetrahydrofuran	197
2-Octanol	174
Cyclohexylamine	111
Cyclohexanol	107
Piperidine	72
2-Butanol	62

peracetate in solvents such as secondary alcohols, primary and secondary amines, and ethers is considerably faster than in solvents that do not induce the decomposition of either acyl or alkyl peroxides. The nature of the induced decomposition of the perester by an α-hydroxyalkyl radical or an α-aminoalkyl radical (having at least one nitrogen-bonded hydrogen) may involve hydrogen atom transfer to the peroxide linkage. However, the two oxygens are not equivalent and the transfer could occur to yield the t-butoxy radical and acetic acid as shown in 10-79 or t-butyl alcohol, carbon dioxide, and methyl radicals

$$R_2\overset{.}{C}OH + O\!-\!OC(CH_3)_3 \rightarrow R_2C\!\!=\!\!O + AcOH + \cdot OC(CH_3)_3$$
$$\underset{\displaystyle CH_3C\!\!=\!\!O}{\big|}$$

$$(10\text{-}79)$$

$$R_2\overset{.}{C}OH + \overset{\displaystyle O}{\overset{\displaystyle \|}{O\!-\!OCCH_3}} \rightarrow R_2C\!\!=\!\!O + (CH_3)_3COH + (\cdot \overset{\displaystyle O}{\overset{\displaystyle \|}{OCCH_3}})$$
$$\underset{\displaystyle (CH_3)_3C}{\big|} \qquad\qquad CH_3\!\cdot + CO_2 \;\;\hookleftarrow$$

$$(10\text{-}80)$$

as shown in 10-80. The formation of acetic acid in these induced decompositions
suggests that transfer of the hydrogen to the oxygen of the acetoxy moiety
rather than the *t*-butoxy moiety must occur to a significant extent. This is not
too surprising in view of the fact that the acetoxy-oxygen is less sterically
hindered toward displacement than the oxygen of the *t*-butoxy moiety.

Induced decompositions of the peresters are observed in ethers, an observation suggesting that stable products might be obtained from decompositions of
peresters in ethers. Reaction of α-alkoxyalkyl radicals with acyl peroxides
indicates that direct displacement on one of the peroxy-oxygens of these
peroxides can occur. Reaction of tetrahydrofuran (**37**) with *t*-butyl peracetate,
on this basis, would be expected to yield the acylal **39** as the result of a similar
displacement on the perester by the radical **38** derived from **37**. This product

$$(10\text{-}81)$$

was not observed, however, whereas 2-*t*-butoxytetrahydrofuran (**40**) was
isolated. The latter need not have been formed by a direct displacement by **38**
on the perester (reaction 10-82), a process which would involve as much steric

$$(10\text{-}82)$$

hindrance as displacement on *t*-butyl peroxide, since reaction of the acylal **39**
with *t*-butyl alcohol yielding acetic acid and **40** does occur readily. The formation of acetic acid as a reaction product would also exclude 10-82 as the reaction

of the ether-derived radical with the perester since the acetoxy radical would be expected to decarboxylate.

$$
\underset{\textbf{39}}{\text{[tetrahydrofuran ring]}}\underset{H}{\overset{O}{\text{OCCH}_3}} + (CH_3)_3COH \rightarrow \underset{\textbf{40}}{\text{[tetrahydrofuran ring]}}\underset{H}{\text{OC(CH}_3)_3} + CH_3\overset{O}{\overset{\|}{C}}OH \tag{10-83}
$$

V. Peroxydisulfate Ions

A. UNIMOLECULAR DECOMPOSITION

The peroxydisulfate ion (**41**) is a water-soluble species and has an oxygen-oxygen linkage similar to those of the organic peroxides. The peroxide linkage of **41** is thermally unstable and the ion can be homolytically cleaved into two sulfate ion radicals (**42**). Peroxydisulfate also decomposes in an acid-catalyzed

$$
S_2O_8^= = \ ^-O\overset{O}{\underset{O}{\overset{\uparrow}{S}}}-O-O-\overset{O}{\underset{O}{\overset{\uparrow}{S}}}-O^- \ \xrightarrow{k_{84}} \ 2 \ ^-O\overset{O}{\underset{O}{\overset{\uparrow}{S}}}O\cdot \ \equiv SO_4^-\cdot \tag{10-84}
$$
$$
\qquad\qquad\qquad \textbf{41} \qquad\qquad\qquad\qquad \textbf{42}
$$

reaction (see next section) and the overall rate law for the decomposition of **41** is given by 10-85. The acid-catalyzed reaction is not an important contributor

$$
\frac{-d[S_2O_8^=]}{dt} = k_{84}[S_2O_8^=] + k_a[H^+][S_2O_8^=] \tag{10-85}
$$

to the overall decomposition in neutral or basic solutions but becomes an important factor only in acidic solutions (pH of 3 or lower).

The rate of the unimolecular decomposition of peroxydisulfate is somewhat slower than that of benzoyl peroxide (see Table 10-6). The activation energy for the unimolecular decomposition is 33.5 kcal/mole.[42] The sulfate ion radicals produced appear to be capable of adding to vinyl monomers and consequently **41** is often used to initiate vinyl polymerization reactions (see Chapter 12).

In the absence of a reactive monomer with which it may react, sulfate ion radicals react with water to produce oxygen. A path that is likely operative for this reaction is shown in 10-86 through 10-90. Hydrogen peroxide is not observed as a reaction product[42, 43] but would decompose to oxygen and water (see Chapter 11).

$$
S_2O_8^= \rightarrow 2SO_4^-\cdot \tag{10-86}
$$

$$SO_4^- \cdot + H_2O \rightarrow HSO_4^- + HO \cdot \qquad (10\text{-}87)$$

$$2HO \cdot \rightarrow H_2O_2 \qquad (10\text{-}88)$$

$$HO \cdot + H_2O_2 \rightarrow H_2O + HOO \cdot \qquad (10\text{-}89)$$

$$2HOO \cdot + H_2O_2 \rightarrow O_2 + \cdot OH + H_2O \qquad (10\text{-}90)$$

B. INDUCED DECOMPOSITIONS

The acid-catalyzed decomposition of peroxydisulfate is a polar reaction and the products of the reaction depend on the acidity of the solution. The intermediate HSO_5^- is stable and accumulates in acidic solutions but in less acidic solutions it decomposes yielding oxygen.[43]

$$H^+ + S_2O_8^{2-} + H_2O \rightarrow SO_4^{2-} + 2H^+ + HSO_5^- \qquad (10\text{-}91)$$

$$HSO_5^- \rightarrow HSO_4^- + \tfrac{1}{2}O_2 \qquad (10\text{-}92)$$

Peroxydisulfate undergoes a marked induced decomposition in primary and secondary alcohols yielding aldehydes and ketones, respectively, as oxidation products.[44] Kinetic studies of the oxidation of isopropyl alcohol with peroxydisulfate indicate the reaction follows the rate law 10-93.[45] This rate law would be consistent with a reaction involving unimolecular decomposition

$$\frac{-d[S_2O_8^-]}{dt} = k'[S_2O_8^-][i\text{-PrOH}]^{1/2} \qquad (10\text{-}93)$$

of $S_2O_8^{2-}$ as the initiation reaction and a cross-termination process as shown in 10-96. The chain-propagating reactions 10-94 and 10-95 are similar to those proposed for the induced decomposition of both alkyl and acyl peroxides as well as peresters by secondary alcohols.

$$S_2O_8^{2-} \rightarrow SO_4^- \cdot$$

$$SO_4^- \cdot + (CH_3)_2CHOH \rightarrow HSO_4^- + (CH_3)_2\dot{C}OH \qquad (10\text{-}94)$$

$$(CH_3)_2\dot{C}OH + S_2O_8^{2-} \rightarrow (CH_3)_2C{=}O + HSO_4^- + SO_4^- \cdot \qquad (10\text{-}95)$$

$$(CH_3)_2\dot{C}OH + SO_4^- \cdot \rightarrow (CH_3)_2C{=}O + HSO_4^- \qquad (10\text{-}96)$$

The rate law for the decomposition of $S_2O_8^{2-}$ in methanol has been known for some time to be three-halves order in $S_2O_8^{2-}$ and thought to be half-order in methanol.[44] Such a rate law would be consistent with a mechanism similar to that shown for 2-propanol but initiated by bimolecular process 10-97 involving $S_2O_8^{2-}$ and methanol. Recent work has shown that the rate law is actually zero order in methanol and three-halves order in $S_2O_8^{2-}$.[46] This rate law is consistent

$$S_2O_8^{2-} + CH_3OH \rightarrow SO_4^- \cdot + \cdot CH_2OH + HSO_4^- \cdot \qquad (10\text{-}97)$$

with a mechanism involving a unimolecular initiation reaction and the bi-molecular termination reaction 10-100.

$$S_2O_8^{2-} \rightarrow SO_4^- \cdot \qquad (10\text{-}84)$$

$$SO_4^- \cdot + CH_3OH \rightarrow HSO_4^- + \cdot CH_2OH \qquad (10\text{-}98)$$

$$\cdot CH_2OH + S_2O_8^{2-} \rightarrow CH_2O + HSO_4^- + SO_4^- \cdot \qquad (10\text{-}99)$$

$$2 \cdot CH_2OH \rightarrow CH_3OH + CH_2O \text{ (or } HOCH_2CH_2OH) \quad (10\text{-}100)$$

It is not immediately obvious why the hydrogen atom transfer reaction 10-99 should be the rate-limiting factor in the chain sequence for methanol oxidation allowing the hydroxymethyl radicals to build up a high enough steady-state concentration to be the sole termination reaction whereas 10-95 is not the limiting reaction in the isopropyl alcohol oxidations. One explanation is that the donor qualities of the methanol-derived radical are less pronounced than those of the isopropyl alcohol-derived radical and, therefore, the hydrogen atom transfer reaction is slower. Another possibility is that electron transfer,

Hydrogen Atom Transfer Mechanism

$$R_2\dot{C}OH + S_2O_8^{2-} \rightarrow$$

$$[R_2C \text{---} O \cdot \cdot H \cdot \cdot O_3^- SO \cdot \cdot OSO_3^-] \rightarrow$$

$$R_2C{=}O + HSO_4^- + SO_4^- \cdot \quad (10\text{-}101)$$

R = H for methanol and R = CH$_3$ for 2-propanol
not hydrogen atom transfer, is involved in the reduction of the peroxydisulfate by the α-hydroxyalkyl radical. In this mechanism a cation results which is

Electron Transfer Mechanism

$$R_2\dot{C}OH + S_2O_8^{2-} \rightarrow R_2\overset{(+)}{\dot{C}}OH + SO_4^{2-} + SO_4^-$$

$$\overset{|}{} {\longrightarrow} R_2C{=}O + H^+ \qquad (10\text{-}102)$$

the protonated carbonyl compound produced as the reaction product. The isopropyl alcohol-derived radical would be expected to react faster than the hydroxymethyl radical in the electron transfer reaction as well.

Aldehydes also induce the decomposition of peroxydisulfate although not to the same extent as secondary alcohols. Table 10-4 lists the half-lives of ammonium peroxydisulfate in the presence of some alcohols and aldehydes. Examination of these data show that the induced decompositions are more pronounced in secondary alcohols than in primary alcohols and somewhat less in aldehydes. Furthermore, aldehydes inhibit the rate of the induced decomposition in secondary alcohols (although the nature of the inhibition is not

TABLE 10-4 Half-Lives (minutes) of $S_2O_8^{2-}$ (0.025 M) at pH 8 in Presence of Alcohols and Aldehydes[47]

Reactant	50°C	70°C
(Peroxydisulfate alone)	11,500	411
iso-Propyl alcohol (0.25 M)	3.2	0.47
iso-Butyl alcohol (0.25 M)	—	24.2
iso-Butyraldehyde (0.25 M)	870	45.5
Propionaldehyde (0.25 M)	730	—
t-Butyl alcohol (0.25 M)	—	182
iso-Propyl alcohol (0.25 M) } iso-Butyraldehyde (0.25 M) }	—	58

completely clear in view of the fact that the alcohol also retards the aldehyde reaction). These data indicate that reaction of aldehydes with the hydrogen abstracting radical is faster than reaction of a secondary alcohol with the hydrogen abstractor. This is not unexpected in terms of the reactivity of aldehydic hydrogens toward abstraction by electron-acceptor radicals. Reaction of the acyl radical with the peroxydisulfate accounts for the induced decomposition of the peroxide but apparently this reaction is slower than the reaction of

$$SO_4^- + RCHO \rightarrow HSO_4^- + R\dot{C}O \qquad (10\text{-}103)$$

$$R\dot{C}O + S_2O_8^{2-} \rightarrow R\overset{(+)}{C}O + SO_4^{2-} + SO_4^-$$
$$\xrightarrow{\;H_2O\;} RCO_2H + H^+ \qquad (10\text{-}104)$$

α-hydroxyalkyl radicals with $S_2O_8^{2-}$. Formation of a carboxylic acid from the aldehyde supports the intermediacy of the acylium ion resulting from the electron-transfer reaction. Part of the anomalous kinetics observed in reactions of primary alcohols with $S_2O_8^{2-}$ is likely the result of interaction with $S_2O_8^{2-}$ of the aldehyde produced in the oxidation of the primary alcohol.

VI. Azo Compounds

Azo compounds decompose thermally yielding free radicals and nitrogen. The rate of the decomposition reaction is markedly influenced by the stabilities of the radicals produced. Table 10-5 lists the activation energies for the thermal

$$R\text{---}N\text{=}N\text{---}R \rightarrow [R\cdot\cdot N\underset{\cdot\cdot}{\text{---}}N\cdot\cdot R] \rightarrow R\cdot + N_2 + R\cdot \qquad (10\text{-}105)$$

decomposition of several azo compounds and the data illustrate the importance of radical stability on the activation energy requirement for the reaction.

TABLE 10-5   Activation Energies for Thermal Decomposition of Azo Compounds[48]

Azo Compound	Radicals Produced	Solvent	$\Delta E^{\ddagger}$
CH_3—N≡N—CH_3	$CH_3 \cdot$	Gas	50.2
$(CH_3)_2CH$—N≡N—$CH(CH_3)_2$	$(CH_3)_2\dot{C}H$	Gas	40.9
$(CH_3)_3C$—N≡N—$C(CH_3)_3$	$(CH_3)_3C \cdot$	Gas	43
C_6H_5CH—N≡N—CHC_6H_5 | | CH_3 CH_3	$C_6H_5\dot{C}H \cdot$ | CH_3	Toluene	32.6
$(C_6H_5)_2CH$—N≡N—$CH(C_6H_5)_2$	$(C_6H_5)_2\dot{C}H$	Toluene	26.6
$(CH_3)_2C$—N≡N—$C(CH_3)_2$ | | CN CN	$(CH_3)_2\dot{C}$ | CN_3	Toluene	31.3

Azobisisobutyronitrile (**43**) is the azo compound most commonly used as an initiator of free-radical reactions. It decomposes readily in the temperature range

$$(CH_3)_2C\text{—N}{=}\text{N—}C(CH_3)_2 \; \rightarrow \; 2(CH_3)_2C \cdot \; + N_2 \qquad (10\text{-}106)$$
$$\underset{\substack{| \\ CN}}{} \qquad \underset{\substack{| \\ CN}}{} \qquad\qquad \underset{\substack{| \\ CN}}{}$$
$$\textbf{43 (AIBN)} \qquad\qquad\qquad \textbf{44}$$

of 60 to 90°C (see Table 10-6) and its decomposition is not induced by solvent-derived radicals. However, it does suffer from the disadvantage that the isobutyronitrile radicals (**44**) are generally less reactive than the radical fragments encountered in the decomposition of most peroxides. Isobutyronitrile radicals add readily to most reactive double bonds and therefore AIBN finds value in initiating vinyl polymerization reactions. The isobutyronitrile radicals often couple to form either tetramethylsuccinonitrile (**45**) or the ketenimine **46**.[49]

$$(CH_3)_2C \cdot \; + \; \cdot C(CH_3)_2 \; \rightarrow \; (CH_3)_2C\text{—}C(CH_3)_2 \qquad (10\text{-}107)$$
$$\underset{\substack{| \\ CN}}{}\;\; \underset{\substack{| \\ CN}}{} \qquad\qquad \underset{\substack{| \;\; | \\ CN \; CN}}{}$$
$$\textbf{44} \qquad\qquad\qquad\qquad \textbf{45}$$

TABLE 10-6 Chemical Initiators

Initiator	Half-Lives							
	50°C	60°C	70°C	85°C	100°C	115°C	130°C	145°C
Azobisisobutyronitrile (AIBN)	—	—	4.8 hr	—	7.2 min	—	—	—
Benzoyl peroxide (Bz_2O_2)	—	—	7.3 hr	1.4 hr	19.8 min	—	—	—
Acetyl peroxide (Ac_2O_2)	158 hr	—	8.1 hr	1.1 hr	—	—	—	—
Decanoyl peroxide	52.7 hr	12.6 hr	3.4 hr	30 min	—	—	—	—
Propionyl peroxide	70.9 hr	—	4.5 hr	42 min	—	—	—	—
Lauroyl peroxide	47.7 hr	12.8 hr	3.5 hr	31 min	—	—	—	—
Peroxydisulfate ($S_2O_8^{2-}$)	191 hr	38.5 hr	8.25 hr	2.1 hr	—	—	—	—
t-Butyl peracetate	—	—	—	88 hr	12.5 hr	1.9 hr	18 min	—
t-Butyl perbenzoate	—	—	—	—	18 hr	3 hr	33 min	—
t-Butyl peroxide	—	—	—	—	218 hr	34 hr	6.4 hr	1.38 hr
Cumyl peroxide	—	—	—	—	—	13 hr	1.7 hr	16.8 min

TABLE 10-6 Chemical Initiators (continued)

Initiator	Comments
Azobisisobutyronitrile (AIBN)	No induced decompositions. Radicals produced are relatively stable. Extensive cage recombination in certain solvents.
Benzoyl peroxide (Bz_2O_2)	Induced decompositions with ethers, alcohols and amines.
Acetyl peroxide (Ac_2O_2)	Induced decompositions with ethers, alcohols and amines.
Decanoyl peroxide	Same as Bz_2O_2 and Ac_2O_2 but may be expected to undergo unimolecular ester formation as well.
Propionyl peroxide	Same as Bz_2O_2 and Ac_2O_2 but may be expected to undergo unimolecular ester formation as well.
Lauroyl peroxide	Same as Bz_2O_2 and Ac_2O_2 but may be expected to undergo unimolecular ester formation as well.
Peroxydisulfate ($S_2O_8^{2-}$)	Induced decompositions in alcohols, amines and aldehydes. Also prone to acid catalyzed ionic reactions.
t-Butyl peracetate	Induced decompositions in alcohols, ethers and amines.
t-Butyl perbenzoate	Induced decompositions in alcohols, ethers and amines.
t-Butyl peroxide	Slow induced decomposition by alcohol and amine derived radicals.
Cumyl peroxide	Slow induced decomposition by alcohol and amine derived radicals.

$$(CH_3)_2\overset{\mid}{\underset{\updownarrow}{C}}\!\!-\!\!C\!\!\equiv\!\!N$$
$$(CH_3)_2C\!\!=\!\!C\!\!=\!\!N\cdot \quad + \cdot C(CH_3)_2 \rightarrow (CH_3)_2C\!\!=\!\!C\!\!=\!\!N\!\!-\!\!C(CH_3)_2 \quad (10\text{-}108)$$
$$\underset{44}{} \qquad\qquad \underset{CN}{\mid} \qquad\qquad\qquad\qquad \underset{CN}{\mid}$$
$$\underset{46}{}$$

Coupling reactions of isobutyronitrile radicals often occur in the solvent cage. When used to initiate the polymerization of styrene, only 65% of the isobutyronitrile radicals escape from the solvent cage to start chain reactions.[50] Decomposition of AIBN in carbon tetrachloride, a substrate having little reactivity toward isobutyronitrile radicals, results in about 90% coupling of the radicals which occurs both in and out of the solvent cage. The presence of butyl mercaptan, a compound reactive toward hydrogen atom abstraction, decreases the amount of the coupling product as the concentration of mercaptan is increased to a limiting amount of about 20% of the coupling product which is probably the result of cage recombination.[51] Liquid bromine appears to be reactive enough toward isobutyronitrile radicals to compete with the cage recombination reaction since no coupling product is observed when AIBN is decomposed thermally in liquid bromine.[52]

VII. Summary

Table 10-6 lists some pertinent rate data of the more commonly encountered and commercially available chemical initiators. Also included in this table are some comments concerning their respective advantages and limitations as initiators of free-radical chain reactions. Some of the compounds (propionyl peroxide, decanonyl peroxide, and lauroyl peroxide) were not discussed in this chapter but can be expected to have chemical properties similar to acetyl peroxide.

References

1. E.g., D. C. Neckers, *Mechanistic Organic Photochemistry*, Reinhold Publishing, New York, 1967.
2. M. S. Kharasch, E. V. Jensen, and W. H. Urry, *J. Org. Chem.*, **10**, 386 (1945); H. C. McBay, O. Tucker, and A. Milligan, *ibid.*, **19**, 1003 (1954).
3. M. S. Kharasch, M. T. Gladstone, *J. Amer. Chem. Soc.*, **65**, 15 (1943); M. S. Kharasch, E. V. Jensen, and W. H. Urry, *J. Org. Chem.*, **10**, 386 (1945); M. S. Kharasch, H. C. McBay, and W. H. Urry, *ibid.*, **10**, 394 (1945).
4. M. F. Ansell, W. J. Hickinbottom, and P. G. Holton, *J. Chem. Soc.*, **1955**, 349; M. S. Kharasch, H. C. McBay, and W. H. Urry, *J. Amer. Chem. Soc.*, **70**, 1269 (1948).

5. A. Rembaum and M. Szwarc, *ibid.*, **77**, 3486 (1955).

6. J. W. Taylor and J. C. Martin, *ibid.*, **88**, 3650 (1966).

7. G. S. Hammond and L. M. Soffer, *ibid.*, **72**, 4711 (1950).

8. P. D. Bartlett and S. G. Cohen, *ibid.*, **65**, 543 (1943).

9. H. Gelissen and P. H. Hermans, *Ber.*, **69**, 662 (1926); P. H. Hermans and J. VanEyk, *J. Polymer Sci.*, **1**, 407 (1946); J. Boeseken and H. Gelissen, *Rec. trav. chim.*, **43**, 869 (1924).

10. Aromatic homolytic substitution is covered in: G. H. Williams, *Homolytic Aromatic Substitution*, Pergamon, London (1960).

11. C. G. Swain, W. H. Stockmayer, and J. T. Clark, *J. Amer. Chem. Soc.*, **72**, 5426 (1950); A. T. Blomquist and A. J. Buselli, *ibid.*, **73**, 3883 (1951).

12. C. Walling, *ibid.*, **66**, 1602 (1944).

13. K. Nozaki and P. D. Bartlett, *ibid.*, **68**, 1686 (1946); **69**, 2299 (1947).

14. W. E. Cass, *ibid.*, **69**, 500 (1947).

15. E. H. Drew and J. C. Martin, *Chem. and Ind.*, **1959**, 925; D. D. Denney and G. Feig, *J. Amer. Chem. Soc.*, **81**, 5322 (1959).

16. W. M. Thomas and M. T. O'Shaughnessy, *J. Polymer Sci.*, **11**, 455 (1953).

17. M. S. Kharasch, J. R. Rowe, and W. H. Urry, *J. Org. Chem.*, **16**, 905 (1951).

18. C. Walling and N. Indictor, *J. Amer. Chem. Soc.*, **80**, 5814 (1958).

19. L. Horner and K. Sherf, *Ann.*, **573**, 35 (1951).

20. E. S. Huyser and J. A. K. Harmony, unpublished results.

21. M. S. Kharasch, J. Kuderna, and W. Nudenberg, *J. Org. Chem.*, **19**, 1283 (1954).

22. J. D. Greene, H. P. Ju..., C. C. Chu, and F. M. Vane, *J. Amer. Chem. Soc.*, **86**, 2080 (1964).

23. N. A. Milas and D. M. Surgenor, *J. Amer. Chem. Soc.*, **68**, 205, 643 (1946); J. H. Raley, F. F. Rust, and W. E. Vaughan, *ibid.*, **70**, 88 (1948).

24. e.g., E. L. Patmore and R. J. Gritter, *J. Org. Chem.*, **27**, 4196 (1962).

25. e.g., F. Lassing and A. W. Tickner, *J. Chem. Phys.*, **20**, 907 (1952).

26. J. H. Raley, F. F. Rust, and W. E. Vaughan, *J. Amer. Chem. Soc.*, **70**, 1336 (1948).

27. E. S. Huyser and R. M. VanScoy, *J. Org. Chem.*, **33**, 3524 (1968).

28. C. Walling and R. Wagner, *J. Amer. Chem. Soc.*, **86**, 3368 (1964).

29. J. H. Raley, F. F. Rust, and W. E. Vaughan, *ibid.*, **70**, 80 (1948).

30. E. R. Bell, F. F. Rust, and W. E. Vaughan, *J. Amer. Chem. Soc.*, **72**, 337 (1950).

31. J. H. Raley, F. F. Rust, and W. E. Vaughan, *ibid.*, **70**, 2767 (1948); M. Flower, L. Batt, and S. W. Benson, *J. Chem. Phys.*, **37**, 2662 (1962).

32. A. M. Hogg and P. Kebarle, *J. Amer. Chem. Soc.*, **86**, 4558 (1964).

33. W. A. Pryor, A. Lee, and C. E. Witt, *ibid.*, **86**, 4229 (1964).

34. E. S. Huyser and C. J. Bredeweg, *ibid.*, **86**, 2401 (1964).

35. E. S. Huyser, C. J. Bredeweg, and R. M. VanScoy, *ibid.*, **86**, 4148 (1964).

36. E. S. Huyser, R. H. S. Wang, and W. T. Short, *J. Org. Chem.*, **33**, 4323 (1968).

37. P. D. Bartlett and R. R. Hiatt, *J. Amer. Chem. Soc.*, **80**, 1398 (1958).

38. P. D. Bartlett and D. M. Simons, *ibid.*, **82**, 1753 (1960).

39. A. T. Blomquist and I. A. Bernstein, *ibid.*, **73**, 5546 (1951).

40. P. D. Bartlett and C. Ruchardt, *ibid.*, **82**, 1756 (1960).

41. L. Hsu, PhD Dissertation, University of Kansas, 1966. [*Diss. Abstr.* **28B**, 2772 (1968)].

42. I. M. Kolthoff and I. K. Miller, *J. Amer. Chem. Soc.*, **73**, 3055 (1951); S. Fronaeus and C. O. Ostman, *Acta. Chem. Scand.*, **9**, 902 (1955).

43. W. K. Wilmarth and A. Haim, "Mechanisms of Oxidation by Persulfate Ion," in *Peroxide Reaction Mechanisms*, J. O. Edwards, Ed., John Wiley, New York, 1962.

44. P. D. Bartlett and J. D. Cotman, *J. Amer. Chem. Soc.*, **71**, 1419 (1949); I. M. Kolthoff, E. J. Meehan, and E. M. Carr, *ibid.*, **75**, 1439 (1953).

45. D. L. Ball, M. M. Crutchfield, and J. O. Edwards, *J. Org. Chem.*, **25**, 1599 (1960).

46. J. O. Edwards, A. R. Gallapo, and E. McIsaac, *J. Amer. Chem. Soc.*, **88**, 3893 (1966).

47. F. L. McMillian, PhD Thesis, University of Kansas, 1965. *Diss. Abstr.*, **XXVII**, 1819-B.

48. For original references, see C. Walling, *Free Radicals in Solution*, John Wiley, New York, 1957, pp. 511–515; W. A. Pryor, *Free Radicals*, McGraw-Hill, New York, 1966, pp. 127–133.

49. G. S. Hammond, C. S. Wu, O. D. Trapp, J. Warkentin, and R. T. Keys, *J. Amer. Chem. Soc.*, **82**, 5394 (1960).

50. J. C. Bevington, *Trans. Faraday Soc.*, **51**, 1392 (1955).

51. G. S. Hammond, J. N. Sen, C. E. Boozer, *J. Amer. Chem. Soc.*, **77**, 3244 (1955).

52. O. D. Trapp and G. S. Hammond, *ibid.*, **81**, 4876 (1959).

Chapter 11

Free-Radical Oxidation-Reduction Reactions

I. Redox Reactions of Organic Functions

A. OXIDATION OF ALCOHOLS

The oxidations of primary and secondary alcohols by peroxides and peresters were discussed in Chapter 10 and reactions of alcohols with hydrogen peroxide are discussed subsequently in this chapter (see Fenton's reagent). This section is concerned with free-radical chain reactions of alcohols with other reagents that result in oxidation of the alcohol function to a carbonyl compound with concurrent reduction of the oxidizing agent.

Oxidations of alcohols with polyhaloalkanes have been known for almost a century. Heating ethanol with carbon tetrabromide at 100°C for 12 hours was reported to yield acetaldehyde, bromoform, and hydrogen bromide.[1] Subsequent work showed that silver halide was formed when either bromoform, methylene iodide, or methyl iodide were heated with ethanolic silver nitrate.[2]

The suggestion that a free-radical chain reaction was involved in these reactions came later when it was shown that the ethanol was oxidized and various polyhalomethanes (CBr_2F_2, CCl_4, CH_2ClBr, $BrCCl_3$) were reduced along with the formation of the silver halide.[3] The mechanism shown in 11-1 through 11-4 accounts for these observations. Reactions 11-1 and 11-2 comprise a free-

$$X_3C\cdot + CH_3CH_2OH \rightarrow HCX_3 + CH_3\dot{C}HOH \qquad (11\text{-}1)$$

$$CH_3\dot{C}HOH + X_4C \rightarrow \underset{\overset{|}{X}}{CH_3CHOH} + X_3C\cdot \qquad (11\text{-}2)$$

$$\underset{\overset{|}{X}}{CH_3CHOH} \rightarrow CH_3CHO + HX \qquad (11\text{-}3)$$

$$HX + Ag^+ \rightarrow AgX + H^+ \qquad (11\text{-}4)$$

radical chain sequence in which the alcohol is halogenated in the α-position yielding the unstable halohydrin.

Razuvaev and his co-workers showed that primary and secondary alcohols are oxidized by carbon tetrachloride but tertiary alcohols are not.[4] The reactions could be initiated chemically with benzoyl peroxide, photochemically and thermally (200°C). The suggestion was made that the oxidation proceeded by a free-radical path similar to that shown in 11-1 and 11-2 except that the halohydrin was not formed. Rather, a complex was proposed that involved the α-hydroxyalkyl radical and the polyhalomethane which yielded the oxidized alcohol, hydrogen halide, and a chain-carrying perhalomethyl radical.[4] Possibly

$$R_2\dot{C}OH + ClCCl_3 \rightarrow (R_2\dot{C}OH\rightarrow ClCCl_3) \rightarrow$$

$$R_2C=O + HCl + \cdot CCl_3 \qquad (11\text{-}5)$$

more likely is that the α-hydroxyalkyl radical reduces the polyhalomethane by a hydrogen atom transfer reaction—a reaction favored by the energetic and polar aspects.

$$R_2\dot{C}OH + XCCl_3 \rightarrow \left[R_2\overset{\delta+}{C}\!-\!\!-\!O\cdot\cdot H\cdot\cdot X\cdot\cdot\overset{\delta-}{C}Cl_3 \right] \rightarrow$$

$$R_2C=O + HX + \cdot CCl_3 \qquad (11\text{-}6)$$

X = Cl			X = Br		
Make	H—Cl	−103 kcal/mole	Make	H—Br	−87 kcal/mole
Make	$\diagdown$C=O	−77 kcal/mole	Make	$\diagdown$C=O	−77 kcal/mole
Break	O—H	+105 kcal/mole	Break	O—H	+105 kcal/mole
Break	Cl_3C—Cl	+73 kcal/mole	Break	Cl_3C—Br	+54 kcal/mole
		−2 kcal/mole			−5 kcal/mole

The peroxide-induced oxidations of alcohols with hexachloroethane proceed by a similar path. Formation of the α-hydroxyalkyl radical in this case occurs by reaction of the alcohol with either a chlorine atom resulting from a β-elimination reaction of the pentachloroethyl radical or by direct interaction of the perhaloalkyl radical with the alcohol.[5]

$$R_2\dot{C}OH + Cl_3CCCl_3 \rightarrow R_2C{=}O + HCl + Cl_2\dot{C}CCl_3 \qquad (11\text{-}7)$$

$$Cl_2\dot{C}CCl_3 + R_2CHOH \rightarrow Cl_2CHCCl_3 + R_2\dot{C}OH \qquad (11\text{-}8)$$

$$Cl_2\dot{C}CCl_3 \rightarrow Cl_2C{=}CCl_2 + Cl\cdot \qquad (11\text{-}9)$$

$$Cl\cdot + R_2CHOH \rightarrow HCl + R_2\dot{C}OH \qquad (11\text{-}10)$$

Trichloromethanesulfonyl chloride (1) does not react with alcohols to form esters but rather causes oxidation of the alcohol. Ethanol and 1 are reported to yield acetaldehyde, sulfur dioxide, and chloroform but no isolable amounts of

$$Cl_3CSO_2Cl + C_2H_5OH \nearrow CH_3CHO + HCCl_3 + SO_2 + HCl \quad (11\text{-}11)$$
$$\underset{\textbf{1}}{} \searrow \underset{\textbf{2}}{Cl_3CSO_2C_2H_5 + HCl} \qquad (11\text{-}12)$$

the expected ester 2.[6] The light-induced reaction of 1 with 2-butanol yielded a variety of products, the formation of which could be explained in terms of a sequence of reactions in which the alcohol is oxidized by 1 in the free-radical chain sequence 11-14 and 11-15. The interaction of the α-hydroxyalkyl radical

$$CH_3CHOHC_2H_5 + \underset{\textbf{1}}{Cl_3CSO_2Cl} \rightarrow$$

$$CH_3COC_2H_5 + HCCl_3 + SO_2 + (CH_3CHClC_2H_5 + H_2O) \quad (11\text{-}13)$$

via

$$Cl_3CSO_2\cdot + CH_3CHOHC_2H_5 \rightarrow Cl_3CSO_2H + CH_3\dot{C}OHC_2H_5 \quad (11\text{-}14)$$

$$CH_3\dot{C}OHC_2H_5 + Cl_3CSO_2Cl \rightarrow CH_3COC_2H_5 + HCl + Cl_3CSO_2\cdot$$
$$(11\text{-}15)$$

$$Cl_3CSO_2H \rightarrow HCCl_3 + SO_2 \qquad (11\text{-}16)$$

$$CH_3CHOHC_2H_5 + HCl \rightarrow CH_3CHClC_2H_5 + H_2O \qquad (11\text{-}17)$$

with 1 is likely a hydrogen atom transfer reaction similar to that suggested for the interaction of α-hydroxyalkyl radicals with the polyhalkanes.[7]

$$R_2\overset{.}{C}OH + ClSCCl_3 \rightarrow \left[R_2\overset{\delta+}{C}\cdots O \cdots H \cdots Cl \cdots \overset{\delta-}{SCCl_3} \right] \rightarrow$$

(with the O groups double-bonded above and below the carbon on the left labeled **3**, and on the right)

$$R_2C{=}O + HCl + \cdot O_2SCCl_3 \quad (11\text{-}18)$$

Secondary alcohols and t-butyl hypochlorite react spontaneously with oxidation of the alcohol to the ketone. The hypochlorite is reduced to t-butyl alcohol and hydrogen chloride is formed as a by-product of the reaction.[8]

$$R_2CHOH + (CH_3)_3COCl \rightarrow R_2C{=}O + HCl + (CH_3)_3COH \quad (11\text{-}19)$$

In some instances, the reaction is performed in the presence of pyridine to remove the hydrogen chloride. The suggestion has been made that this oxidation reaction is ionic and that a pyridine-catalyzed exchange of the hydrogen and chlorine occurs, followed by decomposition of the secondary alkyl hypochlorite to the observed reaction products. This seems unlikely, however, in view of the observation that the oxidation of isopropyl alcohol with t-butyl hypochlorite occurs faster than decomposition of isopropyl hypochlorite to acetone and hydrogen chloride.[9] While the reactions involving secondary alcohols with t-butyl hypochlorite occur spontaneously, the rates of oxidations of primary alcohols are accelerated by light and thus show evidence of being free-radical chain processes. A mechanism for the oxidation is shown in the chain sequence 11-20 and 11-21, the latter being a hydrogen atom transfer reaction which

$$(CH_3)_3CO\cdot + R_2CHOH \rightarrow (CH_3)_3COH + R_2\overset{.}{C}OH \quad (11\text{-}20)$$

$$R_2\overset{.}{C}OH + ClOC(CH_3)_3 \rightarrow \left[R_2\overset{\delta+}{C}\cdots O \cdots H \cdots Cl \cdots \overset{\delta-}{OC(CH_3)_3} \right] \rightarrow$$

$$R_2C{=}O + HCl + \cdot OC(CH_3)_3 \quad (11\text{-}21)$$

Make	H—Cl	-103 kcal/mole
Make	$>$C$=$O	-77 kcal/mole
Break	O—H	$+105$ kcal/mole
Break	O—Cl	$\sim +60$ kcal/mole
		~ -15 kcal/mole

should be quite facile both in terms of the polar factor and the energetics of the process.

The spontaneous nature of the reaction of secondary alcohols with t-butyl hypochlorite as well as the observation that the rate of reaction is faster in polar than in nonpolar solvents are indicative of polar aspects of this reaction. It may be, however, that the oxidation occurs in a free-radical chain sequence that is induced by a bimolecular initiation process involving the hypochlorite and secondary alcohol. If the mechanism of this process involves either an electron transfer reaction from the alcohol to the hypochlorite or a nucleophilic displacement, the influence of the polarity of the solvent on the reaction rate becomes clear.

$$R_2CHOH + (CH_3)_3COCl \rightarrow R_2\dot{C}OH + HCl + (CH_3)_3CO \cdot \quad (11\text{-}22)$$

B. OXIDATION OF NITROGEN-CONTAINING COMPOUNDS

The free-radical oxidations of amines by peroxides were discussed in Chapter 10. The oxidation of nitrogen-containing compounds by other reagents may also proceed by radical paths involving either hydrogen atom or electron transfer reactions.

The Hantzsch ester (3) is oxidized by bromotrichloromethane to the corresponding pyridine with concurrent reduction of the polyhalomethane to

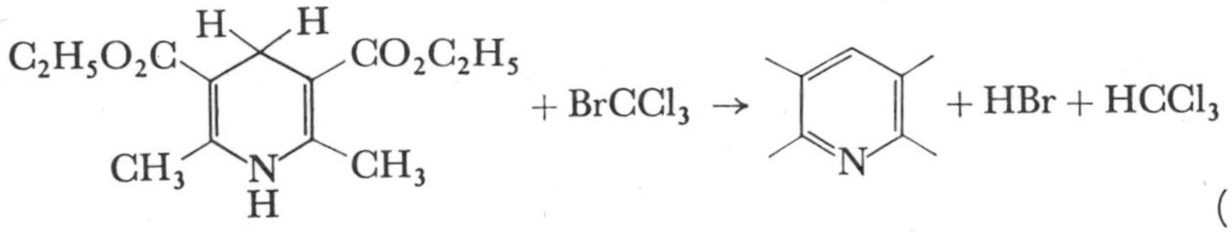

(11-23)

3

chloroform. Similarly, the N-methyl Hantzsch ester (4) is oxidized to the

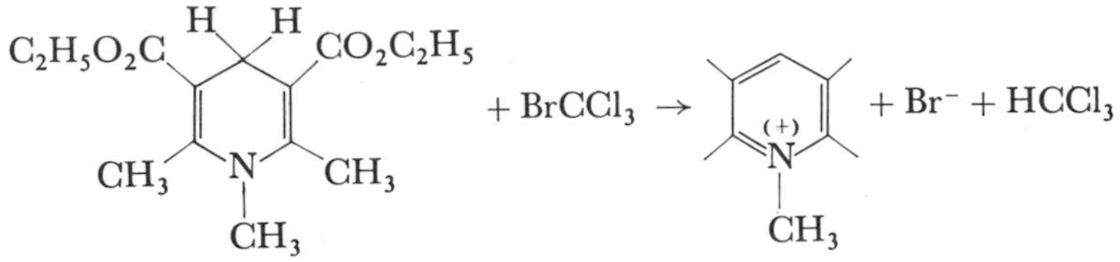

4   **5**

pyridinium cation 5. Both reactions take place in the dark and kinetic analysis of the dark reaction showed a rate law that involved the Hantzsch ester in somewhat greater than first power and, in some reactions, almost three-halves power.[10] These observations are consistent with a mechanism for the reaction that consists of the chain sequence 11-26 and 11-27, a sequence initiated by the bimolecular initiation reaction 11-25. The rates were generally determined in reactions where the polyhalomethane was present in about a 300-fold excess relative to the dihydropyridine making termination by coupling of two trichloromethyl radicals plausible (see Chapter 3). The derived rate law for a

$$\text{BrCCl}_3 + \text{HPyR} \xrightarrow{k_{25}} \text{Br}^- + \text{Cl}_3\text{C}\cdot + \cdot\text{PyR} + \text{H}^+ \quad (11\text{-}25)$$

$$\text{Cl}_3\text{C}\cdot + \text{HPyR} \xrightarrow{k_{26}} \text{HCCl}_3 + \cdot\text{PyR} \qquad (11\text{-}26)$$

$$\cdot\text{PyR} + \text{BrCCl}_3 \xrightarrow{k_{27}} \overset{(+)}{\text{PyR}} + \overset{(-)}{\text{Br}} + \text{Cl}_3\text{C}\cdot \qquad (11\text{-}27)$$

$$\left(\overset{(+)}{\text{PyR}} \rightarrow \text{Py} + \text{R}^+ \text{ when R} = \text{H}\right)$$

$$2\text{Cl}_3\text{C}\cdot \xrightarrow{k_{28}} \text{C}_2\text{Cl}_6 \qquad (11\text{-}28)$$

R = H for **3** and CH$_3$ for **4**

reaction following the sequence of steps shown in 11-25 through 11-28 is given in 11-29. The reduction of the polyhalomethane by the radical derived from **4**

$$\text{Rate} = \left(\frac{k_{25}}{2k_{28}}\right)^{1/2} k_{26}[\text{BrCCl}_3]^{1/2}[\text{HPyR}]^{3/2} \qquad (11\text{-}29)$$

almost certainly is an electron transfer reaction. The reduction of the polyhalomethane by the radical derived from **3** could be either electron transfer or hydrogen atom transfer.

Reductions of the carbonyl functions of pyruvic acid and phenylglyoxalic acid by the Hantzsch ester **3** yield lactic and mandelic acids, respectively, in low yields (5–7%).[11] The available evidence does not indicate that reactions necessarily proceed by a radical path. Reduction of 2-sulfhydrylbenzophenone (**6**) by **3**[12] induced by hydrogen peroxide and ferrous ion (see section on Fenton's reagent) quite possibly is a radical reaction. The α-hydroxyalkyl radical **8** formed in the hydrogen atom transfer 11-33 is able to react in an intramolecular

$$(11\text{-}32)$$

via

$$(11\text{-}33)$$

$$(11\text{-}34)$$

$$(11\text{-}35)$$

hydrogen atom abstraction from the sulfhydryl group yielding the thiyl radical **9** which can react with **3** yielding the product **7** and the chain-carrying mono-hydropyridyl radical.

Other compounds that have been reported to be reduced readily by **3** are maleic anhydride (**10**) and chloranil (**11**).[13] No evidence is available that suggests that these reductions proceed by radical processes and may involve hydride ion transfer reactions from **3** to the substrate. However, in view of susceptibility of **3** toward hydrogen abstraction and the ability of the radical to transfer hydrogen atoms, the free-radical chain reactions shown may be operative.

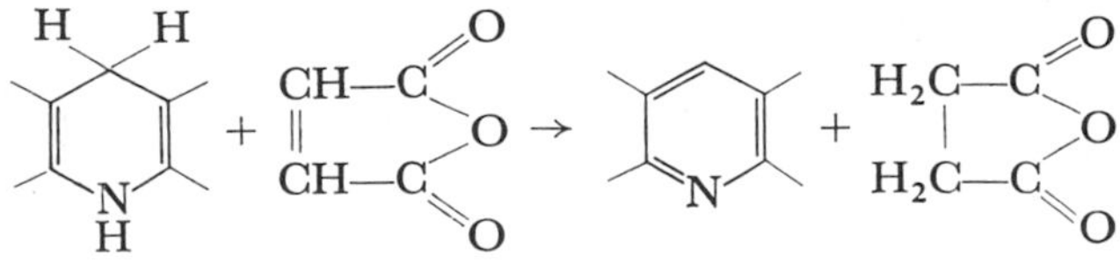

(11-36)

via

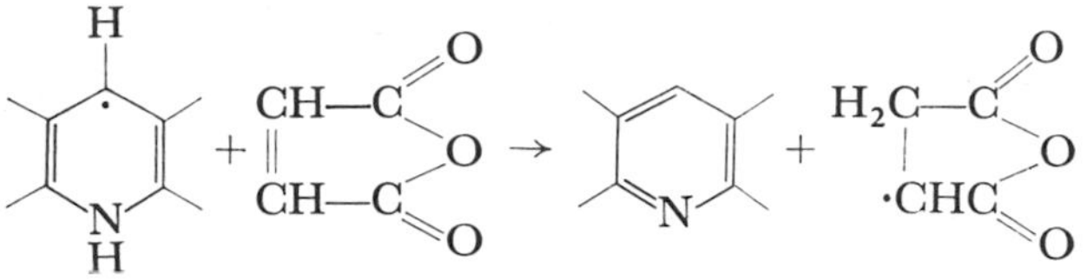

(11-37)

(11-38)

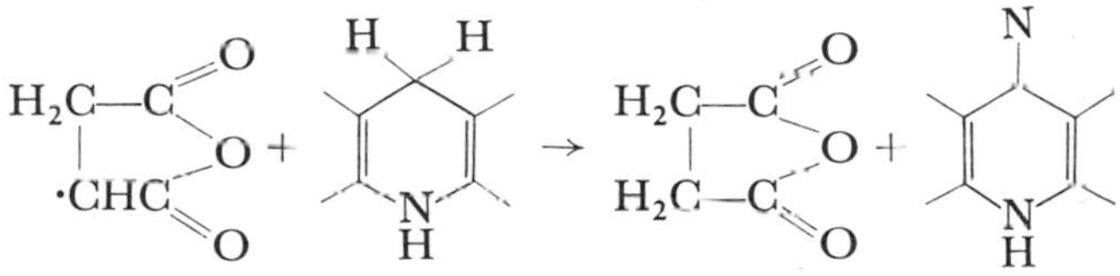

(11-39)

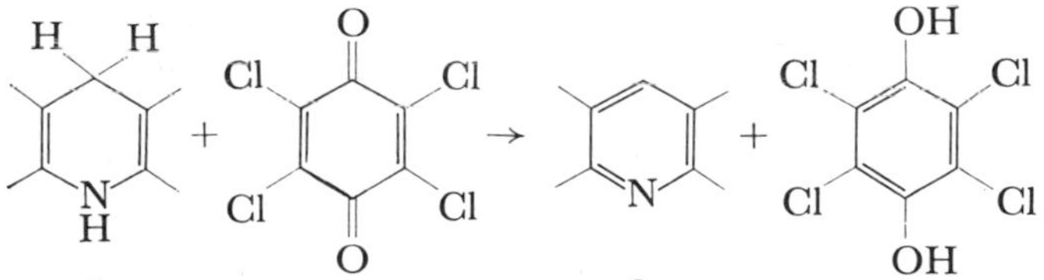

via

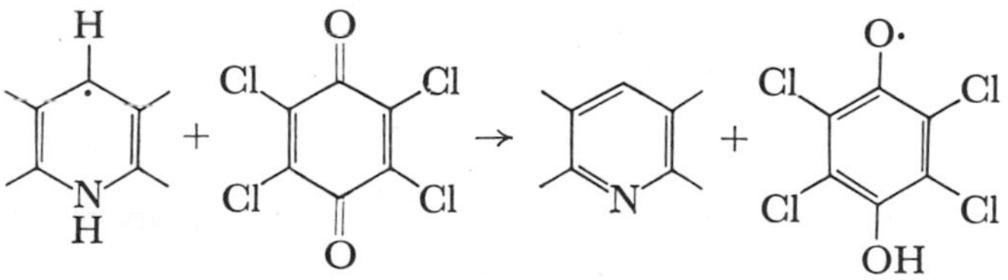

(11-40)

$$(11\text{-}41)$$

1   5

Oxidation of hydrazine (**12**) with carbon tetrachloride yields chloroform and nitrogen among the reaction products.[14] A chain reaction could account for the oxidation of hydrazine to diimide (**13**) with concurrent reduction of the polyhalomethane to chloroform. The diimide, in a subsequent nonradical bimolecular disproportionation reaction, yields nitrogen.[15] The path of the

$$H_2NNH_2 + 2Cl_4C \rightarrow 2HCCl_3 + N_2 + 2HCl \qquad (11\text{-}42)$$
$$\mathbf{12}$$

via $Cl_3C\cdot + H_2NNH_2 \rightarrow HCCl_3 + H\dot{N}NH_2 \qquad (11\text{-}43)$

$$H\dot{N}NH_2 + Cl_4C \begin{cases} \nearrow HCl + HN\!\!=\!\!NH + Cl_3C\cdot \qquad (11\text{-}44) \\ \qquad\qquad \mathbf{13} \\ \searrow Cl_3C\cdot + ClN\dot{H}NH_2 \qquad (11\text{-}45) \\ \qquad\qquad \rightarrow HCl + HN\!\!=\!\!NH \\ \qquad\qquad\qquad \mathbf{13} \end{cases}$$

$$2HN\!\!=\!\!NH \rightarrow H_2NNH_2 + N_2 \qquad (11\text{-}46)$$
$$\mathbf{13} \qquad\qquad \mathbf{12}$$

reduction of the carbon tetrachloride by the hydrazine-derived radical is not known. Either a hydrogen atom or electron transfer reaction may be involved or it may be a two-step process in which a chlorinated intermediate is produced which then decomposes to **13**.[14]

Reactions of phenylhydrazine (**14**) with polyhalomethanes are similar in that phenyldiimide (**15**) appears to be formed in a chain reaction with the polyhalomethane.[16] Subsequent reactions of **15** either with the polyhalomethane or with itself account for the formation of part of the chloroform and the benzene, halobenzene, and nitrogen observed as reaction products (see reactions 11-52 and 11-53). The reaction is accelerated both by light and peroxides, but does occur in the dark and follows the kinetic rate law in 11-47. A free-radical

$$\text{Rate} = k'[C_6H_5NHNH_2][XCCl_3] \qquad (11\text{-}47)$$

chain reaction induced by the bimolecular initiation process 11-48 and terminated by cross-disproportionation reaction 11-51 accounts for both the observed rate law and the formation of the phenyldiimide required for 11-52 and 11-53 as well as part of the chloroform.

$$C_6H_5NHNH_2 + BrCCl_3 \rightarrow C_6H_5\dot{N}NH_2 + HBr + Cl_3C \cdot \quad (11\text{-}48)$$
$$\mathbf{14}$$

$$Cl_3C \cdot + C_6H_5NHNH_2 \rightarrow HCCl_3 + C_6H_5\dot{N}NH_2 \quad (11\text{-}49)$$

$$C_6H_5\dot{N}NH_2 + XCCl_3 \rightarrow C_6H_5N{=}NH + HX + Cl_3C \cdot \quad (11\text{-}50)$$
$$\mathbf{15}$$

$$C_6H_5\dot{N}NH_2 + Cl_3C \cdot \rightarrow C_6H_5N{=}NH + HCCl_3 \quad (11\text{-}51)$$

$$2C_6H_5N{=}NH \rightarrow C_6H_6 + N_2 \quad (11\text{-}52)$$
$$\mathbf{15}$$

$$C_6H_5N{=}NH + XCCl_3 \rightarrow C_6H_5X + N_2 + HCCl_3 \quad (11\text{-}53)$$

The reactions of phenyldiimide are most likely not radical processes. Phenyl-diimide is more selective than a phenyl radical in competition reactions with carbon tetrachloride and bromotrichloromethane. The decomposition of **15** yielding phenyl radicals as shown in 11-54 is not a likely route to the formation of either the halobenzene or benzene. Somewhat more plausible is a direct, nonradical interaction of **15** with either the polyhalomethane yielding nitrogen,

$$C_6H_5N{=}NH \nrightarrow C_6H_5 \cdot + N_2 + II \cdot \quad (11\text{-}54)$$
$$\mathbf{15}$$

chloroform, and the halobenzene as shown in 11-53 or a bimolecular reaction of **15** yielding nitrogen and benzene as shown in 11-52.

C. REDUCTION OF CARBONYL FUNCTIONS

In the previous section, hydrogen atom transfer from a monohydropyridyl radical was proposed to account for the reductions of the keto-functions of pyruvic acid, phenylglyoxalic acids, and 2-sulfhydrylbenzophenone (**6**) by the Hantzsch ester (**3**). Hydrogen atom transfer from α-hydroxyalkyl radicals to ketones has also been observed. The photochemical reduction of benzo-phenone (**16**) to benzpinacol (**17**) by secondary alcohols with concurrent oxidation of the alcohol to a ketone involves a hydrogen atom transfer from the α-hydroxyalkyl radical derived from the alcohol to the ketone. Evidence for the mechanism shown in 11-56 through 11-59 is the observation that one

$$(C_6H_5)_2C{=}O + R_2CHOH \xrightarrow{h\nu} (C_6H_5)_2C{-}C(C_6H_5)_2 + R_2C{=}O \quad (11\text{-}55)$$
$$\mathbf{16} \qquad\qquad\qquad\qquad \underset{\displaystyle \mathbf{17}}{\overset{\displaystyle |\quad\ |}{OH\,OH}}$$

molecule of **17** is produced per quantum of light absorbed and reactions involving optically active secondary alcohols result in no appreciable race-mization of the alcohol during the course of the reaction. The latter observation

$$(C_6H_5)_2C{=}O \xrightarrow[n \to \pi^*]{hv} ((C_6H_5)_2C{=}O)^* \to (C_6H_5)_2\overset{\cdot}{C}{-}O\cdot \quad (11\text{-}56)$$
16

$$(C_6H_5)_2\overset{\cdot}{C}{-}O\cdot + R_2CHOH \to (C_6H_5)_2\overset{\cdot}{C}OH + R_2\overset{\cdot}{C}OH \quad (11\text{-}57)$$

$$R_2\overset{\cdot}{C}OH + (C_6H_5)_2C{=}O \to R_2C{=}O + (C_6H_5)_2\overset{\cdot}{C}OH \quad (11\text{-}58)$$
16

$$2(C_6H_5)_2\overset{\cdot}{C}OH \to (C_6H_5)_2C{-}C(C_6H_5)_2 \quad (11\text{-}59)$$
$$\underset{\overset{|}{OH}\ \overset{|}{OH}}{}$$
17

strongly supports the hydrogen atom transfer reaction 11-58 as the process by which the aliphatic ketone is produced rather than a disproportionation reaction of the α-hydroxyalkyl radicals, a reaction which would result in formation of racemic alcohol.[17]

$$2R_2\overset{\cdot}{C}OH \to R_2C{=}O + R_2CHOH \quad (11\text{-}60)$$

Acetophenone (**18**) is reduced to the substituted glycol **19** by 2-butanol. In this reaction α-hydroxyalkyl radicals are formed by interaction of 2-butanol with t-butoxy radicals resulting from decomposition of t-butyl peroxide.[18] The reductions of phenylcyclopropyl ketone (**20**) and phenylcyclobutyl ketone

$$2C_6H_5COCH_3 + CH_3CHOHC_2H_5 + (CH_3)_3COOC(CH_3)_3 \to$$
18

$$CH_3\ CH_3$$
$$C_6H_5\overset{|}{C}{-}\overset{|}{C}C_6H_5 + CH_3COC_2H_5 + 2(CH_3)_3COH \quad (11\text{-}61)$$
$$\underset{\overset{|}{OH}\ \overset{|}{OH}}{}$$
19

via

$$(CH_3)_3COOC(CH_3)_3 \to 2(CH_3)_3CO\cdot \quad (11\text{-}62)$$

$$(CH_3)_3CO\cdot + CH_3CHOHC_2H_5 \to (CH_3)_3COH + CH_3\overset{\cdot}{C}C_2H_5 \quad (11\text{-}63)$$
$$\underset{\overset{|}{OH}}{}$$

$$\overset{\displaystyle O}{\overset{\displaystyle \|}{}}$$
$$CH_3\overset{\cdot}{C}C_2H_5 + C_6H_5COCH_3 \to CH_3CC_2H_5 + C_6H_5\overset{\cdot}{C}CH_3 \quad (11\text{-}64)$$
$$\underset{\overset{|}{OH}}{} \qquad\qquad\qquad \underset{\overset{|}{OH}}{}$$

$$CH_3\ CH_3$$
$$2C_6H_5\overset{\cdot}{C}CH_3 \to C_6H_5\overset{|}{C}{-}\overset{|}{C}C_6H_5 \quad (11\text{-}65)$$
$$\underset{\overset{|}{OH}}{} \qquad \underset{\overset{|}{OH}\ \overset{|}{OH}}{}$$
19

(21) by secondary alcohols in the peroxide-induced chain sequence 11-66 through 11-68 also involve a hydrogen atom transfer reaction to a carbonyl function.[19]

$$R_2\dot{C}OH + C_6H_5\overset{\overset{O}{\|}}{C}CH(CH_2)_n \rightarrow R_2C=O + C_6H_5\overset{\overset{OH}{|}}{C}{-}\dot{C}H(CH_2)_n \quad (11\text{-}66)$$

$$(\mathbf{20},\ n = 2)$$
$$(\mathbf{21},\ n = 3)$$

$$C_6H_5\overset{\overset{OH}{|}}{\underset{\cdot}{C}}{-}\dot{C}H(CH_2)_n \rightarrow C_6H_5\overset{\overset{OH}{|}}{C}=CH(CH_2)_{n-1}\dot{C}H_2 \quad (11\text{-}67)$$

$$C_6H_5\overset{\overset{OH}{|}}{C}=CH(CH_2)_{n-1}\dot{C}H_2 + R_2CHOH \rightarrow$$

$$R_2\dot{C}OH + C_6H_5\overset{\overset{OH}{|}}{C}=CH(CH_2)_{n-1}CH_3$$

$$\overset{}{\underset{\longrightarrow\ C_6H_5\overset{\overset{O}{\|}}{C}(CH_2)_nCH_3}{\Big\downarrow}} \quad (11\text{-}68)$$

D. MISCELLANEOUS REACTIONS

Trialkyl tin hydrides (22) reduce alkyl halides to the parent alkane. The free-radical chain sequence 11-70 and 11-71 has been suggested as the mechanism for this reaction.[20] The reaction has proven to be valuable in several mechanistic studies since it provides a facile route for the formation of an alkyl radical of known structure.

$$RX + R'SnH \rightarrow RH + R_3'SnX \quad (11\text{-}69)$$
$$\mathbf{22}$$

via

$$R\cdot + R_3'SnH \rightarrow RH + R_3'Sn\cdot \quad (11\text{-}70)$$

$$R_3'Sn\cdot + RX \rightarrow R_3'SnX + R\cdot \quad (11\text{-}71)$$

Reductions of mercaptans by triethyl phosphite (23) yielding hydrocarbon and a trialkyl phosphorothionate (24)[21] proceed by the chain sequence 11-73

$$RSH + P(OC_2H_5)_3 \xrightarrow{\text{AIBN}} RH + (C_2H_5O)_3PS \quad (11\text{-}72)$$

through 11-75. The reaction of the thiyl radical with the trialkyl phosphite yielding the phosphoranyl radical is an addition reaction that involves expansion of the outer valence shell of the phosphorus, a situation similar to the phenomenon observed in the bridged bromine radicals (see Chapter 5). Oxidations

$$R \cdot + RSH \rightarrow RS \cdot + RH \tag{11-73}$$

$$RS \cdot + P(OC_2H_5)_3 \rightarrow RS\dot{P}(OC_2H_5)_3 \tag{11-74}$$
$$\mathbf{23}$$

$$RS\dot{P}(OC_2H_5)_3 \rightarrow R \cdot + (C_2H_5O)_3PS \tag{11-75}$$
$$\mathbf{24}$$

of triethyl phosphite with either *t*-butyl peroxide (**25**) or cumyl peroxide (**26**) follow a path similar to that shown for the mercaptan and phosphite although these oxidations are not chain reactions. However, oxidations of trialkyl

$$ROOR \rightarrow 2RO \cdot \tag{11-76}$$

(**25**, $R = C(CH_3)_3$)

(**26**, $R = C_6H_5C(CH_3)_2$)

$$RO \cdot + P(OC_2H_5)_3 \rightarrow RO\dot{P}(OC_2H_5)_3 \tag{11-77}$$

$$RO\dot{P}(OC_2H_5)_3 \rightarrow R \cdot + (C_2H_5O)_3PO \tag{11-78}$$

$$2R \cdot \rightarrow R_2 \tag{11-79}$$

phosphites with molecular oxygen do involve a chain reaction in which phosphoranyl radicals occur as intermediates. The propagating reactions 11-80 through 11-82 followed by the interaction of the alkoxy radical with the phosphite in the sequence 11-77 and 11-78 would account for the oxidation of two molecules of the trialkyl phosphite to the phosphate and regeneration of the chain-carrying alkyl radical.[21]

$$R \cdot + O_2 \rightarrow ROO \cdot \tag{11-80}$$

$$ROO \cdot + P(OC_2H_5)_3 \rightarrow ROO\dot{P}(OC_2H_5)_3 \tag{11-81}$$

$$ROO\dot{P}(OC_2H_5)_3 \rightarrow RO \cdot + OP(OC_2H_5)_3 \tag{11-82}$$

$$RO \cdot + P(OC_2H_5)_3 \rightarrow RO\dot{P}(OC_2H_5)_3 \tag{11-77}$$

$$RO\dot{P}(OC_2H_5)_3 \rightarrow R \cdot + OP(OC_2H_5)_3 \tag{11-78}$$

The reaction of carbon tetrachloride with triethyl phosphite is catalyzed by

$$Cl_4C + P(OC_2H_5)_3 \xrightarrow{\text{Per. or } h\nu} Cl_3CP(OC_2H_5)_2 + C_2H_5Cl \tag{11-83}$$
$$\downarrow$$
$$O$$

light and by peroxide. The chain sequence shown in 11-84 through 11-86 has been suggested to account for these products.[22]

$$Cl_3C \cdot + P(OC_2H_5)_3 \rightarrow Cl_3C\dot{P}(OC_2H_5)_3 \tag{11-84}$$

$$Cl_3C\dot{P}(OC_2H_5)_3 \rightarrow Cl_3CP(OC_2H_5)_2 + C_2H_5 \cdot \quad (11\text{-}85)$$
$$\downarrow$$
$$O$$

$$C_2H_5 \cdot + Cl_4C \rightarrow C_2H_5Cl + Cl_3C \cdot \quad\quad (11\text{-}86)$$

II. Reactions of Molecular Oxygen

A. AUTOXIDATIONS

Molecular oxygen participates in free-radical chain reactions with many organic substrates yielding hydroperoxides. The hydroperoxide can be isolated in some cases but often undergoes subsequent reactions leading to other reaction products. The chain sequence for the autoxidation of an organic substrate consists of the hydrogen abstraction reaction 11-87 and the coupling reaction

$$ROO \cdot + RH \rightarrow ROOH + R \cdot \quad\quad (11\text{-}87)$$

11-88 which, owing to the diradical nature of molecular oxygen, yields a chain-

$$R \cdot + O_2 \rightarrow ROO \cdot \quad\quad (11\text{-}88)$$

carrying peroxy radical $RO_2 \cdot$. The limiting factor in the chain reaction both from the standpoint of kinetic chain length and the site of reaction of the organic substrate is the hydrogen atom abstraction reaction 11-87.

The peroxy radical $ROO \cdot$ is somewhat more selective than an alkoxy radical as a hydrogen abstractor and displays acceptor qualities. The rate at which it reacts depends on the donor qualities of the substrate from which it abstracts

$$RO_2 \cdot + HR \rightarrow \left[R\overset{\delta-}{O}O \cdot \cdot H \cdot \cdot \overset{\delta+}{R} \right] \rightarrow ROOH + R \cdot \quad (11\text{-}89)$$

a hydrogen. Consequently, autoxidation reactions occur readily with compounds that have donor qualities whereas compounds having only acceptor substrates generally are not as prone to autoxidation.

Alkanes are oxidized to hydroperoxides by molecular oxygen, tertiary hydrogens being more labile than secondary hydrogens which, in turn, are more labile than primary hydrogens toward reaction with the peroxy radical. Methylcyclohexane (27), for example, is oxidized to the tertiary hydroperoxide 28 [23] and n-heptane to a mixture of secondary hydroperoxides.[24] Attack at the

$$(11\text{-}90)$$

27   28

tertiary hydrogen by an alkyl peroxy radical renders a more favorable polar effect to the transition state owing to the more extensive donor qualities of a tertiary alkyl radical relative to a secondary alkyl radical. Likewise, the donor qualities of a secondary alkyl radical being greater than those of a primary alkyl radical result in the secondary hydrogens being more labile than primary hydrogens toward attack by peroxy radicals.

Cumene (29) is converted in high yields to cumene hydroperoxide (30).[25] The greater reactivity of tertiary relative to primary benzylic hydrogens is

$$ \text{(structure 29)} + O_2 \rightarrow \text{(structure 30)} \qquad (11\text{-}91) $$

illustrated by the reaction of p-cymene (31) which yields the tertiary hydroperoxide 32 in greater amounts than the primary hydroperoxide 33.[26]

$$ \text{(structure 31)} + O_2 \rightarrow \text{(structure 32, 80\%)} + \text{(structure 33, 20\%)} \qquad (11\text{-}92) $$

The behavior of unsaturated compounds with molecular oxygen depends on the susceptibility of the double bond toward free-radical addition by a peroxy radical and on the availability of allylic hydrogens for reaction with a peroxy radical. Nonpolymerizable alkenes such as cyclohexene are converted to the allylic hydroperoxides[27] in a chain sequence that involves reaction of the peroxy radical with the allylic hydrogens. The preferential abstraction of allylic hydrogens is dictated both by polar and resonance factors.

Formation of a mixture of products results when the contributors to the hybrid allylic radical are not equivalent. For example, the autoxidation of 1,2-dimethylcyclohexene (34) yields about equivalent amounts of the isomeric

$$(11\text{-}93)$$

$$(11\text{-}94)$$

hydroperoxides **35** and **36**. In this reaction the peroxy radical abstracts an allylic hydrogen from the 3-position yielding a hybrid radical which reacts with molecular oxygen yielding two peroxy radicals and ultimately the hydroperoxides **35** and **36**.[28] Similarly, methyl oleate (**37**) yields a mixture of four

$$(11\text{-}95)$$

34   **35**   **36**

hydroperoxides resulting from interaction of molecular oxygen at the 8-, 9-, 10-, and 11-carbons of the unsaturated ester.[29]

$$CH_3(CH_2)_6CH_2CH\!=\!CHCH_2(CH_2)_6CO_2CH_3$$
37

$$\Big\downarrow O_2$$

$$CH_3(CH_2)_7CH\!=\!CHCH(CH_2)_6CO_2H$$
$$\underset{OOH}{|}$$

$$+\ CH_3(CH_2)_6CHCH\!=\!CH(CH_2)_7CO_2H$$
$$\underset{OOH}{|}$$

$$+\ CH_3(CH_2)_7CHCH\!=\!CH(CH_2)_6CO_2H$$
$$\underset{OOH}{|}$$

$$+\ CH_3(CH_2)_6CH\!=\!CHCH(CH_2)_7CO_2H \qquad (11\text{-}96)$$
$$\underset{OOH}{|}$$

If the unsaturated compound has no allylic hydrogens but a double bond reactive toward free-radical addition, a copolymerization of the compound with oxygen may occur. Thus, styrene (38) and oxygen react in a free-radical chain process yielding, in addition to other products, a polyperoxide.[30] Although the reactions can be initiated with AIBN and other radical sources, initiation can also occur in a bimolecular reaction of the monomer with oxygen.

$$\underset{\overset{|}{C_6H_5}}{\sim CH_2CH\cdot} + O_2 \rightarrow \underset{\overset{|}{C_6H_5}}{\sim CH_2CHOO\cdot} \qquad (11\text{-}97)$$

$$\underset{\overset{|}{C_6H_5}}{\sim CH_2CHOO\cdot} + \underset{\mathbf{38}}{CH_2{=}CHC_6H_5} \rightarrow \underset{\overset{|}{C_6H_5}}{\sim CH_2CHOOCH_2}\underset{\overset{|}{C_6H_5}}{\dot{C}H} \quad (11\text{-}98)$$

The autoxidation of ethers and acetals leads, in some cases, to isolable hydroperoxides. For example, isopropyl ether (39) reacts with molecular oxygen in a chain process yielding first the monohydroperoxide (40) and finally the dihydroperoxide (41).[31] These peroxides are so readily formed that samples of the ether that have been exposed to the atmosphere often contain considerable quantities of the peroxide and, because of the tendency for the peroxide to explode, should be handled with extreme caution. The polar nature of the

$$\underset{\mathbf{39}}{\underset{CH_3}{\overset{CH_3}{\diagdown}}\!\!HCOCH\!\!\underset{CH_3}{\overset{CH_3}{\diagup}}} + \xrightarrow{O_2} \underset{\mathbf{40}}{HOO{-}\underset{CH_3}{\overset{CH_3}{\diagdown}}\!\!COCH\!\!\underset{CH_3}{\overset{CH_3}{\diagup}}} \xrightarrow{O_2} \underset{\mathbf{41}}{HOO\underset{CH_3}{\overset{CH_3}{\diagdown}}\!\!COCOOH\!\!\underset{CH_3}{\overset{CH_3}{\diagup}}} \quad (11\text{-}99)$$

hydrogen abstraction accounts largely for the specificity of the α-hydrogens of the ether toward reaction with the peroxy radical in this chain-propagating reaction although the α-hydrogens in this case are also tertiary hydrogens whereas the remaining hydrogens are primary. Autoxidation of tetrahydrofuran (42) yields only the α-hydroperoxide 43[32] indicating that the polar effect rendered by the oxygen function is the dictating factor in determining the site of hydrogen abstraction by a peroxy radical in this case.

Aldehydes react with molecular oxygen yielding carboxylic acids. This reaction involves formation of an acyl hydroperoxide, which in some cases can be isolated, by the chain sequence shown in 11-102 and 11-103. Interaction of

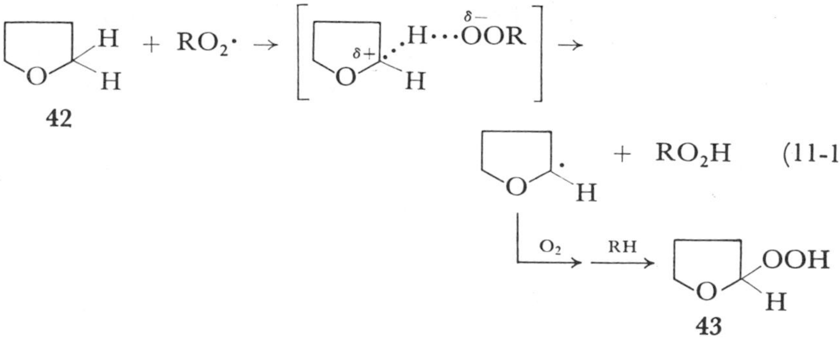

$$\text{(42)} \quad \begin{array}{c}\text{H}\\ \diagdown\\ \text{H}\end{array} + RO_2\cdot \rightarrow \left[\underset{\delta+}{\diagdown}..\overset{\delta-}{H\cdots\dot{O}OR} \right] \rightarrow$$

$$\begin{array}{c}\cdot\\ \diagdown\\ \text{H}\end{array} + RO_2H \qquad (11\text{-}100)$$

$$\xrightarrow{O_2} \xrightarrow{RH} \begin{array}{c}OOH\\ \diagdown\\ \text{H}\end{array}$$

43

the hydroperoxide with unreacted aldehyde yields the carboxylic acid in a process most likely proceeding by a nonradical mechanism. Evidence for

$$2RCHO + O_2 \rightarrow 2RCO_2H \qquad (11\text{-}101)$$

via

$$R\dot{C}O + O_2 \rightarrow RC\diagup^{OO\cdot}_{\diagdown O} \qquad (11\text{-}102)$$

$$RC\diagup^{OO\cdot}_{\diagdown O} + RCHO \rightarrow RC\diagup^{OOH}_{\diagdown O} + R\dot{C}O \qquad (11\text{-}103)$$

$$RC\diagup^{OOH}_{\diagdown O} + RCHO \rightarrow 2RCO_2H \qquad (11\text{-}104)$$

the intermediacy of acyl radicals in this process is found in the observation that at elevated temperatures and low oxygen concentrations, carbon monoxide resulting from decomposition of the acyl radical has been observed.[33]

$$R\dot{C}O \rightarrow R\cdot + CO \qquad (11\text{-}105)$$

The kinetics of free radical autoxidations are informative of certain aspects of chain reactions (see Chapter 3). The mechanism for the chemically initiated autoxidation reaction of a substrate RH yielding the hydroperoxide ROOH involves two different chain-carrying species and hence three different termination reactions are possible. The kinetic rate law for the reaction depends on the termination step that is operative.

$$\text{Init.} \xrightarrow{k_{106}} 2R'\cdot \xrightarrow{O_2} R'OO\cdot \xrightarrow{RH} R'OOH + R\cdot \qquad (11\text{-}106)$$

$$R\cdot + O_2 \xrightarrow{k_{107}} ROO\cdot \tag{11-107}$$

$$ROO\cdot + RH \xrightarrow{k_{108}} ROOH + R\cdot \tag{11-108}$$

$$2ROO\cdot \xrightarrow{k_{109}} ROOR + O_2 \tag{11-109}$$

$$ROO\cdot + R\cdot \xrightarrow{k_{110}} ROOR \tag{11-110}$$

$$2R\cdot \xrightarrow{k_{111}} R_2 \tag{11-111}$$

Rate studies of the autoxidation of ethyl linoleate show that at oxygen pressures above 100 mm (reaction temperature 45°) the rate law 11-112 is

$$\text{Rate} = \frac{-d[O_2]}{dt} = \left(\frac{k_{106}}{2k_{109}}\right)^{1/2} k_{108}[RH][\text{Init.}]^{1/2} \tag{11-112}$$

followed[34] indicating that the termination occurs almost exclusively by the coupling of two peroxy radicals. At lower oxygen pressures, the rate of autoxidation becomes oxygen dependent. The derived rate expressions for autoxidations involving termination by reactions 11-110 and 11-111 are 11-113 and 11-114, respectively.

$$\text{Rate} = \frac{-d[O_2]}{dt} = \left(\frac{k_{106}k_{107}k_{108}}{k_{110}}\right)^{1/2} [RH]^{1/2}[O_2]^{1/2}[\text{Init.}]^{1/2} \tag{11-113}$$

$$\text{Rate} = \frac{-d[O_2]}{dt} = \left(\frac{k_{106}}{2k_{111}}\right)^{1/2} k_{107}[O_2][\text{Init.}]^{1/2} \tag{11-114}$$

The reaction when initiated by a chemical initiator follows the rate law shown in 11-112 but when other means of initiation are employed, the rate law may take the more general form

$$\text{Rate} = \frac{-d[O_2]}{dt} = \left(\frac{R_i}{2k_{109}}\right)^{1/2} k_{108}[RH] \tag{11-115}$$

where R_i is the rate of initiation ($R_i = {}_{106}[\text{Init.}]$ in 11-112). Reactions of polymerizable monomers such as styrene with oxygen involve a bimolecular initiation where

$$R_i = k_i[O_2][CH_2{=}CHR] \tag{11-116}$$

and the observed rate law for the reaction becomes

$$\text{Rate} = \frac{-d[O_2]}{dt} = \left(\frac{k_i}{2k_{109}}\right)^{1/2} k_{108}[CH_2{=}CHR]^{3/2}[O_2]^{1/2} \tag{11-117}$$

It is also significant that the rate is dependent on the reactivity of RH (or $RCH{=}CH_2$ in the case of polyperoxide formation) toward reaction with the peroxy radical. The nature of the termination reaction 11-109, namely the bimolecular process of peroxy radicals in which a peroxide is produced, has been suggested to occur by way of formation of a tetroxide which decomposes to the peroxide and oxygen at least in the case where R is t-butyl.[35]

$$2ROO\cdot \rightarrow ROOOOR \rightarrow ROOR + O_2 \qquad (11\text{-}118)$$

The autoxidation of organic compounds is an important reaction since it can be expected to occur wherever susceptible compounds come in contact with air. The oxidation of many commercial materials is detrimental and there is considerable loss of these substances by their reaction with atmospheric oxygen by the chain reaction leading to the hydroperoxide and, in many cases, to products of further reaction of the hydroperoxide. The autoxidation reaction can be inhibited by use of antioxidants which remove chain-carrying radicals by reacting with them in nonpropagating reactions. In view of the arguments given above, the chain-carrying radical present in relatively greater concentration and most likely to react with the antioxidant is the peroxy radical. Most of the more common antioxidants are aromatics having alkyl, amino, or hydroxy groups. The mechanisms of the interaction of the peroxy radicals with some antioxidants have been investigated and appear to involve either formation of a complex which is too unreactive to propagate the chain or a hydrogen abstraction from the antioxidant producing a free radical that is unreactive.[36]

B. OTHER OXIDATIONS

Free radical chain oxidations involving molecular oxygen that do not lead to formation of a hydroperoxide either as a product or intermediate are also known. The oxidation of perchloroethylene (**44**) to trichloroacetyl chloride (**45**) is one such reaction.[37] The reaction is light-induced and requires a small amount of molecular chlorine as a catalyst. The chain sequence 11-120 through 11-123, which is induced by photolysis of molecular chlorine, is a plausible mechanism for this oxidation. In similar reactions trichloroethylene (**46**) is converted to dichloroacetyl chloride (**47**)[38] and pentachloroethane (**48**) to trichloroacetyl chloride (**49**) with concurrent formation of hydrogen chloride.[39]

$$Cl_2 \xrightarrow{h\nu} 2Cl\cdot \qquad (11\text{-}119)$$

$$Cl\cdot + \underset{\mathbf{44}}{Cl_2C{=}CCl_2} \longrightarrow Cl_3C\dot{C}Cl_2 \qquad (11\text{-}120)$$

$$Cl_3C\dot{C}Cl_2 + O_2 \longrightarrow Cl_3C\overset{\displaystyle Cl}{\underset{\displaystyle Cl}{\overset{|}{\underset{|}{C}}}}OO\cdot \qquad (11\text{-}121)$$

$$2 \; Cl_3C\overset{\overset{\displaystyle Cl}{|}}{\underset{\underset{\displaystyle Cl}{|}}{C}}OO \cdot \;\rightarrow\; O_2 + 2Cl_3C\overset{\overset{\displaystyle Cl}{|}}{\underset{\underset{\displaystyle Cl}{|}}{C}}O \cdot \tag{11-122}$$

$$Cl_3C - \overset{\overset{\displaystyle Cl}{|}}{\underset{\underset{\displaystyle Cl}{|}}{C}} - O \cdot \;\rightarrow\; Cl \cdot + Cl_3CC\overset{\displaystyle O}{\underset{\displaystyle Cl}{\diagup\!\!\!\diagdown}} \tag{11-123}$$

$$\underset{\displaystyle 45}{}$$

In the latter reaction, the pentachloroethyl radical is formed by abstraction of a hydrogen atom from **48** by the eliminated chlorine atom.

$$\underset{\displaystyle 46}{ClCH{=}CCl_2} + \tfrac{1}{2}O_2 \;\rightarrow\; \underset{\displaystyle 47}{Cl_2CH\overset{\displaystyle O}{\overset{\|}{C}}Cl} \tag{11-124}$$

$$\underset{\displaystyle 48}{Cl_3CCHCl_2} + \tfrac{1}{2}O_2 \;\rightarrow\; \underset{\displaystyle 49}{Cl_3CC\overset{\displaystyle O}{\underset{\displaystyle Cl}{\diagup\!\!\!\!\diagdown}}} + HCl \tag{11-125}$$

The reactions of phosphorus trichloride with oxygen and alkanes or alkyl halides follow a different route.[40] The product distributions observed in these reactions indicate that hydrogen abstraction from the organic substrate by a chlorine atom determines the site of the bonding with the phosphorus moiety. Thus, reaction of propane with PCl_3 and oxygen yielded a mixture of 73% secondary and 27% primary alkane phosphonyl chlorides along with $POCl_3$.

$$RH + 2PCl_3 + O_2 \;\rightarrow\; RPOCl_2 + POCl_3 + HCl \tag{11-126}$$

via

$$Cl \cdot + RH \;\rightarrow\; HCl + R \cdot \tag{11-127}$$

$$R \cdot + PCl_3 \;\rightarrow\; R\dot{P}Cl_3 \tag{11-128}$$

$$R\dot{P}Cl_3 + O_2 \;\rightarrow\; \underset{\underset{\displaystyle OO \cdot}{|}}{RPCl_3} \tag{11-129}$$

$$\underset{\underset{\displaystyle OO \cdot}{|}}{RPCl_3} + PCl_3 \diagup\!\!\!\!\diagdown \begin{array}{l} RPOCl_2 + POCl_3 + Cl \cdot \quad (11\text{-}130) \\[2mm] 2 \; POCl_3 + R \cdot \quad\quad\quad\;\; (11\text{-}131) \end{array}$$

Reactions of alkyldichlorophosphines (50) with alkanes and oxygen follow a similar reaction path yielding dialkylphosphonyl chlorides (51).[41]

$$R'PCl_2 + RH + \tfrac{1}{2}O_2 \;\rightarrow\; R'\!\!-\!\!\overset{\displaystyle O}{\underset{\displaystyle R}{\overset{\diagup}{P}Cl}} + HCl \qquad (11\text{-}132)$$

50   51

III. Inorganic Ion Reactions

Many inorganic ions undergo redox reactions that involve loss or gain of a single electron. In this section some reactions of organic substrates with inorganic ions that occur by free-radical mechanisms are discussed. The inorganic species, in some cases, is a reactant that is either oxidized or reduced in the reaction. In other reactions the inorganic ion is a catalyst that is oxidized in one step but reduced to its original oxidation state in another.

A. FERROUS-FERRIC ION REACTIONS

The oxidation of iron(II) (ferrous ion) to iron(III) (ferric ion) by a peroxide is a one electron process and produces a free radical. In the case of hydrogen peroxide, the reaction yields a hydroxy radical,[42] a species capable of interaction with many organic substrates. In the absence of an organic substrate,

$$Fe^{+2} + H_2O_2 \;\rightarrow\; Fe^{+3} + OH^{(-)} + HO \qquad (11\text{-}133)$$

however, the hydroxy radical reacts either with more iron(II) or with peroxide. In the former case, the reaction becomes simply the oxidation of the iron(II) to iron(III) with concurrent reduction of the peroxide to hydroxide ions

$$Fe^{+2} + \cdot OH \;\rightarrow\; Fe^{+3} + {}^-OH \qquad (11\text{-}134)$$

(reactions 11-133 and 11-134). If the hydroxy radical reacts with the peroxide, the result is quite different in that the hydrogen peroxide is decomposed to water and molecular oxygen in a chain reaction. The precise nature of reaction

$$HO\cdot + H_2O_2 \;\rightarrow\; H_2O + HOO\cdot \qquad (11\text{-}135)$$

$$HO\cdot + H_2O + O_2 \;\xleftarrow{\;H_2O_2\;}$$

11-135 is not known and may be either a hydrogen atom transfer or an electron transfer from HOO· to the peroxide linkage as shown in 11-136. The chain

$$HOO\cdot + H_2O_2 \rightarrow HOO^+ + \overset{(-)}{O}H + HO\cdot \qquad (11\text{-}136)$$
$$\downarrow$$
$$ H^+ + O_2$$

reaction shown in 11-135 is long only if the concentration of iron(II) is low compared to that of the peroxide in order to favor the competition of reaction of the latter for the chain-carrying hydroxy radicals.[42]

The hydroxy radical reacts with many organic substrates. In the case of unsaturated species, addition may occur making the iron(II)-hydrogen peroxide system (Fenton's reagent) a hydroxylating reagent[43] as well as an effective initiator system for free-radical vinyl polymerization of water soluble monomers. The reactive nature of the hydroxy radical as a hydrogen abstractor often makes the reaction mixtures resulting from reaction of Fenton's reagent with an organic substrate complex.

An example of the catalytic behavior of an inorganic ion in a free-radical chain process is the iron(II)-iron(III) catalyzed oxidation of primary and secondary alcohols with hydrogen peroxide.[44] A mechanism that accounts for both the products and the catalytic behavior of the iron(II)–iron(III) system is shown in the sequence 11-137 through 11-139. The iron(III) is reduced to the iron(II) in reaction 11-139 by electron transfer from the α-hydroxyalkyl radical.

$$Fe^{+2} + H_2O_2 \rightarrow Fe^{+3} + \overset{(-)}{O}H + HO\cdot \qquad (11\text{-}137)$$

$$HO\cdot + R_2CHOH \rightarrow H_2O + R_2\dot{C}OH \qquad (11\text{-}138)$$

$$R_2\dot{C}OH + Fe^{+3} \rightarrow R_2\overset{(+)}{C}OH + Fe^{+2} \qquad (11\text{-}139)$$
$$\phantom{R_2\dot{C}OH + Fe^{+3} \rightarrow R_2}\downarrow$$
$$\phantom{R_2\dot{C}OH + Fe^{+3} \rightarrow} R_2C{=}O + H^{(+)}$$

Reactions of carboxylic acids with Fenton's reagent yield products that appear to involve a similar chain reaction. The oxidation of hexahydrobenzoic acid (52) with hydrogen peroxide in the presence of iron(II) yields cyclohexanol and cyclohexene among other reaction products. The formation of these products is best explained in terms of a carbonium ion intermediate which would be formed by oxidation of the alkyl radical 53 by the iron(III) as shown in 11-143.[45]

$$52$$

$$+ H_2O_2 \xrightarrow{Fe^{+2}} \quad + \quad + CO_2 + H_2O$$

$$(11\text{-}140)$$

via
$$Fe^{+2} + H_2O_2 \rightarrow Fe^{+3} + \overset{(-)}{OH} + HO\cdot$$

$$HO\cdot + \underset{52}{\text{[cyclohexane-}CO_2H\text{]}} \rightarrow H_2O + \text{[cyclohexane-}CO_2\cdot\text{]} \qquad (11\text{-}141)$$

$$\underset{53}{\text{[cyclohexane-}CO_2\cdot\text{]}} \rightarrow \text{[cyclohexyl}\cdot\text{]} + CO_2 \qquad (11\text{-}142)$$

$$\underset{53}{\text{[cyclohexyl}\cdot\text{]}} + Fe^{+3} \rightarrow \text{[cyclohexyl}^{(+)}\text{]} + Fe^{+2} \qquad (11\text{-}143)$$

$$\text{[cyclohexyl}^{(+)}\text{]} \xrightarrow{H_2O} \text{[cyclohexanol-}OH\text{]} + H^+ \qquad (11\text{-}144a)$$

$$\text{[cyclohexyl}^{(+)}\text{]} \searrow \text{[cyclohexene]} + H^+ \qquad (11\text{-}144b)$$

Reactions of alkyl hydroperoxides with iron(II) appear to be similar in that the peroxide linkage is reduced to an anion and a free radical. The radical formed generally is the alkoxy radical and, in the presence of appropriate organic substrates, reacts by either hydrogen abstraction or addition to an unsaturated linkage. For example, reactions of alkyl hydroperoxides with

$$ROOH + Fe^{+2} \rightarrow RO\cdot + \overset{(-)}{OH} + Fe^{+3} \qquad (11\text{-}145)$$

butadiene in the presence of iron(II) yield octadiene derivatives. This product is formed by the dimerization of the adduct radicals formed by addition of alkoxy radicals to the diene.[46]

$$RO\cdot + CH_2{=}CHCH{=}CH_2 \rightarrow ROCH_2\dot{C}HCH{=}CH_2 \qquad (11\text{-}146)$$
$$\updownarrow$$
$$ROCH_2CH{=}CH\dot{C}H_2$$

$$2ROCH_2CH{=}CH\dot{C}H_2 \rightarrow$$

$$ROCH_2CH{=}CHCH_2CH_2CH{=}CHCH_2OR + \text{isomers} \qquad (11\text{-}147)$$

B. COPPER ION REACTIONS

Copper(I) (cuprous ion) catalyzes reactions of diacyl and dialkyl peroxides that yield products different from those observed in the thermal decomposition of the peroxides. t-Butyl peroxide and benzaldehyde in the presence of a catalytic amount of copper(I), yield t-butyl benzoate (54).[47] A mechanism that

$$(CH_3)_3COOC(CH_3)_3 + C_6H_5CHO \xrightarrow{Cu^+}$$

$$(CH_3)_3COH + C_6H_5CO_2C(CH_3)_3 \quad (11\text{-}148)$$
$$\underset{54}{}$$

accounts for the product shown includes reduction of the peroxide linkage in a

$$Cu^+ + ROOR \rightarrow Cu^{+2} + RO\cdot + RO^{(-)} \quad (11\text{-}149)$$

$$RO\cdot + C_6H_5CHO \rightarrow ROH + C_6H_5\overset{.}{C}O \quad (11\text{-}150)$$

$$C_6H_5\overset{.}{C}O + Cu^{+2} \rightarrow Cu^+ + C_6H_5\overset{(+)}{C}O \quad (11\text{-}151)$$
$$\xrightarrow[]{ROH} C_6H_5CO_2R$$
$$\underset{54}{}$$

$$R = C(CH_3)_3$$

one electron transfer from the copper(I) (11-149) and oxidation of the acyl radical to the acylium ion by the copper(II) in reaction 11-151. The latter reaction apparently is fast since most of the benzoyl radicals undergo oxidation to the cation rather than add to unreacted benzaldehyde yielding the α-benzoyoxybenzyl radicals which are formed in reactions of benzaldehyde and t-butyl peroxide in the absence of copper ions (see Chapter 6).

Cumene (55) reacts with benzoyl peroxide in the presence of copper ions yielding cumyl benzoate (56) in a reaction that involves a mechanism similar to that proposed for the benzaldehyde-t-butyl peroxide reaction. Copper ion catalyzed reactions of peroxides and peresters with hydrocarbons are of synthetic

$$\underset{55}{\overset{\displaystyle CH_3}{\underset{\displaystyle CH_3}{C_6H_5\overset{|}{\underset{|}{C}}H}} + Bz_2O_2 \xrightarrow[30\%]{Cu^+} \underset{56}{\overset{\displaystyle CH_3}{\underset{\displaystyle CH_3}{C_6H_5\overset{|}{\underset{|}{C}}OBz}} + BzOH \quad (11\text{-}152)$$

value since the reaction allows for introduction of a functional group in the molecule.[48] An example of the use of this reaction is the synthesis of 7-t-butoxy-norbornadiene (59) from t-butyl perbenzoate (57) and norbornadiene (58) in

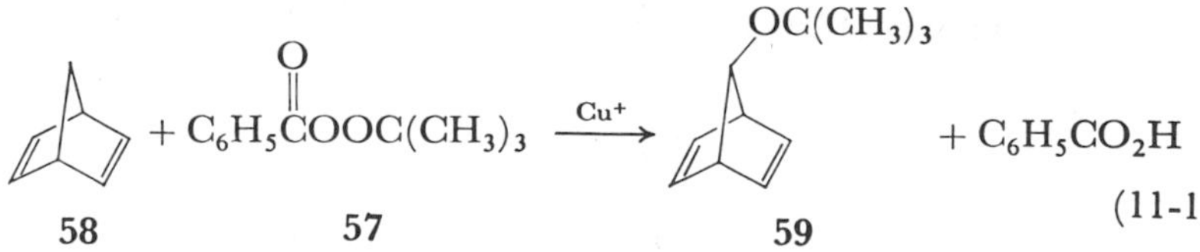

$$\text{58} + C_6H_5\overset{\overset{\displaystyle O}{\|}}{C}OOC(CH_3)_3 \xrightarrow{Cu^+} \text{59} + C_6H_5CO_2H \tag{11-153}$$

58   57   59

via

$$Cu^+ + C_6H_5\overset{\overset{\displaystyle O}{\|}}{C}OOC(CH_3)_3 \rightarrow Cu^{+2} + (CH_3)_3CO\cdot + C_6H_5\overset{\overset{\displaystyle O}{\|}}{C}O^- \tag{11-154}$$

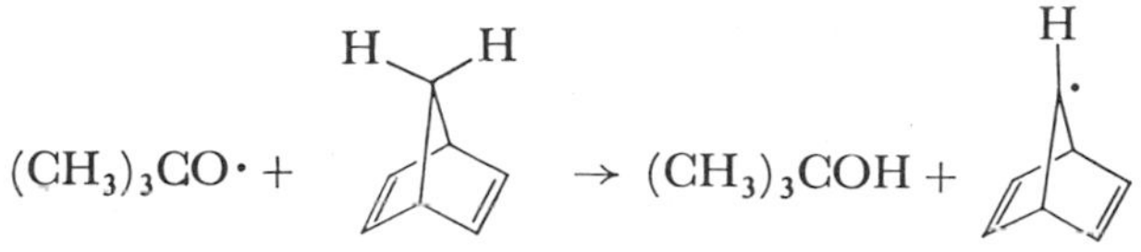

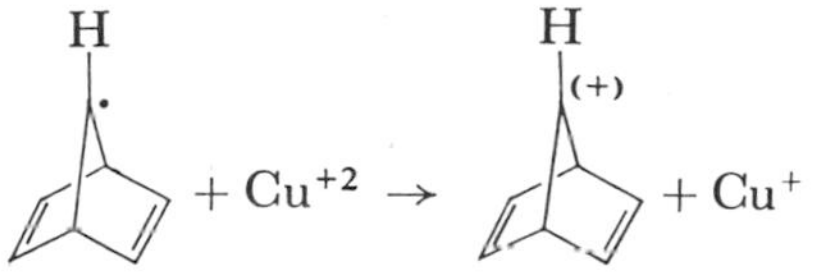

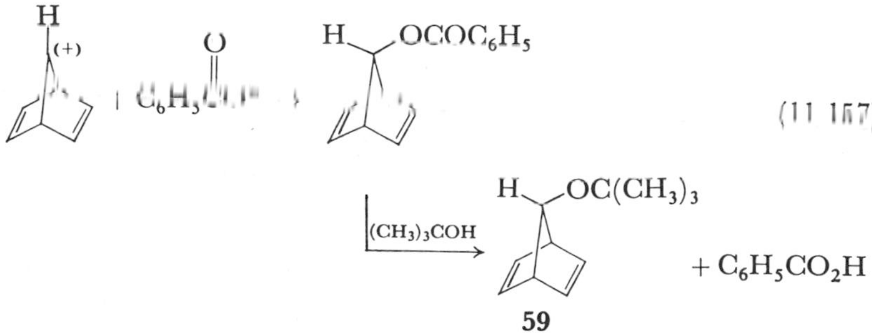

$$(CH_3)_3CO\cdot + \quad \rightarrow (CH_3)_3COH + \tag{11-155}$$

$$+ Cu^{+2} \rightarrow + Cu^+ \tag{11-156}$$

$$+ C_6H_5CO^- \rightarrow \tag{11-157}$$

$$\xrightarrow{(CH_3)_3COH} \text{59} + C_6H_5CO_2H$$

59

a copper ion catalyzed reaction.[49, 50] The benzoate ester is the only isolated product in some cases and may be an intermediate for the formation of 59.

Decompositions of diacyl peroxides induced by copper ions apparently are free-radical chain reactions. Evidence for the oxidation of the alkyl radicals by the copper(II) to carbonium ions comes from the nature of the products resulting from the alkyl moieties of the peroxide. In most cases an alkene resulting from fragmentation of the carbonium ion is formed. For example, the reaction of isovaleryl peroxide (60) with copper ions yields isovaleric acid, carbon dioxide, and isobutylene by the chain sequence of reactions shown in 11-158 and 11-159. Similarly, 1-butene is formed in the reaction of n-valeryl

$$((CH_3)_2CHCH_2CO_2)_2 + Cu^+ \rightarrow$$
$$\mathbf{60}$$

$$(CH_3)_2CHCH_2CO_2^- + Cu^{+2} + (CH_3)_2CHCH_2\cdot + CO_2 \quad (11\text{-}158)$$

$$(CH_3)_2CHCH_2\cdot + Cu^{+2} \rightarrow Cu^+ + (CH_3)_2CH\overset{(+)}{C}H_2 \quad (11\text{-}159)$$

$$\phantom{(CH_3)_2CHCH_2\cdot + Cu^{+2} \rightarrow} \longrightarrow H^+ + (CH_3)_2C{=}CH_2$$

peroxide with copper ions. Compelling evidence for the oxidation of the alkyl radical to the carbonium ion by the copper(II) ion comes from the observation of a rearrangement of the carbon skeleton of the alkyl moiety in the copper-catalyzed reaction of *bis-t*-butylacetyl peroxide (**61**) in acetic acid. The neopentyl radical (**62**) does not rearrange whereas the carbonium ion produced by the oxidation of the radical can rearrange to the *t*-amyl carbonium ion. The formation of the 2-methyl-1-butene, 2-methyl-2-butene, and *t*-amyl acetate as reaction products can readily be explained in terms of reactions of this carbonium ion.[51]

$$((CH_3)_3CCH_2CO_2)_2 + Cu^+ \rightarrow$$
$$\mathbf{61}$$

$$Cu^{+2} + (CH_3)_3CCH_2CO_2 + (CH_3)_3CCH_2\cdot + CO_2 \quad (11\text{-}160)$$
$$\phantom{Cu^{+2} + (CH_3)_3CCH_2CO_2 + (CH_3)_3CCH_2\cdot} \mathbf{62}$$

$$(CH_3)_3CCH_2\cdot + Cu^{+2} \rightarrow (CH_3)_3C\overset{(+)}{C}H_2 + Cu^+ \quad (11\text{-}161)$$

$$(CH_3)_3C\overset{(+)}{C}H_2 \rightarrow (CH_3)_2\overset{+}{C}CH_2CH_3$$

$$\overset{-H^+}{\swarrow} \quad \overset{-H^+}{\downarrow} \quad \overset{HOAc}{\searrow}$$

$$\underset{CH_2{=}CHC_2H_5}{\overset{CH_3}{|}} \quad (CH_3)_2C{=}CHCH_3 \quad \underset{(CH_3)_2CC_2H_5}{\overset{OAc}{|}} \quad (11\text{-}162)$$

The catalytic effect of copper salts on the addition of carbon tetrachloride to alkenes is another example of a redox reaction involving the copper(I)–copper(II) system intervening in the course of a familiar free-radical reaction. Because of the comparatively low chain-transfer constants of carbon tetrachloride with polymerizable monomers such as styrene, acrylonitrile, and ethyl acrylate, high molecular weight telomers are formed in free-radical addition reactions of the polyhalomethane to these monomers. However, in the presence of copper salts, good yields of the 1:1 addition product are formed.[52] The mechanism shown in 11-163 through 11-165 offers an explanation for the fact that the transfer reaction appears to proceed more readily in the presence of copper salts. The displacement reaction of the adduct radical leading to the formation of the 1:1 addition product occurs on the copper(II) chloride which apparently is more reactive toward reaction with the adduct radical than is the polyhalomethane. The abstraction reaction is, however, a reduction process since it generates the copper(I) species that reacts with carbon tetrachloride reducing this species to the trichloromethyl radical and generating the copper(II) chloride. The addition of chloroform to 1-octene in the presence of copper

$$Cu^+ + Cl_4C \rightarrow Cu^{II}Cl + Cl_3C \cdot \tag{11-163}$$

$$Cl_3C \cdot + CH_2{=}CHR \rightarrow Cl_3CCH_2\dot{C}HR \tag{11-164}$$

$$Cl_3CCH_2\dot{C}HR + Cu^{II}Cl \rightarrow Cl_3CCH_2\underset{\underset{Cl}{|}}{C}HR + Cu^+ \tag{11-165}$$

provides support for this mechanism. The free-radical addition of chloroform to terminal alkenes generally results in formation of a 1,1,1-trichloroalkane as

$$HCCl_3 + CH_2{=}CHR \xrightarrow{Per.} Cl_3CCH_2CH_2R \tag{11-166}$$

the addition product (reaction 11-166), a product indicative of addition of a trichloromethyl radical to the unsaturated linkage (see Chapter 6). In the presence of copper, however, 1,1,3-trichlorononane (63) is produced in the reaction of chloroform with 1-octene. Copper(I) reacts with chloroform producing, as the adding radical, the dichloromethyl radical and the copper(II) chloride which reacts with the adduct radical producing the observed reaction product and regenerating copper(I).

$$HCCl_3 + CH_2{=}CHC_6H_{13}{-}n \xrightarrow{Cu^+} Cl_2CHCH_2\underset{\underset{Cl}{|}}{C}HC_6H_{13}{-}n \tag{11-167}$$
$$\textbf{63}$$

via

$$Cu^+ + HCCl_3 \longrightarrow Cu^{II}Cl + H\dot{C}Cl_2 \tag{11-168}$$

$$H\dot{C}Cl_2 + CH_2{=}CHC_6H_{13}{-}n \longrightarrow Cl_2CHCH_2\dot{C}HC_6H_{13}{-}n \tag{11-169}$$

$$Cl_2CHCH_2\dot{C}HC_6H_{13}{-}n + Cu^{II}Cl \longrightarrow Cl_2CHCH_2\underset{\underset{Cl}{|}}{C}HC_6H_{13}{-}n + Cu^+$$
$$\textbf{63}$$

$$\tag{11-170}$$

Other reactions involving the copper(I)–copper(II) system that probably proceed by free-radical chain mechanisms are the reactions of copper(I) salts with aryl diazonium cations. For example, the benzene diazonium cation reacts with halide ions or cyanide ions in the presence of the corresponding copper(I) salts (Sandmeyer reaction) yielding the halobenzene or benzonitrile, respectively. The reaction may proceed by a free-radical chain process similar

$$C_6H_5N^+ + X^- \xrightarrow{Cu^+} C_6H_5X + N_2 \qquad (11\text{-}171)$$

to those outlined above in that the copper(I) acts as a reducing agent converting the diazonium cation to the free radical and nitrogen. Interaction of the phenyl radical with the ligand of the copper(II)-complex yields the product of the reaction and regenerates the copper(I).

$$C_6H_5N_2^+ + Cu^IX_n \rightarrow C_6H_5\cdot + N_2 + Cu^{II}X_n \qquad (11\text{-}172)$$

$$C_6H_5\cdot + Cu^{II}X_n^- \rightarrow C_6H_5X + Cu^IX_{n-1} \qquad (11\text{-}173)$$

Other evidence of the intermediacy of the aryl radical is that reaction of aryl diazonium cations comes from their reduction to the hydrocarbon by either hypophosphorous acid[53] (64) or methanol.[54] The former reaction is catalyzed by copper(I) salts and likely involves the chain sequence 11-175 and 11-176

$$Cu^+ + ArN_2^+ \rightarrow Cu^{+2} + Ar\cdot \qquad (11\text{-}174)$$

$$Ar\cdot + H_3PO_2 \rightarrow ArH + H_2\dot{P}O_2 \qquad (11\text{-}175)$$
$$\mathbf{64}$$

$$H_2\dot{P}O_2 + ArN_2^+ \rightarrow \overset{(+)}{H_2PO_2} + Ar\cdot + N_2 \qquad (11\text{-}176)$$
$$\downarrow{\scriptstyle H_2O} \quad H_3PO_3 + H^+$$

which can be induced by interaction of copper(I) with the aryl diazonium cation. The reduction by alcohols likely involves a very similar chain sequence, the reducing radical in this case being the α-hydroxyalkyl radical.

$$Ar\cdot + CH_3OH \rightarrow ArH + \cdot CH_2OH \qquad (11\text{-}177)$$

$$\cdot CH_2OH + ArN_2^+ \rightarrow \overset{(+)}{CH_2OH} + Ar\cdot + N_2 \qquad (11\text{-}178)$$
$$\downarrow \quad CH_2O + H^+$$

C. SILVER ION REACTIONS

Silver(I) can be oxidized to either silver(II) or silver(III) by peroxydisulfate ions.[55] Although salts of silver(III) have been isolated and characterized, most of the reactions of silver in higher oxidation states in solution can be considered to be those of silver(II) since disproportionation of silver(I) and silver(III) would be expected to occur. Silver(II) is a strong oxidizing agent and is capable

$$S_2O_8^{2-} + Ag^+ \rightarrow 2SO_4^{2-} + Ag^{+3} \tag{11-179}$$

$$Ag^{+3} + Ag^+ \rightarrow 2Ag^{+2} \tag{11-180}$$

$$2Ag^+ + S_2O_8^{2-} \rightarrow 2Ag^{+2} + 2SO_4^{-2} \tag{11-181}$$

of oxidizing water to molecular oxygen. A likely mechanism is that given in

$$4Ag^{+2} + 2H_2O \rightarrow 4Ag^+ + 4H^+ + O_2 \tag{11-182}$$

11-183 through 11-185 which involves oxidation of water by silver(II) to

$$Ag^{+2} + H_2O \longrightarrow Ag^+ + H^+ + HO\cdot \tag{11-183}$$

$$2HO\cdot \longrightarrow H_2O_2 \tag{11-184}$$

$$(2H_2O_2 \xrightarrow{HO\cdot} 2H_2O + O_2) \tag{11-185}$$

hydroxy radicals which couple to form hydrogen peroxide, the latter decomposing by a free-radical chain mechanism involving hydroxy radicals similar to that outlined previously for the Fenton reaction.

In the presence of catalytic amounts of silver(I), peroxydisulfate can effect certain oxidative cleavage reactions that do not occur in the absence of silver ions. For example, pinacol (65) is oxidized to acetone in good yields by peroxydisulfate if a silver(I) is present.[56] Other metal ions do not display a similar

$$(CH_3)_2C\!\!-\!\!C(CH_3)_2 + S_2O_8^{2-} \xrightarrow{Ag^+} 2SO_4^{2-} + 2H^+ + 2(CH_3)_2C\!\!=\!\!O$$
$$\underset{OH\ OH}{|\quad\ |}$$
$$\mathbf{65} \tag{11-186}$$

catalytic effect suggesting that silver(II) may be a reaction intermediate. A mechanism that accounts for the reaction products and the catalytic effect of the silver is given in 11-187 through 11-191. Reaction 11-187 is a bimolecular

$$S_2O_8{}^{2-} + Ag^+ \rightarrow Ag^{+2} + SO_4{}^{\bar{\cdot}} + SO_4{}^{2-} \tag{11-187}$$

$$Ag^{+2} + (CH_3)_2\underset{HO}{\overset{}{C}}\!\!-\!\!\underset{OH}{\overset{}{C}}(CH_3)_2 \rightarrow Ag^+ + H^+ + (CH_3)_2\underset{\cdot O}{\overset{}{C}}\!\!-\!\!\underset{OH}{\overset{}{C}}(CH_3)_2 \tag{11-188}$$

$$(CH_3)_2\underset{:O}{\overset{\curvearrowright}{C}}\!\!-\!\!\underset{OH}{\overset{}{C}}(CH_3)_2 \rightarrow (CH_3)_2C\!\!=\!\!O + \cdot\underset{OH}{\overset{}{C}}(CH_3)_2 \tag{11-189}$$

$$\cdot\underset{OH}{\overset{}{C}}(CH_3)_2 + S_2O_8{}^{2-} \rightarrow {}^{(+)}\underset{OH}{\overset{}{C}}(CH_3)_2 + SO_4{}^{2-} + SO_4{}^{\bar{\cdot}} \tag{11-190}$$
$$\llcorner\!\!\rightarrow H^+ + (CH_3)_2C\!\!=\!\!O$$

$$SO_4^- \cdot + Ag^+ \rightarrow SO_4^{2-} + Ag^{+2} \qquad (11\text{-}191)$$

initiation reaction producing two chain-carrying radicals (Ag^{2+} and $SO_4^- \cdot$). The chain sequence of reactions 11-188 through 11-191 involves oxidation of the glycol by silver(II) and ultimately regeneration of this species by oxidation of silver(I) by the sulfate ion radical. The other radical-propagating reactions in the chain sequence, namely the fragmentation of the alkoxy radical (reaction 11-189) and induced decomposition of the peroxydisulfate by the α-hydroxy-alkyl radical (reaction 11-190), have been discussed previously in Chapters 8 and 10, respectively.

Other systems can also be oxidized by peroxydisulfate in silver-catalyzed reaction. For example, pyruvic acid (66) undergoes oxidative decarboxylation in a chain reaction.[57]

$$CH_3COCO_2H + S_2O_8{}^{2-} + H_2O \xrightarrow{\ Ag^+\ } CH_3CO_2H + CO_2 + 2H^+ + 2SO_4{}^{2-}$$
$$\underset{\textbf{66}}{}$$
$$\qquad (11\text{-}192)$$

via

$$Ag^{+2} + \underset{\textbf{66}}{CH_3COCO_2H} \rightarrow CH_3COCO_2 \cdot + Ag^+ + H^+ \quad (11\text{-}193)$$

$$CH_3COCO_2 \cdot \rightarrow CH_3\dot{C}O + CO_2 \qquad (11\text{-}194)$$

$$CH_3\dot{C}O + S_2O_8{}^{2-} \rightarrow CH_3\overset{(+)}{C}O + SO_4{}^{2-} + SO_4^- \quad (11\text{-}195)$$
$$\phantom{CH_3\dot{C}O + S_2O_8{}^{2-} \rightarrow}\big\downarrow{\scriptstyle H_2O} \quad CH_3CO_2H + H^+$$

$$SO_4^- + Ag^+ \rightarrow SO_4{}^{2-} + Ag^{+2} \qquad (11\text{-}196)$$

D. OTHER INORGANIC ION REACTIONS

Several other inorganic species undergo one electron reactions with organic substrates but do not participate readily in chain reactions. A few examples of such reactions are given in this section to illustrate the processes that occur in these reactions.

1. Cerium(IV) Oxidations. The cerium(IV) ion is a strong oxidizing agent and is reduced to cerium(III) in its reactions with many organic substrates. This is obviously a single electron transfer process and it is not unexpected that the reactions bear many of the earmarks of free-radical reactions. Oxidations of primary and secondary alcohols by ceric sulfate,[58-60] for example, involve removal of an α-hydrogen from the alcohol in the rate-determining step as evidenced by a deuterium isotope effect ($k_H/k_D = 1.9$) in the oxidations of cyclohexanol and α-deuterated cyclohexanol with cerium(IV). A mechanism has been suggested for these oxidations which involves abstraction of the

α-hydrogen from the alcohol with concurrent reduction of the cerium(IV) in a cyclic process involving the complexed ion [59, 60] (reaction 11-197). Subsequent

$$\text{(cyclic Ce}^{IV}\text{ complex)} \rightarrow \text{(Ce}^{III}\text{ complex)} \quad \rightleftarrows \quad Ce^{III} + R_2\dot{C}OH + HSO_4^- \qquad (11\text{-}197)$$

$$R_2\dot{C}OH + Ce^{IV} \rightarrow R_2\overset{(+)}{C}OH + Ce^{III} \qquad (11\text{-}198)$$
$$\qquad\qquad \downarrow$$
$$\qquad R_2C{=}O + H^+$$

reaction of the α-hydroxyalkyl radical—either complexed or free of the Ce(III) —with cerium(IV) results in oxidation of the radical yielding the carbonyl function.

Oxidative cleavage of glycols by cerium(IV) is a process that has overtones of certain free-radical chain-propagating reactions. As in the alcohol oxidations, formation of a complex between cerium(IV) may be involved, however, it is not essential that both hydroxy groups be involved. This is evidenced by the fact that ethylene glycol (67) and 2-methoxyethanol (68) are oxidized at the same rate.[58] The reaction is different in that alcohol functions in the diol are

$$\begin{array}{cc} \underset{\overset{|}{OH}}{CH_2}-\underset{\overset{|}{OH}}{CH_2} & \underset{\overset{|}{OH}}{CH_2}\underset{\overset{|}{OCH_3}}{CH_2} \\ 67 & 68 \end{array}$$

not oxidized to the carbonyl functions but carbon-carbon cleavage likely occurs in the rate-determining step. Thus, while 2,3-butanediol (69) is oxidized to acetaldehyde, acetoin (70) is oxidized to biacetyl (71) but neither 70 nor 71

$$\begin{array}{ccc} CH_3CHCCH_3 & CH_3CHCOCH_3 & CH_3COCOCH_3 \\ \;\;|\;\;| & \;\;| & \\ \;\;|\;\;OH & \;\;OH & \\ OH & & \\ 69 & 70 & 71 \end{array}$$

are observed as reaction products of the oxidation of 69. Pinacol, which cannot undergo oxidation of a single hydroxy function without carbon-carbon bond rupture, is oxidized readily by cerium(IV) to acetone. A mechanism for the reaction consistent with these observations is one in which the cerium(IV) oxidizes the alcohol function to the carbonyl with concurrent rupture of the carbon-carbon bond. The α-hydroxyalkyl reacts with another cerium(IV) ion.

$$\begin{array}{c} -\overset{|}{\underset{|}{C}}-OH \\ -\overset{|}{\underset{|}{C}}-OH \end{array} + Ce^{IV} \rightarrow \quad \overset{|}{\underset{-\overset{|}{C}-OH}{C}}\overset{O}{\diagup}Ce^{IV} \quad \rightarrow \quad \begin{array}{c} -\overset{|}{C}=O + Ce^{III} \\ + \\ -\overset{|}{\underset{|}{\dot{C}}}-OH \end{array} \qquad (11\text{-}199)$$

$$\overset{\diagup}{\diagdown}\dot{C}OH + Ce^{IV} \rightarrow {}^{(+)}\overset{\diagup}{\diagdown}C-OH + Ce^{III} \qquad (11\text{-}200)$$

$$\overset{L}{\longrightarrow} \overset{\diagup}{\diagdown}C=O + H^+$$

2. Cobalt(III) Oxidations. Oxidations of alcohols with cobalt(III) ions, in which the cobalt(III) is reduced to cobalt(II), have free-radical character.[61] Reactions of tertiary alcohols with cobalt(III) yield mixtures of products resulting from carbon-carbon cleavage processes much like the fragmentation reactions of t-alkoxy radicals. For example, t-amyl alcohol is oxidized to acetone and presumably ethanol by cobalt(III). The reaction may involve the formation of the alkoxy radical which then fragments (11-202), or the carbon-carbon bond cleavage may be concurrent with the electron transfer to the cobalt(III) (11-203). The eliminated alkyl radical is oxidized to the corresponding carbonium ion by another cobalt(III) ion (11-204).

$$\begin{array}{c} CH_3 \\ | \\ C_2H_5\overset{|}{C}OH + Co^{+3} \rightarrow \\ | \\ CH_3 \end{array}$$

$$\left[\begin{array}{c} CH_3 \\ | \\ C_2H_3-\overset{|}{C}-O\diagdown Co^{III} \\ | \\ CH_3 \;\; + H^+ \end{array} \right] \rightarrow \begin{array}{c} CH_3 \\ | \\ C_2H_5\overset{|}{C}O\cdot + Co^{+2} + H^+ \\ | \\ CH_3 \end{array} \qquad (11\text{-}201)$$

$$\begin{array}{c} CH_3 \\ | \\ C_2H_5\overset{|}{C}O\cdot \rightarrow C_2H_5\cdot + (CH_3)_2C=O \\ | \\ CH_3 \end{array} \qquad (11\text{-}202)$$

or

$$\left[\begin{array}{c} CH_3 \\ | \\ CH_3-\overset{|}{C}\diagdown O\diagdown Co^{III} \\ \overset{|}{\underset{}{C_2H_5}} \end{array} \right] \rightarrow \begin{array}{c} CH_3 \\ | \\ CH_3\overset{|}{C}=O + C_2H_5\cdot + Co^{+2} \end{array} \qquad (11\text{-}203)$$

$$C_2H_5 \cdot + Co^{+3} \rightarrow C_2H_5^{(+)} + Co^{+2}$$

$$\underset{H_2O}{\big\lfloor} \rightarrow C_2H_5OH + H^+$$

(11-204)

Secondary alcohols are apparently oxidized in a similar manner in that an alkoxy-like intermediate is involved. In much the same manner as described for the tertiary alcohol, fragmentation occurs by elimination of either a hydrogen atom or an alkyl radical. Although the latter is essentially the same process as the fragmentation of tertiary alkoxy radicals, the removal of the α-hydrogen likely resembles a displacement reaction of some yet unspecified nature. It is

$$\rightarrow \left[\overset{H}{\underset{O}{R_2C \diagdown Co^{III}}} \right] \rightarrow R_2C{=}O + (Co^{II}H) \quad (11\text{-}205)$$

$$\overset{H}{\underset{|}{R_2COH}} + Co^{+3} \rightarrow \overset{H}{\underset{|}{R_2C}}{-}O{-}Co^{III}$$

$$\rightarrow \left[\underset{R}{\overset{H}{\underset{|}{RC}}} \,O{-}Co^{III} \right] \rightarrow RCHO + R \cdot + Co^{II} \quad (11\text{-}206)$$

interesting that the oxidative cleavage reaction competes well in many instances with the oxidation of the secondary alcohol to the corresponding ketone. The ease of β-elimination of an alkyl radical appears to be related to the stability of the alkyl radical as evidenced by the relative amounts of oxidative cleavage and oxidation observed for the following secondary alcohols.[62]

$$\underset{C_3H_7\text{-}n}{\overset{|}{CH_3CHOH}} + Co^{+3} \rightarrow \underset{C_3H_7\text{-}n}{\overset{|}{CH_3C}}{=}O + CH_3CHO \quad (11\text{-}207)$$
$$\phantom{CH_3CHOH + Co^{+3} \rightarrow CH_3C=O}51\% \qquad 49\%$$

$$\underset{C_3H_7\text{-}i}{\overset{|}{CH_3CHOH}} + Co^{+3} \rightarrow \underset{C_3H_7\text{-}i}{\overset{|}{CH_3C}}{=}O + CH_3CHO \quad (11\text{-}208)$$
$$\phantom{CH_3CHOH + Co^{+3} \rightarrow CH_3C=O}2\% \qquad 98\%$$

$$(C_2H_5)_2CHOH + Co^{+3} \rightarrow (C_2H_5)_2C{=}O + C_2H_5CHO \quad (11\text{-}209)$$
$$\phantom{(C_2H_5)_2CHOH + Co^{+3} \rightarrow}15\% \qquad 85\%$$

$$(C_2H_5)_2CDOH + Co^{+3} \rightarrow (C_2H_5)_2C{=}O + C_2H_5CHO \quad (11\text{-}210)$$
$$5\% \qquad\qquad 95\%$$

Oxidations of carboxylic acids by cobalt(III) yield carbon dioxide and alkyl radicals and appear to be processes in which carbon-carbon bond rupture

$$RCO_2H + Co^{+3} \rightarrow R\cdot + CO_2 + Co^{+2} + H^+ \quad (11\text{-}211)$$

occurs concurrently with the electron transfer.[63] The ease of oxidation of the acid by cobalt(III) is related to the stability of the eliminated alkyl radical. Phenylacetic, isobutyric, and pivalic acids yield the resonance-stabilized benzyl, isopropyl, and t-butyl radicals, respectively, and are decarboxylated readily. On the other hand, β-phenylpropionic and acetic acid, which yield the less stable β-phenylethyl and methyl radicals, respectively, are oxidized less readily.

3. Other Inorganic Ionic Species. Oxidations similar to those outlined above are effected by other inorganic species. Oxidation of primary and secondary alcohols by vanadium(V) in acid-catalyzed reactions involve removal of an α-hydrogen in the rate-determining reaction.[64] The mechanism possibly involves formation of the tetracoordinated vanadium(V) species resulting from interaction of the pervanadyl ion (VO_2^+) with the alcohol and water in the acidic medium. In some instances, carbon-carbon cleavage occurs

$$VO_2{}^+ + R_2CHOH + H_3O^+ \rightarrow$$

$$\rightarrow R_2\dot{C}OH + (V^{IV}O)^{2+} + 2H_2O \quad (11\text{-}212)$$

more readily than abstraction of the α-hydrogen. For example, reaction of β-phenylethanol (**72**) with vanadium(V) yields benzaldehyde by a route in which a benzyl radical is eliminated rather than an α-hydrogen abstracted in

$$C_6H_5CH_2CH_2OH + VO_2{}^+ + H_3{}^+O \rightarrow$$
$$\textbf{72}$$

$$C_6H_5\dot{C}H_2 + CH_2O + (V^{IV}O)^{2+} + 2H_2O \quad (11\text{-}213)$$

$$C_6H_5\dot{C}H_2 \xrightarrow[H_2O]{VO_2^{2+}} C_6H_5CH_2OH \xrightarrow[H_3O^+]{VO_2^{2+}} C_6H_5CHO \quad (11\text{-}214)$$

the oxidation.[65] Oxidative cleavage of tertiary glycols by pervanadyl ions likely proceeds by a very similar process.

$$+ (V^{IV}O)^{2+} + H_2O \qquad (11\text{-}215)$$

Manganese(III) acetate oxidizes pinacol readily in a reaction that is mechanistically similar to the oxidative cleavage of tertiary diols by cerium(IV) and vanadium(V) complexes. Oxidations of α-hydroxy acids by manganese(III) pyrophosphosphate possibly proceed in a similar manner.[66]

$$+ H_3P_2O_7 \qquad (11\text{-}216)$$

$$\rightarrow R_2\underset{HO}{C} + CO_2 + Mn^{II} + 2H_3P_2O_7$$

References

1. T. Bolas and C. E. Groves, *J. Chem. Soc.*, **24**, 773 (1871).
2. A. Tchakirian, *Compt. rend.*, **196**, 1026 (1933); F. C. Whitmore, E. L. Whittle, and H. H. Papkin, *J. Amer. Chem. Soc.*, **61**, 1506 (1939).
3. J. W. Heberling, Jr., and W. W. McCormack, *ibid.*, **78**, 5433 (1955).
4. G. A. Razuvaev, *Vistas in Free-Radical Chemistry*, W. A. Waters, Ed., Pergamon Press, New York, N.Y., 1959, pp. 226–228.
5. G. A. Razuvaev, B. N. Moryganov, and H. E. Kronan, *J. Gen. Chem., U.S.S.R.*, **26**, 2485 (1956).
6. M. M. Battegay and W. Kern, *Bull. soc., chim.*, **41**, 34 (1927).
7. E. S. Huyser, Abstr. 140th ACS Meeting, Sept. 1960, p. 6P.
8. M. Anbar and D. Ginsberg, *Chem. Rev.*, **54**, 925 (1954); C. A. Grob and H. J. Schmid, *Helv. Chim. Acta.*, **36**, 1763 (1963).

9. C. Walling and M. J. Mintz, *J. Amer. Chem. Soc.*, **89**, 1515 (1967).

10. J. L. Kurz, R. F. Hutton, and F. H. Westheimer, *J. Amer. Chem. Soc.*, **83**, 584 (1961).

11. D. Mauzerall and F. H. Westheimer, *ibid.*, **77**, 2261 (1955); R. Abeles and F. H. Westheimer, *ibid.*, **80**, 459 (1958).

12. K. A. Schellenberg and F. H. Westheimer, *J. Org. Chem.*, **30**, 1859 (1965).

13. E. A. Braude, J. Hannah, and R. Linstead, *J. Chem. Soc.*, **1960**, 3249, 3257, 3268.

14. J. Wolinsky and T. Schultz, *J. Org. Chem.*, **30**, 3980 (1965).

15. P. K. C. Huang and E. M. Kosower, *J. Amer. Chem. Soc.*, **89**, 3911 (1967).

16. E. S. Huyser and R. H. S. Wang, *J. Org. Chem.*, **33**, 3901 (1968).

17. J. N. Pitts, Jr., R. Letsinger, R. Taylor, S. Patterson, G. Recktenwald, and R. Martin, *J. Amer. Chem. Soc.*, **81**, 1068 (1959).

18. E. S. Huyser and D. C. Neckers, *ibid.*, **85**, 3641 (1963).

19. D. C. Neckers, A. P. Schaap, and J. Hardy, *ibid.*, **88**, 1265 (1966); D. C. Neckers, J. Hardy, and A. P. Schaap, *J. Org. Chem.*, **31**, 622 (1966).

20. L. W. Menapace and H. G. Kuivila, *J. Amer. Chem. Soc.*, **86**, 3047 (1964); H. G. Kuivila and L. W. Menapace, *J. Org. Chem.*, **28**, 2165 (1963).

21. C. Walling and R. Rabinowitz, *J. Amer. Chem. Soc.*, **79**, 5326 (1957); **81**, 1243 (1959).

22. C. E. Griffen, *Chem. and Ind.*, **1958**, 415.

23. K. I. Ivanov and V. K. Savinova, *Doklady Akad. Nauk S.S.S.R.*, **59**, 493 (1948).

24. W. Pritzkow and K. A. Muller, *Ann.*, **597**, 12 (1956).

25. G. G. Joris, U.S. Patent 2,681,936 (1954).

26. G. S. Serif, C. F. Hunt, and A. N. Bourns, *Can. J. Chem.*, **31**, 1229 (1953).

27. R. Criegee, *Ann.*, **522**, 75 (1935); R. Criegee, H. Pilz, and H. Flygare, *Ber.*, **B72**, 1799 (1939).

28. E. H. Farmer and D. A. Sutton, *J. Chem. Soc.*, **1946**, 10.

29. J. Ross, A. I. Gibhart, and J. F. Gerecht, *J. Amer. Chem. Soc.*, **71**, 282 (1949).

30. A. A. Miller and F. R. Mayo, *ibid.*, **78**, 1017 (1956).

31. K. I. Ivanov, V. K. Savinova, and Mikhailova, *J. Gen. Chem. U.S.S.R.*, **16**, 65, 1003, 1015 (1946).

32. A. Robertson, *Nature*, **762**, 153 (1948).

33. P. Thuring and A. Perret, *Helv. Chim. Acta*, **36**, 13 (1953).

34. L. Bateman, *Quart. Revs.*, **8**, 147 (1954).

35. P. D. Bartlett and T. G. Traylor, *J. Amer. Chem. Soc.*, **85**, 2407 (1963).

36. C. E. Boozer and G. S. Hammond, *ibid.*, **76**, 3861 (1954); C. E. Boozer, G. S. Hammond, C. E. Hamilton, and J. N. Sen, *ibid.*, **77**, 3233, 3238 (1955); C. E. Boozer, G. S. Hammond, C. E. Hamilton, and C. Peterson, *ibid.*, **77**, 3380 (1955).

37. R. A. Dickinson and J. A. Leermakers, *ibid.*, **54**, 3852 (1932).

38. K. L. Muller and H. J. Schumacker, *Z. physik. Chem.*, **B37**, 365 (1937).
39. H. J. Schumacker and K. L. Muller, *ibid.*, **A184**, 183 (1941).
40. (a) L. Z. Soborovskii, Y. M. Zinov'ev, and M. A. Englin, *Doklady Akad. Nauk S.S.S.R.*, **67**, 293 (1949); **73**, 333 (1950);
    (b) F. R. Mayo, L. J. Durham, and K. S. Griggs, *J. Amer. Chem. Soc.*, **85**, 3156 (1963).
41. L. Z. Soborovskii and Y. M. Zinov'ev, *Zhur. Obshchei Khim.*, **24**, 516 (1954).
42. N. Uri, *Chem. Rev.*, **50**, 375 (1952).
43. J. H. Baxendale, M. G. Evans, and G. S. Park, *Trans. Faraday Soc.*, **42**, 155 (1946).
44. I. M. Kolthoff and A. I. Medelia, *J. Amer. Chem. Soc.*, **71**, 3777 (1949).
45. C. Matazack and E. S. Huyser, unpublished results. See also R. F. Brown, S. E. Jamison, U. K. Pandit, J. Pinkus, G. R. White, and H. P. Braendlin, *J. Org. Chem.*, **29**, 146 (1964).
46. M. S. Kharasch, F. S. Arimoto, and W. Nudenberg, *ibid.*, **16**, 1556 (1951).
47. M. S. Kharasch and A. Fono, *J. Org. Chem.*, **23**, 324 (1958); **24**, 606 (1959).
48. G. Sosnovsky and N. C. Yang, *ibid.*, **25**, 899 (1960).
49. M. S. Kharasch, G. Sosnovsky, and N. C. Yang, *J. Amer. Chem. Soc.*, **81**, 5819 (1959).
50. P. R. Story, *J. Org. Chem.*, **26**, 287 (1960).
51. J. K. Kochi, *J. Amer. Chem. Soc.*, **85**, 1958 (1963); see also *ibid.*, **78**, 4815 (1956); **79**, 2942 (1957); *Tetrahedron*, **18**, 483 (1962).
52. M. Asscher and D. Vofsi, *Chem. and Ind.*, **1961**, 209.
53. N. Kornblum, G. D. Cooper, and J. E. Taylor, *J. Amer. Chem. Soc.*, **72**, 3013 (1950).
54. D. F. DeTar and M. N. Turetsky, *ibid.*, **77**, 1745 (1955).
55. J. A. McMillan, *Chem. Rev.*, **62**, 65 (1962).
56. F. P. Greenspan and H. M. Woodburn, *J. Amer. Chem. Soc.*, **76**, 6345 (1954).
57. D. D. Mishra and S. Ghosh, *Jour. Ind. Chem. Soc.*, **41**, 397 (1964).
58. J. S. Littler and W. A. Waters, *J. Chem. Soc.*, **1960**, 2767; S. M. Chou and S. V. Gorbachev, *Zh. Fiz. Khim.*, **32**, 635 (1958).
59. J. S. Littler, *J. Chem. Soc.*, **1959**, 4135.
60. F. H. Westheimer, *Chem. Rev.*, **61**, 265 (1961).
61. D. G. Hoare and W. A. Waters, *J. Chem. Soc.*, **1964**, 2552.
62. W. A. Waters and J. S. Littler, Chap. III, *Oxidation in Organic Chemistry*, K. B. Wiberg, Ed., Academic Press, New York (1965), pp. 202–203.
63. A. A. Clifford and W. A. Waters, *J. Chem. Soc.*, **1965**, 2796.
64. J. R. Jones and W. A. Waters, *ibid.*, **1962**, 2068.
65. J. R. Jones and W. A. Waters, *ibid.*, **1960**, 2772.
66. P. Levesley and W. A. Waters, *ibid.*, **1955**, 217.

Chapter 12

Free-Radical Vinyl Polymerization

I. Introduction

Few reactions, if any, have been as extensively investigated as the conversion of certain unsaturated compounds into high molecular weight polymers. One of the more effective means of accomplishing the polymerization of many terminally unsaturated compounds (vinyl monomers) into polymers is by a free-radical chain reaction. Various aspects of the vinyl polymerization reaction have been covered earlier in this book (e.g., steady-state rate laws, chain-transfer constants with solvents, reversibility of the addition reaction, resonance, polar and steric effects in chain-propagating reactions) and will be familiar in terms of the polymerization reaction. It should not be inferred that because this subject appears in the last chapter of this book that an appreciation of many of these concepts of vinyl polymerization resulted from an understanding of the reactions discussed in the previous chapters. Indeed, the opposite is true in that many of the basic ideas concerning the free-radical reactions already discussed originated in investigations of vinyl polymerization reactions.

The following discussion is quite limited and deals only with basic kinetic aspects of free-radical polymerizations that, for the most part, are pertinent to the concepts of other free-radical chain reactions. The reader is urged to refer to the references (both recently and not so recently published) in order to be informed more completely about other intriguing aspects of polymers and polymerization reactions.

The reaction to be discussed in this chapter can be illustrated in its simplest form by considering the conversion of a vinyl monomer such as styrene to polystyrene in a reaction initiated with benzoyl peroxide. The polymer produced in an effective polymerization of the monomer may have hundreds and maybe thousands of the monomer units per molecule of polymer. The mechanism for the reaction is shown in 12-1 through 12-4. This mechanism implies that the only chain-propagating reactions are addition reactions and

$$Bz_2O_2 \rightarrow 2BzO\cdot \tag{12-1}$$

$$BzO\cdot + CH_2\text{—}CHC_6H_5 \rightarrow BzOCH_2\overset{\cdot}{C}H \tag{12-2}$$
$$\underset{\displaystyle C_6H_5}{|}$$

$$BzOCH_2\overset{}{C}H\cdot + CH_2\text{=}CHC_6H_5 \rightarrow BzOCH_2CHCH_2\overset{\cdot}{C}H \tag{12-3}$$
$$\underset{\displaystyle C_6H_5}{|} \qquad\qquad\qquad \underset{\displaystyle C_6H_5}{|}\ \ \underset{\displaystyle C_6H_5}{|}$$

$$2BzO\left(\underset{\displaystyle C_6H_5}{\overset{\displaystyle CH_2CH}{|}}\right)_n \overset{\displaystyle CH_2CH\cdot}{\underset{\displaystyle C_6H_5}{|}} \rightarrow$$

$$BzO\left(\underset{\displaystyle C_6H_5}{\overset{\displaystyle CH_2CH}{|}}\right)_n \overset{\displaystyle C_6H_5}{\overset{|}{\underset{\displaystyle C_6H_5}{CH_2CHCHCH_2}}} \left(\underset{\displaystyle C_6H_5}{\overset{\displaystyle CHCH_2}{|}}\right)_n OBz \tag{12-4}$$

that the product of the reaction, namely the polymer molecule, is produced in the termination reaction 12-4. Only one molecule of polymer is formed for each molecule of peroxide that decomposes into benzoyloxy radicals (12-1) starting two chains (12-2). High molecular weight radicals result from repetition of the chain-propagating reaction 12-3.

A set of symbols is used to represent the components of this reaction. The initiator In yields the initiator fragments $R\cdot$ which add to the monomer M yielding a monomer-derived radical $M\cdot$. Two of the latter couple (or undergo disproportionation) yielding the polymer. Furthermore, the reaction rate constants for each step in the sequence are labelled appropriately to designate the particular reaction; k_d for the decomposition of the initiator, k_i for initiation of the chain sequence, k_p for the propagation reaction, and k_t for the termination reaction. Thus, the mechanism for the vinyl polymerization reaction of any monomer using any initiator can be represented in the abbreviated form 12-5 through 12-8.

$$In \xrightarrow{\ k_d\ } 2R\cdot \tag{12-5}$$

$$R\cdot + M \xrightarrow{\ k_i\ } M\cdot \tag{12-6}$$

$$M\cdot + M \xrightarrow{\ k_p\ } M\cdot \tag{12-7}$$

$$2M\cdot \xrightarrow{\ k_t\ } \text{Polymer} \tag{12-8}$$

II. Steady-State Rate Relationships

A. CHEMICALLY INITIATED POLYMERIZATIONS

A rate law for the polymerization of a vinyl monomer to a high molecular weight polymer by the mechanism outlined above can be derived using the principles outlined in Chapter 3. The reaction of interest is the rate at which the monomer M is consumed. Two reactions involve consumption of M, the initiation of the chain process by reaction of an initiator fragment with M and the propagation reaction in which the monomer-derived radical $M\cdot$ adds to M, a reaction that may repeat itself many times. The rate of reaction M, therefore, becomes

$$\frac{-d[M]}{dt} = k_i[R\cdot][M] + k_p[M\cdot][M] \tag{12-9}$$

The first term of 12-9 makes only a very small contribution to the overall consumption rate of the monomer if the chain process represented by 12-7 is long. Since this is necessarily the case if a high molecular weight polymer is produced, the rate of consumption of the monomer, or the rate of polymerization, R_p, can be represented by the second term of 12-9. A workable rate law for the polymerization reaction can be derived by invoking the steady-state

$$\frac{d[M]}{dt} = R_p = k_p[M\cdot][M] \tag{12-10}$$

assumption concerning the concentration of $M\cdot$, namely

$$\frac{d[M\cdot]}{dt} = 0 = k_i[R\cdot][M] - 2k_t[M\cdot]^2 \tag{12-11}$$

and

$$[M\cdot] = \left(\frac{k_i[R\cdot][M]}{2k_t}\right)^{1/2} \tag{12-12}$$

Substituting this value for $[M\cdot]$ in 12-10 gives 12-13 for the rate law. A steady-

$$R_p = k_p\left(\frac{k_i[R\cdot][M]}{2k_t}\right)^{1/2}[M] \tag{12-13}$$

state concentration for $[R\cdot]$ can also be assumed since $R\cdot$ is formed only in the slow thermal decomposition of the initiator and is consumed in the rapid radical-propagating addition reaction to the monomer. The "f" in the term representing the rate at which $R\cdot$ is produced is the efficiency factor relating the effectiveness of the initiator decomposition rate with its ability to produce

$$\frac{d[R\cdot]}{dt} = 0 = k_d f[\text{In.}] - k_i[R\cdot][M] \tag{12-14}$$

and

$$[R\cdot] = \frac{k_d f[\text{In.}]}{k_i[M]} \tag{12-15}$$

radicals that start chains. Reactions other than formation of a pair of radicals that may start chains are, as pointed out in Chapter 10, available to many initiators (e.g., induced decompositions and cage recombinations). Substituting the expression for $[R\cdot]$ in 12-15 into 12-13 gives a rate law for the polymerization reaction that has no radical concentrations.

$$R_p = k_p\left(\frac{k_d f}{2k_t}\right)^{1/2}[M][\text{In.}]^{1/2} \tag{12-16}$$

This rate law for polymerization can be derived more simply by assuming that there is a steady-state concentration for all radicals and that the rate of initiation (R_i) is equal to the rate of termination (R_t). An expression for $[M\cdot]$

$$R_t = 2k_t[M\cdot]^2 = R_i \tag{12-17}$$

and

$$[M\cdot] = \left(\frac{R_i}{2k_t}\right)^{1/2} \tag{12-18}$$

in terms of R_i can be obtained which can be substituted in 12-10 giving the

$$R_p = k_p\left(\frac{R_i}{2k_t}\right)^{1/2}[M] \tag{12-19}$$

general rate law for the rate of polymerization shown in 12-19. In the case of the chemically initiated polymerization,

$$R_i = k_d f [\text{In.}] \tag{12-20}$$

and substitution in the general rate law gives the same rate law shown in 12-16.

The rate law 12-16 predicts that the rate of polymerization is first order in monomer and half order in the initiator. This derived rate law is amply verified experimentally. Table 12-1 list the initial rates of polymerization of some monomers at various concentrations of the reactants along with the calculated values of $R_p/[M][In]^{1/2}$ which, according to equation 12-16, should be a constant equal to $(k_d f/2k_t)^{1/2} k_p$. The observation that a constant relationship is found supports assumptions made in the derivation of the rate law 12-16.

B. LIGHT-INDUCED POLYMERIZATIONS

Consider the mechanism for a polymerization of a monomer M in which the initiation process is the formation of a pair of chain-carrying radicals resulting from absorption of light by a molecule of the monomer. Absorption of a quantum of light yields an excited monomer molecule $M*$ which produces a pair of

$$M \xrightarrow{h\nu} M* \xrightarrow{(M?)} 2R \cdot \text{ (or } 2M' \cdot] \tag{12-21}$$

$$R \cdot \text{ (or } M' \cdot) + M \rightarrow M \cdot \tag{12-22}$$

$$M \cdot + M \rightarrow M \cdot \tag{12-23}$$

$$2M \cdot \rightarrow \text{polymer} \tag{12-24}$$

radicals that become involved in the chain-propagating reaction 12-23. The precise nature of the photochemical process by which the initiation is accomplished is open to speculation. At the outset it might appear inviting to propose the formation of a triplet-diradical in which both radical sites react with the monomer. The difficulty with this suggestion is that reaction incorporating two monomer units would result in a diradical that undergoes intramolecular coupling to yield a cyclohexane derivative faster than it could react with

$$CH_2{=}CHX \xrightarrow{h\nu} (CH_2{=}CHX)* \rightarrow \dot{C}H_2{-}\dot{C}HX$$

$$\dot{C}H_2{-}\dot{C}HX \xrightarrow{CH_2{=}CHX} \dot{C}H_2CHXCH_2\dot{C}HX \tag{12-26}$$

TABLE 12-1 Polymerization Rates of Vinyl Monomers

Monomer	$[M]$ (moles/liter)	$[In]$ (moles/liter)	R_p (mole/liter/sec)	$R_p/[M][In]^{1/2}$	Reference
	(AIBN initiated at 77°C)				
$CH_2{=}CCO_2CH_3$ $\;\;\mid$ $\;\;CH_3$	8.63	2.06×10^{-4}	1.70×10^{-4}	13.7×10^{-4}	1
$CH_2{=}CCO_2CH_3$ $\;\;\mid$ $\;\;CH_3$	4.75	1.92×10^{-4}	9.37×10^{-5}	14.3×10^{-4}	1
$CH_2{=}CCO_2CH_3$ $\;\;\mid$ $\;\;CH_3$	2.07	2.11×10^{-4}	4.15×10^{-5}	13.8×10^{-4}	1
	(Bz$_2$O$_2$ initiated at 50°C)				
$CH_2{=}CCO_2CH_3$ $\;\;\mid$ $\;\;CH_3$	7.55	4.13×10^{-2}	1.3×10^{-4}	8.6×10^{-5}	2
$CH_2{=}CCO_2CH_3$ $\;\;\mid$ $\;\;CH_3$	3.78	4.13×10^{-2}	6.74×10^{-5}	8.8×10^{-5}	2
$CH_2{=}CCO_2CH_3$ $\;\;\mid$ $\;\;CH_3$	1.89	4.13×10^{-2}	3.34×10^{-5}	8.1×10^{-5}	2
$C_6H_5CH{=}CH_2$	8.35	4×10^{-2}	2.55×10^{-5}	4.8×10^{-5}	3
$C_6H_5CH{=}CH_2$	5.85	4×10^{-3}	1.73×10^{-5}	4.7×10^{-5}	3
$C_6H_5CH{=}CH_2$	3.34	4×10^{-3}	9.3×10^{-6}	4.4×10^{-5}	3

another monomer unit. Fragmentation of the excited monomer yielding two radical fragments $R' \cdot$ or the rapid reaction of the excited monomer with another monomer molecule in the ground state yielding a pair of monomer-derived radicals $M' \cdot$ are reactions that are more likely involved in the photo-initiation reaction. In either case, two chains would be initiated by the absorption of a single quantum of energy.

$$CH_2\!=\!CHX \xrightarrow{h\nu} (CH_2\!=\!CHX)^* \nearrow^{2R\cdot \hspace{3em} (12\text{-}27)}_{\searrow_{(CH_2\!=\!\overset{\cdot}{C}X + CH_3\overset{\cdot}{C}HX)\,?}}$$

$$(12\text{-}28)$$

In the case of a photo-initiated reaction, R_i is dependent on the number of quanta of energy actually absorbed by the monomer. The light absorbed (I_{abs}) by a species is determined by the intensity of the incident light (I_o), the extinction coefficient (ϵ) of the compound (that is, the efficiency with which the light is absorbed) and, if only part of the incident light is absorbed, the concentration of the compound. Thus, in the case of a light-induced vinyl polymerization,

$$R_i = I_o\,\epsilon f[M] \qquad (12\text{-}29)$$

where f again refers to the efficiency of the initiation process in producing radicals that start chain reactions. Substituting this value for R_i in the general kinetic equation 12-19 leads to the rate expression which predicts that the

$$R_p = \left(\frac{I_o\,\epsilon f}{2k_t}\right)^{1/2} k_p[M]^{3/2} \qquad (12\text{-}30)$$

rate of polymerization is three-halves order in the monomer and depends on the square root of the intensity of the incident light.

If all of the light should be absorbed, the rate of initiation would no longer be dependent on the extinction coefficient nor on the concentration of the

$$R_i = I_o f \qquad (12\text{-}31)$$

monomer. The rate law for the polymerization reaction is still dependent on

$$R_p = \left(\frac{I_o f}{2k_t}\right)^{1/2} k_p[M] \qquad (12\text{-}32)$$

the square root of the intensity of the illumination but since $[M]$ is not a factor in determining R_i, the rate is only first order in M.

C. THERMAL POLYMERIZATION

Styrene displays the interesting property of undergoing vinyl polymerization simply by being heated to temperatures in the range of 75 to 150°C. Methyl methacrylate also undergoes thermal polymerization but somewhat less readily than styrene. Since no initiator or other means of inducing radical formation (e.g., light or high-energy radiation) is required for these polymerizations, the formation of radicals in the initiation reaction must result from some reaction of the monomer molecules. The kinetic aspects of the thermal polymerization reaction provide some insight into the process of radical formation from monomers such as styrene and methyl methacrylate.

Thermal polymerizations follow a rate law (12-33) which indicates that the reaction is second order in monomer. It appears reasonable to assume once radicals are introduced into the system that the chain-propagating reaction 12-7 is identical to that encountered in the chemical and light-induced reactions. From the general rate equation 12-19, it follows that the initiation reaction

$$R_p = k'[M]^2 = k_p \left(\frac{k_i}{2k_t} \right)^{1/2} [M]^2 \tag{12-33}$$

if

$$R_i = k_i[M]^2 \tag{12-34}$$

must be second order in M to account for the observed overall second-order rate law 12-33 for thermal polymerization.

Although the mechanism of the initiation reaction is not yet completely understood, the activation parameters for the initiation process provide some insight concerning the nature of the process. The calculated $\Delta E^{\ddagger}$ for thermal polymerization of styrene is 21 kcal/mole and $\Delta E^{\ddagger}$ for the initiation reaction is 29 kcal/mole. Thus, whatever the initiation reaction may be, its enthalpy cannot exceed 29 kcal/mole. The bimolecular coupling reaction yielding the diradical, although energetically feasible, does not seem likely since it would result in

$$2CH_2{=}CHC_6H_5 \; \rightarrow \; C_6H_5\dot{C}HCH_2CH_2\dot{C}HC_6H_5$$

$$\tag{12-35}$$

formation of cyclic products (see previous section). An initiation process in which two monoradicals are formed such as 12-36, a reaction which is also energetically possible in terms of the limitations set above, seems more likely

$$2CH_2{=}CHC_6H_5 \; \rightarrow \; CH_3\dot{C}HC_6H_5 + CH_2{=}\dot{C}C_6H_5 \tag{12-36}$$

in view of the propensity of diradicals toward intramolecular coupling.

The initiation process has a small pre-exponential factor (PZ of the Arrhenius rate equation $k = PZe^{-\Delta E^{\ddagger}/RT}$, where P is the probability of reaction occurring if molecules having the energy $E^{\ddagger}$ collide with the frequency Z). In the case of styrene, PZ is about 4×10^4 and for methyl methacrylate PZ is less than unity. These values contrast markedly with the PZ values generally encountered for bimolecular reactions, values that are most often in the range of 10^9 to 10^{14}. Such extremely small PZ values indicate that the probability of radicals being produced from a collision having sufficient energy to cause reaction is small. Indeed, in the case of methyl methacrylate, one about one collision in 10^{13} is effective in producing radicals. However, each time an effective collision does occur, the event results in the ultimate reaction of many molecules of the monomer owing to the nature of the subsequent reaction of the radicals formed. The chain-propagating reaction, therefore, serves as a sensitive detector of a reaction that has an extremely low probability of occurring.

A possible mode of reaction that is worthy of consideration for the formation of two monoradicals (12-36) is the formation of a Diels-Alder adduct that may

$$2CH_2{=}CHC_6H_5 \rightarrow \text{Diels-Alder Adduct} \qquad (12\text{-}37)$$

$$CH_3\dot{C}HC_6H_5 + CH_2{=}\dot{C}C_6H_5 \qquad (12\text{-}38)$$

Diels-Alder Adduct $_{CH_2{=}CHC_6H_5}$
$(C_{16}H_{16})$

$$CH_3\dot{C}HC_6H_5 + C_{16}H_{15}\cdot \qquad (12\text{-}39)$$

either fragment into a pair of free radicals or react by means of a hydrogen atom transfer reaction with another molecule of monomer producing a pair of free radicals.

Some studies indicate that the initiation reaction may be a termolecular process $(R_p = k'[M]^{5/2})$. If so, the low pre-exponential factors can be explained only in part by a decrease in the collision frequency Z. A low probability factor is still the governing factor in the initiation reaction.

Whatever the mechanism of the initiation reaction, it is noteworthy that a reactive double bond conjugated with another unsaturated linkage may be necessary. Vinyl acetate and vinyl chloride do not undergo any detectable amounts of thermal polymerization at $100°$. Although a reactive double bond may be necessary, it is not sufficient for thermal polymerization since neither methyl vinyl ketone nor acrylonitrile undergo thermal polymerization when heated to $200°$ but rather form stable Diels-Alder products.

D. AUTO-ACCELERATION

The kinetic derivations for the vinyl polymerization reactions discussed up to this point are based on the assumption that a steady-state concentration of free radicals is achieved in the reaction mixture. In view of the correlation between the derived and experimentally observed rate laws, this assumption appears to be valid. When the derived rate laws are not observed experimentally, determination of the reason for the deviation is warranted. One of the deviations that is informative, in that it illustrates some basic principles of free-radical chain reactions, is the auto-acceleration of the polymerization rate of certain vinyl monomers.

The rate law for the polymerization of a monomer M, given by 12-19,

$$R_p = \left(\frac{R_i}{2k_t}\right)^{1/2} k_p[M]$$

predicts that the rate is first order in monomer and half order in the initiator. Although this rate law is observed experimentally when the polymerization is performed in an inert solvent, when the undiluted monomer is polymerized, the derived rate law applies only at the outset of the reaction and deviation from it becomes apparent as the reaction progresses. The interesting aspect of this deviation is that the rate of polymerization increases. Since vinyl polymerization is an exothermic reaction (see section on heats of polymerization in this chapter), the rate increase is often accompanied by evolution of considerable amounts of heat. In some cases, the rate can become so rapid that the polymerization mixture explodes. Methyl acrylate and acrylic acid are particularly notorious in this respect.

It might appear attractive to attribute auto-acceleration to the inability of the neat polymerization mixture to dissipate the heat evolved in the chain-propagating reaction causing the rapid rate increase as the result of an increased reaction temperature. Although this certainly would have the effect of increasing the reaction rate, it is not the reason for auto-acceleration since an increase in rate is observed when isothermal conditions for the reaction are maintained.

Examination of the rate law 12-19 shows that the increase in rate must be attributed to either an increase in k_p or a decrease in k_t since the concentrations of the reactants cannot increase nor, in isothermal reactions, would the rate of initiation be expected to increase. A reasonable explanation for auto-acceleration in terms of the rate constants k_p and k_t can be arrived at if the effect on the reaction mixture of converting simple monomer molecules into polymer molecules is considered. One change is that the viscosity of the polymerizing mixture increases markedly. This is the result of introducing extremely long, and for the most part, linear molecules into the reaction medium. The mobility of the long, growing polymer radical $M\cdot$ decreases as the viscosity of the medium in which it finds itself increases. Monomer molecules will also find their mobility somewhat retarded but certainly not to the same extent as the long polymer radicals. As a consequence, collisions between polymer radicals $M\cdot$ and the monomer molecules M occur at essentially the same frequency in both viscous and nonviscous media. Since the number of effective collisions is comparatively small in chain-propagating reactions (about one in 10^9 compared to about one in 10^4 for the termination reaction), the rate of the chain-propagating reaction is not markedly affected by an increase in viscosity. Its rate certainly would not be increased as would have to be the case if a rate change in the propagating reaction were responsible for auto-acceleration.

The situation for the termination reaction is different. The mobility of both reactants of this process, namely two polymer radicals $M\cdot$, is decreased as the viscosity of the medium is increased. Consequently, fewer collisions occur between radical centers in the more viscous media. Since the collisions between polymer radicals resulting in termination are more effective than those of the propagating reaction (see above), any decrease in the number of collisions can be expected to have a significant effect on the rate of the termination reaction causing it to be slower. The rate of termination is no longer given by 12-17

$$R_t = 2k_t[M\cdot]^2$$

in such a case since the rate constant k_t is no longer activation controlled but becomes almost entirely diffusion controlled. Thus, as the viscosity of the medium increases, the value of k_t becomes smaller and, as a result, the rate of the polymerization increases.

Viewed from a somewhat different standpoint, the rate of termination in an auto-accelerating polymerization becomes slower than the rate of initiation. Since $d[M\cdot]/dt \neq 0$ but $[M\cdot]$ is constantly increasing in such a situation, it follows that R_p must also increase.

E. EVALUATION OF RATE CONSTANTS

An interesting consequence of steady-state kinetics is that, even with all of the experimental information available concerning reaction rates, rates of initiation, and reactant concentrations, the best that we can determine numerically is a value for a combination of reaction rate constants. In the case of vinyl polymerization, reliable values for R_p, R_i, and $[M]$ can be found. Rearrangement of the general rate law 12-19 allows for determination of k_p^2/k_t. Values for the individual rate constants can be determined by relating

$$\frac{k_p^2}{k_t} = \frac{2R_p^2}{R_i[M]^2} \tag{12-40}$$

k_p and k_t to some parameter other than R_p. The details for doing so are outside our scope and the reader is referred to the references listed at the end of this chapter. Two methods, both involving light-induced polymerizations, have been used with some degree of success for obtaining k_p and k_t in terms of some kinetic aspect of the vinyl polymerization reaction other than R_p. In one case, the rate of decay of the polymerization rate is determined immediately after illumination has been removed. This halts any further initiation and k_p and k_t are related to the nonsteady-state polymerization rate. In another method, the rate of polymerization is determined for a reaction that is illuminated intermittently (rotating sector method) and the rate constants are related to τ_s, the average lifetime of a radical chain sequence in a steady-state polymerization. Both methods are subject to experimental problems far greater than those encountered in determining R_p at the steady-state.

Table 12-2 lists the rate constants and the activation energies for the chain-propagating and termination reactions of some of the more common monomers (see Table 3-1). Compared to most second-order reactions, the chain-propagating reactions have comparatively high rate constants. The termination reaction

TABLE 12-2   Absolute Rate Constants[a]

Monomer	k_p			$k_t \times 10^{-7}$		
	30°C	60°C	E (kcal)	30°C	60°C	E (kcal)
Styrene	49	145	7.3	0.24	0.3	2.4
Methyl methacrylate	350	705	6.3	1.5	1.8	2.8
Vinyl acetate	990	2300	7.3	2.0	2.9	5.2
Acrylonitrile	—	1960	—	—	60	—
Vinyl chloride	6800	12300	3.7	1200	2300	—

[a] Reference 2, Chapter 3.

TABLE 12-3   Kinetic Parameters of Vinyl Acetate Polymerization (Light-Induced at 25°C)[a]

	High Intensity	Low Intensity	Units
Rate of initiation (R_i)	7.29×10^{-9}	1.11×10^{-9}	Moles liter^{-1} sec^{-1}
Rate of polymerization (R_p)	1.19×10^{-4}	4.5×10^{-5}	Moles liter^{-1} sec^{-1}
Average chain lifetime (τ_s)	1.5	4.0	Seconds
Propagation rate constant (k_p)	1.01×10^3	0.94×10^3	Liter mole^{-1} sec^{-1}
Termination rate constant (k_t)	3.06×10^7	2.83×10^7	Liter mole^{-1} sec^{-1}
Steady-state radical concentration [$M\cdot$]	0.54×10^{-8}	0.44×10^{-8}	Moles liter^{-1}

[a] Reference 2, Chapter 3.

rate constants are very much higher than those of the propagating reactions. Although the activation energy requirements for both types of reaction are low, it is significant that finite activation energies are observed for the very rapid termination reactions.

Pertinent data concerning the polymerization of vinyl acetate using the rotating sector method are found in the data in Table 12-3. The values found for the average lifetimes of the chain reactions and the steady-state concentrations of the radicals are indicative of the nature of free-radical chain reactions.

III. Molecular Weight Relationships

A. KINETIC CHAIN LENGTH AND DEGREE OF POLYMERIZATION

Among the most interesting questions that can be asked about a vinyl polymerization reaction are those concerning the molecular weight of the product formed in the reaction. This particular concern is interesting to us because the molecular weight is intimately related to the fundamental free-radical chain aspects of the reaction.

Examination of the mechanism for the polymerization reaction given 12-5 through 12-8 indicates that the size of the growing polymer radical increases by one monomer unit for each chain-propagating reaction 12-7. The growing radical continues to increase in size until it becomes involved in the termination reaction 12-8. The molecular weight of the polymer is, therefore, directly related to the rate-ratio of these reactions, namely R_p/R_t. This ratio is the kinetic chain length (v) of the reaction and, invoking the steady-state assumption, v can be

$$v = \frac{R_p}{R_t} \tag{12-41}$$

$$v = \frac{R_p}{R_i} \tag{12-42}$$

related to the rate of initiation (R_i) as well. The relationship of the kinetic chain length and the average degree of polymerization $(\bar{P})$, which is the average number of monomer units per molecule of polymer, depends on the particular mode of termination that may be operative. Termination by both coupling and

$$\bar{P} = 2v \text{ for coupling} \tag{12-43}$$

$$\bar{P} = v \quad \text{for disproportionation} \tag{12-44}$$

disproportionation are encountered in vinyl polymerization reactions although coupling appears to be the more common route.

Examination of equation 12-41 gives some insight into the control available in determining the size of the polymer molecules formed in an ideal polymerization, that is, one in which other factors that may have an effect on the molecular weight are not involved. The rate expressions 12-10 and 12-17 can be substituted

$$v = \frac{k_p[M\cdot][M]}{2k_t[M\cdot]^2} = \frac{2k_t[M]}{k_p[M\cdot]} \tag{12-45}$$

in 12-41 giving 12-45. Rearranging 12-10 results in 12-46, an expression for

$$[M\cdot] = \frac{R_p}{k_p[M]} \tag{12-46}$$

$[M\cdot]$ that can be substituted in 12-45 producing an expression for v having no

$$v = \frac{k_p^2[M]^2}{2k_t R_p} \tag{12-47a}$$

radical concentrations. This expression for v points out an important principle of vinyl polymerization, namely that the molecular weight of a polymer is inversely proportional to the rate of the polymerization reaction in which it was formed. The rate law 12-16 can be substituted in 12-47a giving 12-47b which

$$v = \frac{k_p[M]}{(2k_t k_d f[\text{In.}])^{1/2}} \tag{12-47b}$$

indicates that the molecular weight of the polymer is inversely proportional to the initiator concentration, an experimental parameter readily controlled in a polymerization reaction.

The equation for the kinetic chain length of a thermal polymerization can be derived by substituting the rate law for thermal polymerization in 12-47a.

$$v = \frac{k_p^2[M]^2}{2k_t(k_i/2k_t)^{1/2}k_p[M]^2} = \frac{k_p}{(2k_t k_i)^{1/2}} \tag{12-48}$$

The kinetic chain length and hence degree of polymerization of a thermal polymerization is a constant and can be varied only by varying the temperature of the reaction. The actual degree of polymerization for a thermal polymerization is, however, influenced to some extent by chain transfer reactions involving both the monomer and polymer (see next section).

B. CHAIN TRANSFER

1. Solvent. Real polymerization reactions are somewhat more complex than those described so far in that other chain-propagating reactions occur that affect both the molecular weight and, to some extent, the structure of the polymer. A chain-transfer reaction with the solvent, a reaction in which the growing polymer radical $M\cdot$ is converted to a polymer molecule, will occur if the solvent is reactive to displacement by the growing polymer radical. A new chain is started by addition of the solvent-derived radical to a molecule of the monomer. A classic example of chain transfer with a solvent occurs in the polymerization of styrene in carbon tetrachloride. Using the symbol S for the solvent (Cl_4C in this case) and $S\cdot$ for the solvent-derived radical ($Cl_3C\cdot$ in this case) a general scheme for chain transfer with a solvent can be represented as shown in 12-49 through 12-51. The molecular weight of the polymer

$$M\cdot \quad + \quad S \quad \xrightarrow{k_{tr}} \quad \text{Polymer} \quad + \quad S\cdot \qquad (12\text{-}49)$$
$$\sim CH_2\dot{C}H \quad Cl_4C \qquad \qquad \sim CH_2CHCl \quad Cl_3C\cdot$$
$$\qquad\quad | \qquad\qquad\qquad\qquad\qquad\quad |$$
$$\qquad C_6H_5 \qquad\qquad\qquad\qquad\quad C_6H_5$$

$$M\cdot \quad + \quad M \quad \xrightarrow{k_p} \quad M\cdot \qquad\qquad (12\text{-}50)$$
$$\sim CH_2\dot{C}H \quad CH_2{=}CHC_6H_5 \qquad \sim CH_2CHCH_2\dot{C}H$$
$$\quad\;\; | \qquad\qquad\qquad\qquad\qquad\quad | \qquad\quad |$$
$$\quad C_6H_5 \qquad\qquad\qquad\qquad\quad C_6H_5 \quad C_6H_5$$

$$S\cdot \quad + \quad M \quad \xrightarrow{k_a} \quad M\cdot$$
$$Cl_3C\cdot \quad CH_2{=}CHC_6H_5 \qquad Cl_3CCH_2\dot{C}H_2$$
$$\qquad\qquad\qquad\qquad\qquad\qquad\quad |$$
$$\qquad\qquad\qquad\qquad\qquad\quad C_6H_5$$

$$(12\text{-}51)$$

produced will depend on the relative rates of reactions 12-49 and 12-50. The chain-transfer constant with a solvent (C_s), is defined as the ratio of the rate constants of these reactions. Solvents with high chain-transfer constants have

$$C_s = \frac{k_{tr}}{k_p} \qquad\qquad (12\text{-}52)$$

a marked effect on the molecular weight of the polymer if present in a sufficient concentration to allow the rate of the chain-transfer reaction 12-49 to compete with 12-50, the reaction that results in growth of the polymer radical. If C_s is high enough for a given solvent-monomer system and the solvent is present in a sufficiently high concentration, chain-transfer will occur to the exclusion of the polymerization reaction and a 1:1 addition product of the solvent and monomer will result (see Chapter 6). The factors relating chain-transfer constants to the structure of solvents and monomers can be summarized making use of the chain-transfer constants listed in Table 12-4. A more detailed discussion of this subject can also be found in Chapter 6.

1. Monomers that yield resonance-stabilized radicals (e.g., styrene and methyl methacrylate) have lower chain-transfer constants with a given solvent than those that yield less stabilized and hence more reactive monomer-derived radicals (e.g., vinyl acetate).

2. Solvents that yield resonance-stabilized radicals (e.g., toluene and ethyl benzene) have higher chain-transfer constants with a given monomer than those that yield radicals having little resonance stabilization (e.g., cyclohexane and benzene).

TABLE 12-4   Solvent Chain-Transfer Constants at 60°C ($C_s \times 10^6$)

Solvent	Monomers		
	Styrene	Vinyl Acetate	Methyl Methacrylate
Cyclohexane	2.4	6.6	1
Benzene	1.8	300	7.5
Toluene	12.5	2100	52.5
Ethyl benzene	67	5500	135
Cumene	82	—	—
t-Butyl benzene	6	—	—
Methylene chloride	15	—	—
Chloroform	50	—	—
Carbon tetrachloride	9×10^3	10^6	240
Carbon tetrabromide	1.36×10^6	3.9×10^7	3.3×10^5
n-Butyl mercaptan	2.2×10^7	4.8×10^7	6.7×10^5

3. Solvents that display acceptor qualities in a polar transition state (e.g., carbon tetrachloride and mercaptans) generally have higher chain-transfer constants with monomers having donor properties (e.g., styrene) than with monomers having acceptor qualities (e.g., methyl methacrylate).

4. Conversely, solvents having donor qualities (e.g., toluene, ethyl benzene) have higher chain-transfer constants with monomers that display acceptor behavior in the transition state than with monomers that are donors.

The determination of chain-transfer constants for a solvent-monomer system is based on the effect that the concentration of the solvent relative to that of the monomer has on the molecular weight of the polymer produced. In the case of solvent-monomer systems having small chain-transfer constants, polymer molecules are formed both in reaction 12-49 and in the termination reaction 12-8. The degree of polymerization $\bar{P}$ in this case is the ratio of the rate of the chain-propagating reaction 12-10 with respect to the combined rates of the two reactions which stop the further growth of the polymer radical, namely 12-8 and 12-49.

$$\bar{P} = \frac{k_p[M\cdot][M]}{k_{tr}[M\cdot][S] + 2k_t[M\cdot]^2} \tag{12-53}$$

or

$$\frac{1}{\bar{P}} = \frac{k_{tr}[S][M\cdot] + 2k_t[M\cdot]^2}{k_p[M][M\cdot]} \tag{12-54}$$

Rearranging 12-54 gives

$$\frac{1}{\bar{P}} - \frac{2k_t[M\cdot]}{k_p[M]} = C\frac{[S]}{[M]} \tag{12-55}$$

Substitution of suitable values for $[M\cdot]$ in 12-55 results in expressions that can be used for determining chain-transfer constants of a solvent. For example, the value for $[M\cdot]$ given in 12-46 substituted in 12-55 results in 12-56. The second

$$\frac{1}{\bar{P}} - \frac{2k_t R_p}{k_p^2[M]^2} = C\frac{[S]}{[M]} \tag{12-56}$$

term on the left of 12-56 is the reciprocal of the degree of polymerization in the

$$\frac{1}{\bar{P}_o} = \frac{2k_t R_p}{k_p^2[M]^2} \tag{12-57}$$

absence of chain-transfer, namely, $\bar{P}_o$. Thus, 12-56 can be represented by 12-58,

$$\frac{1}{\bar{P}} - \frac{1}{\bar{P}_o} = C\frac{[S]}{[M]} \tag{12-58}$$

a linear equation in which C_s can be found by determining $\bar{P}_o$ for the monomer at certain standard conditions and then finding $\bar{P}$ at various concentration ratios of the solvent and monomer. One particularly useful application of this method is the determination of chain-transfer constants for various solvents with monomers that are capable of thermal polymerization. In such cases, $\bar{P}_o$ is a constant, as shown in 12-48, and the chain-transfer equation 12-58 becomes a comparatively simple expression which indicates that a plot of $1/\bar{P}$ against $[S]/[M]$ will be linear with a slope of C_s and the intercept being $1/\bar{P}_o$. This particular approach is limited in that it can be used only with monomers that undergo thermal polymerization. It has been used extensively for determination of chain-transfer constants of many solvents with styrene.

If the solvent has a large chain-transfer constant, the polymer molecules are mainly produced in the transfer reaction 12-49 and only few, if any, are formed in the termination reaction 12-8. The chain-transfer constant for such a system can be determined by assuming that $\bar{P}$ is the ratio of the rate at which the monomer reacts relative to that of the solvent.

$$\bar{P} = \frac{d[M]}{d[S]} = \frac{k_p[M][M\cdot] + k_a[M][S\cdot]}{k_{tr}[S][M\cdot]} \tag{12-59}$$

or

$$\bar{P} = \frac{k_p}{k_{tr}}\frac{[M]}{[S]} + \frac{k_a[M][S\cdot]}{k_{tr}[S][M\cdot]} \tag{12-60}$$

Assuming a steady-state concentration for $[S\cdot]$, namely

$$k_{tr}[M\cdot][S] = k_a[S\cdot][M], \tag{12-61}$$

then

$$[S\cdot] = \frac{k_{tr}[M\cdot][S]}{k_a[M]} \ . \tag{12-62}$$

Substitution of this term for $[S\cdot]$ in 12-60 gives 12-63. If the chain-transfer

$$\bar{P} = \frac{1}{C}\frac{[M]}{[S]} + 1 \tag{12-63}$$

constant is small, the equation simplifies to

$$\bar{P} \simeq \frac{1}{C}\frac{[M]}{[S]} \tag{12-64}$$

The appearance of the second term on the right of 12-63, namely the "1," results from consideration of the consumption of monomer by the addition of the solvent-derived radical. This reaction becomes important only if the chain-transfer constant is large and $\bar{P}$ is small.

2. Monomer. Unreacted monomer may participate in a chain-transfer reaction in much the same way that a solvent does. The chain-transfer constant for the monomer, C_m, is a constant peculiar to the monomer and introduces a limiting effect on the molecular weight that the polymer from a given monomer

$$M\cdot + M \begin{array}{c} \overset{k_{tr}}{\nearrow} M'\cdot + \text{polymer} \qquad (12\text{-}65) \\ \\ \underset{k_p}{\searrow} M\cdot \qquad\qquad\qquad (12\text{-}66) \end{array}$$

can attain. If C_m has a large value, the degree of polymerization can never

$$C_m = \frac{k_{tr(m)}}{k_p} \tag{12-67}$$

exceed C_m since most of the polymer molecules are formed in the transfer reaction 12-65. Monomers that yield high molecular weight polymers generally have small chain-transfer constants and the kinetic principles relating the molecular weight with the kinetic chain length pertain since few polymer molecules are formed in the transfer reaction but are mostly formed in the termination reaction 12-8.

Table 12-5 lists the chain-transfer constants of some of the more common monomers along with those of some unsaturated compounds that are of limited value as monomers for synthesizing high molecular weight polymers. Compounds having allylic hydrogens which are reactive toward displacement by the growing polymer radical fall into this category. It is interesting to note that the more important monomers apparently have no reactive hydrogens as evidenced by their low chain-transfer constants. One exception is vinyl acetate

TABLE 12-5   Chain-Transfer   Constants   of Monomers (C_m)

Monomer	C_m	Temperature (°C)
Styrene	6×10^{-5}	60
Methyl acrylate	$7\text{--}40 \times 10^{-6}$	60
Vinyl chloride	2×10^{-5}	60
Vinyl acetate	2×10^{-3}	60
Allyl chloride	1.6×10^{-1}	80
Allyl acetate	7×10^{-2}	80

which reacts readily with the growing polymer radical in a transfer reaction on the methyl group of the acetate moiety. The transfer reaction is both exothermic and has a favorable polar effect.

$$\sim CH_2CH\cdot + CH_3CO_2CH{=}CH_2 \xrightarrow{k_{tr(m)}}$$
$$\underset{OAc}{|}$$

$$\left[\sim CH_2\overset{\delta+}{CH}\cdots H\cdots \overset{\delta-}{CH_2}CO_2CH{=}CH_2 \right] \rightarrow$$
$$\underset{OAc}{|}$$

$$\sim CH_2CH_2 + \cdot CH_2CO_2CH{=}CH_2 \quad (12\text{-}68)$$
$$\underset{OAc}{|} \qquad\qquad M'\cdot$$

3. Polymer. After the polymerization reaction has proceeded to a certain extent, another reagent becomes available on which chain-transfer may occur, namely the polymer itself. Consider the situation encountered in the poly-

$$\sim CH_2\dot{C}H + \sim CH_2CHCH_2CH\sim \xrightarrow{k_{tr(p)}} \sim CH_2CH_2 + \sim CH_2\dot{C}CH_2CH\sim$$
$$\underset{C_6H_5}{|} \qquad \underset{C_6H_5}{|}\ \underset{C_6H_5}{|} \qquad\qquad \underset{C_6H_5}{|} \qquad \underset{C_6H_5}{|}\ \underset{C_6H_5}{|}$$

$$(12\text{-}69)$$

$$\qquad\qquad\qquad\qquad\qquad\qquad CH_2\dot{C}HC_6H_5$$
$$\qquad\qquad\qquad\qquad\qquad\qquad\qquad |$$
$$\sim CH_2\dot{C}CH_2CH\sim + CH_2{=}CHC_6H_5 \xrightarrow{k_a} \sim CH_2CCH_2CH\sim$$
$$\underset{C_6H_5}{|}\ \underset{C_6H_5}{|} \qquad\qquad\qquad\qquad\qquad \underset{C_6H_5}{|}\ \underset{C_6H_5}{|}$$

$$(12\text{-}70)$$

merization of styrene in the presence of polystyrene. The chain-transfer constant for the polymer (C_p) should be similar to that of cumene $(C_s = 8.2 \times 10^{-5}$ at $60°)$ since the chain-transfer reactions in each case involve attack by the monomer-derived radical on similar structures. Having a chain-transfer constant as small as this makes it obvious that chain transfer on the polymer does not become important until a considerable amount of the monomer has been converted to polymer giving rise to a concentration of benzylic hydrogens large enough to allow the transfer reaction to compete with addition to the monomer.

Chain transfer on the polymer alters the nature of the product formed in that the polymer is not a linear arrangement of monomer units but is composed of "branches" along the polymer backbone. These branches may well be composed of as many monomer units as the backbone of the polymer itself.

As is the case with polystyrene, most polymers resulting from vinyl monomers have small chain-transfer constants. Examination of the process for the styrene polymerization indicates a feature that is common to many vinyl monomers with regard to chain transfer on their respective polymers, namely the chain-transfer reaction is one in which the radical produced is very similar in structure to the growing polymer radical. Although the process is slightly exothermic

$$\sim CH_2\dot{C}H + \sim CH_2CHCH_2CH\sim \rightarrow \sim CH_2CH_2 + \sim CH_2\dot{C}CH_2CH\sim \quad (12\text{-}71)$$
$$\underset{X}{|} \qquad \underset{X}{|} \quad \underset{X}{|} \qquad \underset{X}{|} \qquad \underset{X}{|} \quad \underset{X}{|}$$

(two additional β-hydrogens contributing by means of hyperconjugation to the hybrid of the polymer derived radical), there are no inmobile additional resonance or polar factors that lower the activation energy of the transfer reaction. Very near the end of a polymerization reaction when the polymer concentration is high and the monomer concentration is low, extensive chain-transfer on the polymer occurs and extensive branching results. In the case of polymers resulting from polymerization of vinylidene monomers (1,1-disubstituted ethylenes), chain transfer on the polymer might be expected to be even less important because the reaction is generally endothermic.

$$\underset{Y}{\overset{X}{\sim CH_2\dot{C}\cdot}} + \underset{Y}{\overset{X}{\sim CH_2C}}-CH_2-\underset{Y}{\overset{X}{C\sim}} \rightarrow \underset{Y}{\overset{X}{\sim CH_2C}}-H + \underset{Y}{\overset{X}{\sim CH_2C}}-\underset{Y}{\dot{C}H}-C\sim$$

$$(12\text{-}72)$$

Chain transfer on the polymer can be a significant factor if the growing polymer radical is able to attack a reactive site on the polymer. Such is the case with vinyl acetate where $C_p = 3 \times 10^{-3}$ at $40°$.

$$\underset{\substack{|\\ \text{OAc}}}{\sim\text{CH}_2\text{CH}\cdot} + \underset{\substack{|\\ \text{OCOCH}_3}}{\sim\text{CH}_2\text{CHCH}_2\sim} \rightarrow \underset{\substack{|\\ \text{OAc}}}{\sim\text{CH}_2\text{CH}_2} + \underset{\substack{|\\ \text{OCO}\dot{\text{C}}\text{H}_2}}{\sim\text{CH}_2\text{CHCH}_2\sim}$$

$$(12\text{-}73)$$

Free-radical polymerization of ethylene produces polymer having a considerable amount of branching. In contrast to the branching that results from chain transfer on the polymer followed by incorporation of many monomer units at this radical site, the branches in polyethylene are generally short, being either n-butyl or n-amyl groups. The explanation for this particular phenomenon is that the growing polymer radical undergoes chain transfer on itself. The growing polymer radical can assume conformations that allow for abstraction of a hydrogen atom from either the 5- or 6-carbon. The new radical site that results can react with ethylene to continue the formation of the polymer backbone leaving the relatively short branch which is unable to react further.

$$(12\text{-}74)$$

The amount of this type of branching is comparatively high in polyethylene not only because of the reactivity of the primary alkyl radical but also because the effective concentration of hydrogens available for transfer is large even at the outset of the polymerization.

4. Initiator. Another reagent present in many polymerizations is the initiator and it is also subject to chain transfer by the growing polymer radical. Table 12-6 lists the chain-transfer constants of some initiators with styrene. The

TABLE 12-6   Chain-Transfer Constants of Initiators (C_i) with Styrene

Initiator	C_i	Temperature $(°C)$
Benzoyl peroxide	5.5×10^{-2}	60
Benzoyl peroxide	7.5×10^{-2}	70
n-Propyl peroxide	8.4×10^{-4}	60
Cumene hydroperoxide	1×10^{-1}	70
Bis(p-chlorobenzoyl) peroxide	2.1×10^{-1}	70

chain-transfer constants for the initiators are appreciably larger than those of most solvents. However, the initiator concentration in most polymerizations is generally very small and consequently chain transfer on the initiator is insignificant. If chain transfer on the initiator does occur, the previously derived relationships between molecular weight and initiator concentration are no longer valid. This is observed in the polymerization of styrene with bis(2,4-dichlorobenzoyl) peroxide in that the degree of polymerization very markedly decreased as the concentration of the initiator is increased. Polymerization of styrene with initiators having lower chain-transfer constants show the inverse square root dependency of the initiator concentration on the degree of polymerization as predicted by 12-47b.

IV. Copolymerization

A. COPOLYMERIZATION COMPOSITION EQUATION

When two reactive monomers, M_1 and M_2, are allowed to polymerize in the presence of each other, a copolymer consisting of units of each may result. The copolymer can have properties that incorporate the features of each of the homopolymers (a polymer consisting of a single type of monomer unit) but it may also have properties that are peculiar to itself. The composition of the copolymer formed in such a reaction is of interest since it depends not only on the relative amounts of the two monomers present and, therefore, available for reaction, but also on the reactivities of each toward free-radical addition.

Consider a copolymerization of M_1 and M_2. In this reaction, addition of a radical to M_1 produces the M_1-derived radical $M_1 \cdot$ and addition of a radical to M_2 produces the M_2-derived radical $M_2 \cdot$. The chain-propagating reactions that are operative in converting the two monomers to a macromolecule are summarized in 12-75 through 12-78. The reaction rate constants k_{11} and k_{22}

$$M_1 \cdot + M_1 \xrightarrow{k_{11}} M_1 \cdot \qquad (12\text{-}75)$$

$$M_1 \cdot + M_2 \xrightarrow{k_{12}} M_2 \cdot \qquad (12\text{-}76)$$

$$M_2 \cdot + M_1 \xrightarrow{k_{21}} M_1 \cdot \qquad (12\text{-}77)$$

$$M_2 \cdot + M_2 \xrightarrow{k_{22}} M_2 \cdot \qquad (12\text{-}78)$$

are the rate constants for the homopolymerization of M_1 and M_2, respectively. The rate constants k_{12} and k_{21} are for the reactions that involve incorporation of a given monomer unit by a radical derived from the other monomer. The assumption is made that the previous monomer units of the growing radical have no effect on the reaction rate constants. The composition of the copolymer is the rate at which one monomer is consumed relative to the other, namely $d[M_1]/d[M_2]$. Since M_1 is consumed in reactions 12-75 and 12-77, its rate of reaction is given by 12-79. M_2 reacts in 12-76 and 12-78 and its rate of reaction

$$\frac{-d[M_1]}{dt} = k_{11}[M_1\cdot][M_1] + k_{21}[M_2\cdot][M_1] \tag{12-79}$$

is 12-80. Dividing 12-79 by 12-80 results in an expression for the composition

$$\frac{d[M_2]}{dt} = k_{12}[M_1\cdot][M_2] + k_{22}[M_2\cdot][M_2] \tag{12-80}$$

$$\frac{d[M_1]}{d[M_2]} = \frac{k_{11}[M_1\cdot][M_1] + k_{21}[M_2\cdot][M_1]}{k_{12}[M_1\cdot][M_2] + k_{22}[M_2\cdot][M_2]} \tag{12-81}$$

of the copolymer. Invoking the steady-state assumption, namely

$$\frac{d[M_1\cdot]}{dt} = \frac{d[M_2\cdot]}{dt} = 0 = k_{12}[M_1\cdot][M_2] - k_{21}[M_2\cdot][M_1] \tag{12-82}$$

expressions for $[M_1\cdot]$ or $[M_2\cdot]$ can be found and substituted in 12-81 as shown in 12-83 (substitution of value for $[M_2\cdot]$). After appropriate cancellations and

$$\frac{d[M_1]}{d[M_2]} = \frac{k_{11}[M_1\cdot][M_1] + k_{21}(k_{12}[M_1\cdot][M_2]/k_{21}[M_1\cdot])[M_1]}{k_{12}[M_1\cdot][M_2] + k_{22}(k_{12}[M_1\cdot][M_2]/k_{21}[M_1\cdot])[M_2]} \tag{12-83}$$

rearrangements, the equation 12-84 results. The ratios of the rate constants in 12-84 are particularly interesting in that they are the relative reactivities of the

$$\frac{d[M_1]}{d[M_2]} = \frac{[M_1]}{[M_2]}\left(\frac{k_{11}/k_{12}[M_1] + [M_2]}{[M_1] + k_{22}/k_{21}[M_2]}\right) \tag{12-84}$$

two monomers toward addition by each of the radicals encountered in the reaction. Thus, k_{11}/k_{12}, referred to as r_1, is the relative reactivity of M_1 with respect to M_2 toward addition by radical $M_1\cdot$ and k_{22}/k_{21}, referred to as r_2, is the relative reactivity of M_2 with respect to M_1 toward addition by radical $M_2\cdot$. The copolymer composition equation, therefore, can be represented by 12-85.

$$\frac{d[M_1]}{d[M_2]} = \frac{[M_1]}{[M_2]}\cdot\left(\frac{r_1[M_1] + [M_2]}{r_2[M_2] + [M_1]}\right) \tag{12-85}$$

The value of equation 12-85 lies in the fact that r_1 and r_2 can be obtained by determining the composition of the copolymer ($d[M_1]/d[M_2]$) at various monomer "feed ratios" ($[M_1]/[M_2]$). A certain amount of care must be taken in these determinations since 12-85 is not an integrated equation and, if the monomers are consumed at different rates, the feed ratio changes. Copolymerization constants r_1 and r_2 can be determined from reactions that are allowed to proceed until only a few percent of the monomers have reacted and consequently only little change in the feed ratio has occurred. The purified copolymer can be analyzed by various techniques to determine its composition. Knowing the composition at two feed ratios is all that is necessary to find values for r_1 and r_2. The reader is referred to the general references at the end of this chapter for more details concerning the procedures employed for making these determinations as well as exposure to an integrated form of the composition equation.

B. EVALUATION OF COPOLYMERIZATION DATA

Copolymerization constants r_1 and r_2 are of value in making predictions concerning feed ratios and copolymer compositions. They have also proven valuable in elucidating various relationships between structure and reactivity of both free radicals and unsaturated linkages in the chain-propagating addition reaction. Since most of the principles relating structure and reactivity in chain-propagating reactions (many of which originated from copolymerization data) have been outlined earlier in this book, the subject is treated in a more general nature in this chapter. The reader is referred to the general sources listed at the end of this chapter for more extensive discussions of these data in terms of the copolymerization reaction.

1. Effects of r_1 and r_2 on Copolymer Composition. The customary manner of examining copolymer composition in terms of r_1 and r_2 is to plot the composition of the copolymer against the feed ratio as predicted by the values of r_1 and r_2 for the system. In practice, the mole fraction of either monomer in the feed is plotted against the mole fraction of this monomer in the copolymer. The situation is similar in many regards to that of a plot of vapor composition against liquid composition of a binary mixture.

An "ideal" copolymerization is defined as one in which the two monomers show exactly the same reactivities toward each of the two free radicals. In an ideal copolymerization, $r_1 = 1/r_2$ (or $r_1 r_2 = 1$). Some composition plots for ideal copolymerizations are shown in Figure 12.1. If M_1 is far more reactive toward addition by a free radical than M_2, r_1 and r_2 will be large and small, respectively, and the copolymer will be largely composed of M_1 units at all feed ratios except when the concentration of M_1 is small. Conversely, if M_2 is more reactive than M_1, r_1 and r_2 will have small and large values, respectively and the copolymer will consist mainly of M_2 units except when the concentration of M_2 is small. If both monomers are equal in reactivity toward both radicals $(r_1 = r_2 = 1)$, the composition of the copolymer is the same as that of the feed at all feed ratios. Obviously, the more similar r_1 and r_2 are in an ideal copolymerization, the more the composition of the copolymer will resemble that of the feed.

Mole fraction of M_2 in feed

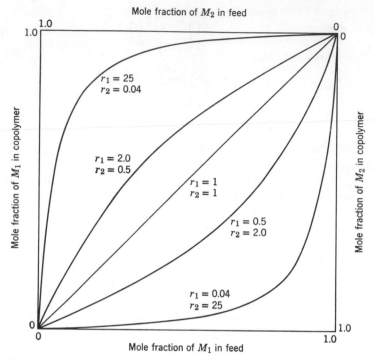

Fig. 12.1.   Feed-copolymer composition plots for ideal copolymerizations.

Another situation encountered in copolymerization is one in which both r_1 and r_2 are less than unity. In such cases, the radical derived from one monomer always prefers to add to the other monomer rather than to one of its own kind, generally because of polar effects that are operative in the addition reaction (see next section). If both r_1 and r_2 are very small, an alternating copolymer $(\sim M_1 M_2 M_1 M_2 M_1 M_2 \sim)$ is formed. Although the product $r_1 r_2$ will be small in such a case, a low value for $r_1 r_2$ is not sufficient to predict formation of an alternating copolymer since only one of the two parameters needs to be small in order for $r_1 r_2$ to be small, but both must be small if an alternating copolymer is to be formed.

If both r_1 and r_2 are either greater than or less than unity, a feed ratio exists that will produce a copolymer having the same composition as the feed. Such a reaction is referred to as an azeotropic copolymerization. The precise feed ratio necessary to have an azeotropic copolymerization can be determined from the values of r_1 and r_2. In such a copolymerization,

$$\frac{d[M_1]}{d[M_2]} = \frac{[M_1]}{[M_2]}$$

(12-86)

Substituting in 12-85 gives 12-87

$$1 = \frac{r_1[M_1] + [M_2]}{r_2[M_2] + [M_1]} \qquad (12\text{-}87)$$

which on rearranging results in

$$r_1[M_1] - [M_1] = r_2[M_2] - [M_2] \qquad (12\text{-}88)$$

or

$$\frac{[M_1]}{[M_2]} = \frac{(r_2 - 1)}{(r_1 - 1)} \qquad (12\text{-}89)$$

From 12-89 the feed ratio that will result in formation of a copolymerization having the same composition can be determined from r_1 and r_2.

Figure 12.2 shows feed-copolymer composition plots for several azeotropic copolymerization situations. If r_1 and r_2 are both greater than unity, a situation not yet encountered in free-radical polymerizations, the reaction involves a pair of monomers that prefer to homopolymerize leading to formation of a "block-copolymer." A block-copolymer is one in which the same monomer unit is repeated several times along the polymer chain, namely,

$$(\sim M_1\, M_1\, M_1\, M_1\, M_1\, M_2\, M_2\, M_2 \sim) \cdot$$

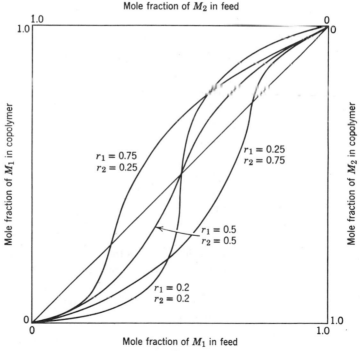

Mole fraction of M_2 in feed

$r_1 = 0.75$
$r_2 = 0.25$

$r_1 = 0.25$
$r_2 = 0.75$

$r_1 = 0.5$
$r_2 = 0.5$

$r_1 = 0.2$
$r_2 = 0.2$

Mole fraction of M_1 in copolymer

Mole fraction of M_2 in copolymer

Mole fraction of M_1 in feed

Fig. 12.2.   Feed-copolymer composition plots for azeotropic copolymerizations.

Most of the common vinyl monomers tend to copolymerize with each other in reactions that are either close to being ideal copolymerizations or that show an alternating effect.

2. Structure and Reactivity. Table 12-7 lists the r_1 and r_2 values for some of the more common monomers both with each other and with other unsaturated

TABLE 12-7 Copolymerization Data (60°C)

Item	M_1	M_2	r_1	r_2	$r_1 r_2$
(1)	Styrene	Acrylonitrile	0.41	0.04	0.016
(2)	Styrene	Butadiene	0.78	1.39	1.09
(3)	Styrene	Dimethyl fumarate	0.21	0.025	0.005
(4)	Styrene	Dimethyl maleate	8.5	0.03	0.3
(5)	Styrene	Vinylidene cyanide[a]	0.005	0.001	5×10^{-6}
(6)	Styrene	Vinyl acetate	55	0.02	1.0
(7)	Styrene	Vinyl chloride	17	0.02	0.34
(8)	Styrene	*cis*-1,2-Dichloroethylene	210	0	—
(9)	Styrene	*trans*-1,2-Dichloroethylene	37	0	—
(10)	Styrene	Vinylidene chloride	1.85	0.085	0.16
(11)	Styrene	Trichloroethylene	16	0	—
(12)	Styrene	Tetrachloroethylene	185	0	—
(13)	Styrene	Maleic anhydride[b]	0.04	0	0
(14)	Methyl methacrylate	Styrene	0.46	0.52	0.24
(15)	Methyl methacrylate	Vinyl acetate	20	0.015	0.30
(16)	Methyl methacrylate	Acrylonitrile	1.35	0.18	0.24
(17)	Vinyl acetate	Diethyl fumarate	0.011	0.44	0.004
(18)	Vinyl acetate	Diethyl maleate	0.17	0.043	0.007
(19)	Vinyl acetate	Isopropenyl acetate[c]	1.0	1.0	1.0
(20)	Vinyl acetate	Acrylonitrile	0.061	4.05	0.25
(21)	Vinyl acetate	Vinyl chloride	0.23	1.68	0.38
(22)	Vinyl acetate	*cis*-1,2-Dichloroethylene	6.3	0.018	0.11
(23)	Vinyl acetate	*trans*-1,2-Dichloroethylene	0.99	0.086	0.085
(24)	Acrylonitrile	Butadiene	0.25	0.33	0.08
(25)	Acrylonitrile	Vinyl chloride	3.28	0.02	0.06
(26)	Acrylonitrile	Vinylidene chloride	0.91	0.37	0.34
(27)	Acrylonitrile	Trichloroethylene	67	0	—
(28)	Acrylonitrile	Tetrachloroethylene	470	0	—
(29)	Acrylonitrile	1-Hexene	12.2	0	—
(30)	Acrylonitrile	Butyl acrylate	1.0	1.0	1.0

[a] At 75°C.
[b] At 80°C.
[c] At 75°C.

compounds. Generalizations concerning the factors that relate the structures of the reactants in a free-radical addition reaction with their reactivities can be made from these data.

(a) *Resonance.*   Monomers that yield resonance-stabilized radicals are more reactive toward addition than those that do not. Styrene, for example, is about 50 times more reactive than vinyl acetate (item 6) toward addition of both styrene- and vinyl acetate-derived radicals. When only resonance plays a significant role in determining the relative reactivities of the monomer, ideal copolymerizations ($r_1 r_2 = 1$) generally result. Such is the case when the monomers are both donors or both acceptors and the steric factors for both are similar. Thus, the donor monomer styrene undergoes a very nearly ideal copolymerization with butadiene (item 2) which is also a donor and vinyl acetate does so with isopropenyl acetate (item 19). The resonance factor is also evident in the lower reactivity of vinyl acetate with respect to acrylonitrile toward addition by both monomer-derived radicals (item 20). Acrylonitrile and butyl acrylate are both acceptor species and participate in an ideal copolymerization (item 30).

(b) *Polar Factors.*   Much of the early evidence that led to the recognition and, to some extent, the elucidation of the nature of polar effects in free-radical reactions came from copolymerization data. The formation of alternating copolymers can best be explained in terms of polar factors. Examination of the data in Table 12-8 shows that an alternating effect is observed in copolymerizations of the donor monomer styrene with the acceptor monomers acrylonitrile (item 1), dimethyl fumarate (item 3), vinylidene cyanide (item 5), and maleic anhydride (item 13). (The problem of dimethyl maleate (item 4) will be discussed in the section concerning steric effects). Very nearly perfect alternating copolymers are formed in the reactions of styrene with vinylidene cyanide and maleic anhydride as evidenced by the small values of both r_1 and r_2 in each case. The acceptor monomer acrylonitrile and the donor monomer butadiene also display a degree of alternation (item 24).

(c) *Steric Effects.*   The reactivities of the chloroethylenes toward addition of styrene-, vinyl acetate-, and acrylonitrile-derived radicals illustrate both resonance and steric effects in radical addition reactions. Comparison of the reciprocals of the r_1 values for a given monomer M_1 with other monomers serves as a relative reactivity series of these monomers toward addition by $M_1 \cdot$. The reactivities of the chloroethylenes toward addition of various monomer-derived radicals are summarized in Table 12-8. The steric effect toward radical addition rendered by a 1,2-disubstituted ethylene is obvious in the lower reactivities of the 1,2-dichloroethylenes relative to vinyl chloride. The additional resonance contribution rendered by a second chlorine bonded to the carbon having the unpaired electron is illustrated by the higher reactivity of vinylidene chloride relative to vinyl chloride and trichloroethylene relative to the 1,2-dichloroethylenes. The very low reactivity of tetrachloroethylene is indicative of extensive steric hindrance, but, when compared to the reactivity of cis-1,2-dichloroethylene, resonance contribution of the additional chlorines is evident.

TABLE 12-8   Reactivities of Chloroethylenes toward Free-Radical Addition

	Radical		
Chloroethylene	$CH_2CH\cdot$ | C_6H_5	$CH_2CH\cdot$ | OAc	$CH_2CH\cdot$ | CN
Vinyl chloride	0.059[a]	4.6[b]	0.305[c]
Vinylidene chloride	0.54	—	1.1
cis-1,2-Dichloroethylene	0.0048	0.16	—
trans-1,2-Dichloroethylene	0.027	1.01	—
Trichloroethylene	0.063	—	0.015
Tetrachloroethylene	0.0054	—	0.0002

[a] Relative to addition to styrene.
[b] Relative to addition to vinyl acetate.
[c] Relative to addition to acrylonitrile.

The greater reactivity of *trans*-1,2-dichloroethylene relative to the *cis*-isomer is due to the higher ground-state energy of the thermodynamically less stable *trans*-isomer. Both compounds result ultimately in the formation of the same adduct radical and, if the structure of this adduct radical makes a significant contribution to the transition state of each addition reaction, the more exothermic of the two reactions, namely addition to the *trans*-isomer, should be faster.

Examination of the r_1 and r_2 values in Table 12-7 for the fumarate and maleate esters with styrene (items 3 and 4) and with vinyl acetate (items 17 and 18) would contradict this explanation for the difference in reactivities of the *cis*- and *trans*-1,2-dichloroethylenes. Other factors must be operative in the reactions of these esters. One explanation is that the carbonyl functions of the maleate esters are not coplanar with the unsaturated linkage. They are unable, therefore, to make significant contributions to the resonance stabilization of the transition state of the addition reaction. The fumarate esters, although thermodynamically more stable than the maleate esters, are more reactive toward addition because the carbonyl function can make its full resonance contribution to the transition state of the reaction.

$$(12\text{-}90)$$

$$(12\text{-}91)$$

V. Thermodynamics of Vinyl Polymerization

A. HEATS OF POLYMERIZATION

Vinyl polymerization is an exothermic reaction in which essentially all of the heat changes occur in the chain-propagating reaction if a high molecular weight polymer is produced. A chain-terminating coupling reaction, although very exothermic, occurs only $1/P$ times to P times that the chain propagating reaction takes place. The energy requirement for decomposition of a chemical initiator is also small compared to the energy evolved in the propagation reaction if $\bar{P}$ is large. Heats of polymerization, therefore, serve well to describe the energy changes that occur in the chain-propagating additions of monomer-derived radicals to the monomer.

Employing the bond dissociation energies given in Chapter 4, the heat of reaction of the addition of a primary alkyl radical to ethylene is exothermic to the extent of about 22 kcal/mole. This reaction, of course, is the chain-propagating step in the free-radical polymerization of ethylene. The calculated heat of polymerization of ethylene, which is close to the observed value, can be regarded as a standard for the vinyl polymerization reaction. Deviations from this value for the heats of polymerizations of other monomers must be accounted for in terms of either the structure of the monomer or of the polymer formed. The heats of polymerization of some of the more common monomers listed in Table 12-9 show that there are appreciable variations in the heats of polymerization although none are greater than that of ethylene. These deviations can be accounted for partly in terms of the resonance energies of the unsaturated

$$\sim CH_2\dot{C}H_2 + CH_2{=}CH_2 \rightarrow \sim CH_2CH_2{-}CH_2\dot{C}H_2 \qquad (12\text{-}92)$$

$$\text{Make} \quad C{-}C = \sim{-}80 \text{ kcal/mole}$$
$$\text{Break} \quad C{=}C = \sim{+}58 \text{ kcal/mole}$$

$$\Delta H_p = \sim{-}22 \text{ kcal/mole}$$

TABLE 12-9   Energetics of Vinyl Polymerization Reactions[a]

Monomer	ΔH_p (obs.)	Estimated Resonance Energy[b]	ΔH_p (calc.)[c]	ΔH_p (calc.)– ΔH_p (obs.)
Ethylene	22.3	0	22	0
Vinyl acetate	21.3	1.8	20.5	0.8
Methyl acrylate	18.7	2	20	1
Isoprene	17.9	4.4	17.9	0
Acrylonitrile	17.3	4	18	0
Styrene	16.7	4.4	17.9	1.2
Vinylidene chloride	14.4	3	19.3	5
Methyl methacrylate	13.0	3.1	19.2	6.2
Isobutylene	12.8	4.6	17.7	4.9
α-Methylstyrene	9.0	4	18	9

[a] All values in kcal/mole.
[b] Based on zero resonance energy for ethylene. Estimates made from heats of hydrogenation of the monomer or an analogous compound.
[c] ΔH_p (calc.) = ΔH_p (obs.) for ethylene minus the estimated resonance energy.

linkage when part of a conjugated system and partly as the result of steric factors that exist in the polymer formed in the polymerization reaction.

The estimated resonance energies of the unsaturated linkage when conjugated with other functionalities in the monomer are also listed in Table 12-9. This resonance energy is lost when the monomer is converted to polymer and consequently does not appear in the observed heats of polymerization. The calculated heats of polymerization take the resonance energy into consideration and, for the most part, are in good agreement with the observed heats of polymerization for vinyl monomers.

There is little agreement between the calculated and observed heats of polymerization in the reactions of the vinylidene monomers (1,1-disubstituted ethylenes). In each case, the observed heat of polymerization is considerably lower than would be expected. The discrepancy between the calculated and observed heats of polymerization (ΔH_p(calc.) – ΔH_p(obs.)) can be attributed largely to steric repulsions that exist in the polymer. Examination of the structure of a polymer produced from a vinylidene monomer shows that steric repulsions resulting from interactions of the substituents with each other cannot be eliminated by allowing the polymer to assume proper conformations as readily as in the case of polymers formed from vinyl monomers. The size of the two substituents plays a significant role in determining the extent of the steric repulsion and becomes considerable in poly-α-methylstyrene.

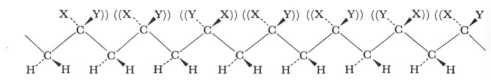

B. REVERSIBILITY OF POLYMERIZATION

The chain-propagating step in vinyl polymerization is a reversible process and certain of the basic details concerning all β-eliminations can be deduced by examining the character of this particular reaction. The equilibrium involved in the reaction of a monomer-derived radical with the monomer is shown in 12-93 where K_p is k_p/k_{-p}. Since the reaction is exothermic in the propagating

$$M \cdot + M \underset{k_{-p}}{\overset{k_p}{\rightleftarrows}} M \cdot \qquad (12\text{-}93)$$

reactions encountered in polymerization of vinyl and vinylidene monomers, the addition reaction has a lower activation energy than the elimination reaction. At temperatures in the range where most vinyl polymerizations are performed, the addition reaction is very much faster than the elimination reaction for most monomers. Increasing the temperature of the reaction has a greater effect on increasing k_{-p} than k_p because of the higher activation energy requirement of the former (see Figure 12-3). The temperature at which the rate of the addition reaction is equal to the rate of the elimination reaction for a given system is called the ceiling temperature (T_c) of the polymerization. Since T_c depends on the rate of the addition reaction, it is dependent on $[M]$, increasing with increasing monomer concentration.

The equilibrium situation in reaction 12-93 can be treated in terms of a simple thermodynamic equilibrium. Thus,

$$-RT \ln K_p = \Delta F_p \qquad (12\text{-}94)$$

where ΔF_p is the free energy of the polymerization reaction. At the ceiling temperature,

$$\Delta F_p = 0 \qquad (12\text{-}95)$$

and

$$\Delta H_p - T_c \Delta S_p = 0 \qquad (12\text{-}96)$$

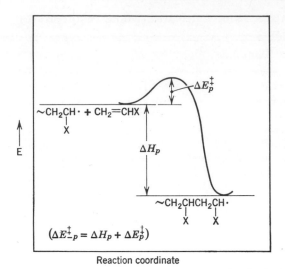

Reaction coordinate

Fig. 12.3. Energy profiles for chain-propagating addition reaction in vinyl polymerization.

or

$$T_c = \frac{\Delta H_p}{\Delta S_p} \qquad (12\text{-}97)$$

Having values for ΔH_p (see Table 12-9), T_c can be calculated if the entropy of the polymerization (ΔS_p) is known. The value of ΔS_p determined for styrene polymerization is -27 e.u. (cal/mole). The primary reason for the entropy difference is a decrease in the translational degrees of freedom in the addition reaction because of the decrease in the number of components. Assuming ΔS_p is similar for other polymerizations as well, an assumption having some validity since the primary entropic changes are very nearly the same, ceiling temperatures for various polymerization reactions can be calculated (Table 12-10). Although only approximate owing to the assumptions made concerning ΔS_p, these calculated ceiling temperatures indicate that vinylidene monomers, because of steric repulsions in the polymer, have considerably lower ceiling temperatures than vinyl monomers. In the case of α-methylstyrene, the ceiling temperature for the polymerization reaction is in the temperature range for decomposition of many chemical initiators.

A polymer can be depolymerized if heated above its ceiling temperature provided a radical center is produced (possibly by thermal fission of a bond along the polymer backbone). This reaction does introduce a serious limitation in the utility of a material such as poly-α-methylstyrene as a high molecular weight polymer.

Investigations of copolymerization reactions performed above the ceiling

TABLE 12-10   Calculated Ceiling Temperatures[a]

Monomer	H_p	T_c (°K)	t_c (°C)
Styrene	16.7	619	346
Ethylene	22.3	826	553
Vinyl acetate	21.3	790	517
Acrylonitrile	17.3	640	367
Methyl acrylate	18.7	693	420
Methyl methacrylate	13.0	483	210
Isobutylene	12.8	473	200
Vinylidene chloride	14.4	534	261
α-Methylstyrene	9.0	334	61

[a] Calculated on the assumption that $S_p = -27$ e.u. in each case.

temperature of α-methylstyrene give some insight into the nature of the steric problems encountered in a vinylidene polymer.[6] No more than two α-methylstyrene units are found bonded consecutively in the copolymer if the reaction is performed above the ceiling temperature for the polymerization of α-methylstyrene even if the concentration of α-methylstyrene is appreciably higher than that of the other monomer. Apparently conformations are available if only two α-methylstyrene units are bonded consecutively that allow for reduction of steric repulsions caused by interactions of the phenyl and methyl groups. Incorporation of a third α-methylstyrene unit in the polymer evidently introduces steric problems that cannot be diminished by any conformational changes in the backbone of the polymer. However, incorporation of a vinyl monomer unit following the two α-methylstyrene units apparently does not introduce serious steric repulsions. The radical having three consecutive α-methylstyrene units likely does form but undergoes fragmentation faster than it adds to another monomer unit since the addition would result in formation of polymeric structure having a considerable amount of steric energy.

The ceiling temperature concept can be used to explain, in part, the reversible additions and fragmentation reactions discussed in Chapter 8. When fragmentation results in formation of two particles, there will be a marked increase in entropy. If the enthalpy of the addition reaction is less than $T\Delta S$, elimination will be preferred over the addition. This becomes an important factor when the enthalpy of the reaction is comparatively low because the "ceiling temperature" of the reaction may be in the temperature range of the reaction. This likely is the case in the reactions of bromine atoms and thiyl radicals with alkenes. Additions of alkyl radicals to alkenes are exothermic and have ceiling temperatures well above the temperature range where most radical reactions in solution

$$\sim M_1\text{-}M_1\text{-}M_1\text{\textperiodcentered} \xrightarrow{M_1} \begin{array}{l} \sim M_1M_1M_1M_1\text{\textperiodcentered} \\ \text{(Have steric} \\ \text{repulsions)} \end{array}$$

$$\sim M_1\text{-}M_1\text{\textperiodcentered} \xrightarrow[\text{fast}]{M_1} \sim M_1\text{-}M_1\text{-}M_1\text{\textperiodcentered} \xrightarrow{M_2} \sim M_1M_1M_1M_2\text{\textperiodcentered}$$

$$\text{(12-98)}$$

$$\sim M_1\text{-}M_1\text{\textperiodcentered} \xrightarrow{M_2} \sim M_1M_1M_2\text{\textperiodcentered} \xrightarrow{M_1} \begin{array}{l} \sim M_1M_1M_2M_1\text{\textperiodcentered} \\ \text{(Have no serious} \\ \text{steric repulsions)} \end{array}$$

$$\sim M_1M_1M_2\text{\textperiodcentered} \xrightarrow{M_2} \sim M_1M_1M_2M_2$$

$M_1 = \alpha$-Methylstyrene, $M_2 =$ Vinyl monomer

are performed. Reactions of alkyl radicals with carbonyl functions are comparatively less exothermic because the bond strength of the carbonyl π-bond is greater than that of an olefinic bond. Consequently, α-alkoxyalkyl and alkoxy radicals fragment more readily than alkyl radicals. In the β-elimination reactions that involve ring opening, the increase in the entropy results from the increase in the number of rotational degrees of freedom possessed by the open-chain system than by the ring system. The entropy change likely is not as great in a ring-opening reaction as it is in a fragmentation reaction leading two separate species with a concurrent increase in the number of translational degrees of freedom. As a consequence, fragmentation reactions that result in ring opening may require higher temperatures than those that yield two species unless the ring-opening reaction results in a significant relief of strain. If the latter is the case, the enthalpy of the addition reaction will necessarily be low and the ceiling temperature would be low even if the entropy change in the reaction is not great.

References

Since the topic of vinyl polymerization was treated in a more general manner in this book than were some of the other areas of free-radical chain reactions, the material was not referenced in the same manner. Almost all of the data cited in this chapter can be found in the general references listed below, often with more detailed discussions than found here. Some other references that pertain to specific subjects discussed in this chapter are also listed.

1. P. J. Flory, *Principles of Polymer Chemistry*, Cornell University Press, Ithaca, New York, 1953.

2. C. Walling, *Free Radicals in Solution*, John Wiley and Sons, New York, 1957.
3. R. W. Lenz, *Organic Chemistry of Synthetic High Polymers*, Interscience Publishers, New York, 1967.
4. T. E. Ferington, "Kinetics of Polymer Formation," *J. Chem. Ed.*, **36**, 174 (1959).
5. F. R. Mayo and C. Walling, "Copolymerization," *Chem. Revs.*, **46**, 191 (1950).
6. G. G. Lowry, "The Effect of Depropagation on Copolymer Composition," *J. Polymer Science*, **42**, 463 (1960).

Author Index

Numbers in parentheses are reference numbers found on the page(s) indicated, The complete reference can be found at the end of the chapter in which the work is cited.

Subject Index

Abstraction reactions, definition, 8
Acceptor species, 81
  Table, 82
Acetals, addition reactions, 158
  autoxidation, 309
  fragmentation, 29, 226
Acetoacetic ester, addition reactions, 26, 157
  polar effect, 83
Acetoin, oxidation by Ce(IV), 324
Acetophenone, reduction by 2-butanol, 303
Acetoxy radical, fragmentation, 259
Acetyl peroxide, 5
  decomposition, 258, 287
  homolytic atomic substitution with, 195
  induced decompositions, 86
    alcohols, 264
    amines, 265
  reactions with, cumene, 36
    Hantzsch ester, 267
Acetylene, additions to, n-butyl mercaptan, 174, 176
  iodotrifluoromethane, 175
Acetylenes, addition to, 174
  stereochemistry of addition, 204
Acids, addition reactions, 156
  oxidative dimerization, 258
Acrylic acid, addition of alcohols, 155
  auto-correlation in polymerization, 340
Acrylonitrile, additions to, 144
  rate constants for polymerization, 339
  thermal reaction, 342
Activation energies, chain-propagating reactions, 93, 105, 342
  chain-terminating reactions, 342
  decomposition reactions, azo-compounds, 386
    peroxy-compounds, 271, 278, 282
  thermal polymerization, 338
Active methylene compounds, addition reactions, 26, 156
Acyl groups, 1,2-shift of, 250
Acyl peroxides, ester formation, 267

reactions with, amines, 265
  alcohols, 264
  ethers, 262
  unimolecular decomposition, 258
Acyl radicals, 25
  decarbonylation, 153, 310
  polar effect, 80
  reaction with oxygen, 310
    peroxydisulfate, 12, 285
Addition reactions, 146
  definition, 10
  mechanism of free-radical, 22
  stereochemistry of, 199
Alcohols, addition reactions, 25, 154
  chain-transfer constants, 155
  oxidation by, acyl peroxides, 264
    t-butyl peroxide, 34, 57, 274
    Ce(IV), 323
    oxygen, 309
    polyhaloalkanes, 128, 293
    Vanadium(V), 327
  polar effects, 80
  rate law for t-butyl peroxide oxidation, 57
  solvation of chlorine atoms, 111
Aldehydes, addition reactions, 25, 152
  polar effects, 80
  reaction with, carbon tetrachloride, 128
    peroxydisulfate, 284
Aldrin, addition to, 201
Allene, additions to, 176
Alkalie metals, 1
Alkanes, reactions with, N-bromosuccinimide, 126
  carbon tetrachloride, 84
  chloroethylenes, 214
  oxygen, 306
Alkenes, additions to, 22, 140–169
  allylic bromination, 21, 124, 130
Alkoxy radicals, fragmentation, 11, 28, 60, 85, 229, 365
  1,5-interactions, 236
a-Alkoxyalkyl radicals, electron transfer from, 12